Chemistry in Motion

Chemistry in Motion: Reaction–Diffusion Systems for Micro- and Nanotechnology

Bartosz A. Grzybowski
Northwestern University, Evanston, USA

WILEY
A John Wiley and Sons, Ltd, Publication

This edition first published 2009

Registered office
John Wiley & Sons Ltd, The Atrium, Southern Gate, Chichester, West Sussex, PO19 8SQ, United Kingdom

For details of our global editorial offices, for customer services and for information about how to apply for permission to reuse the copyright material in this book please see our website at www.wiley.com.

Library of Congress Cataloging-in-Publication Data

Grzybowski, Bartosz A.
Chemistry in motion : reaction-diffusion systems for micro- and nanotechnology / Bartosz A. Grzybowski.
p. cm.
Includes bibliographical references and index.
ISBN 978-0-470-03043-1 (cloth : alk. paper)
1. Reaction mechanism. 2. Reaction-diffusion equations.
3. Microtechnology–Mathematics. 4. Nanotechnology–Mathematics. I. Title.
QD502.5.G79 2009
541′.39–dc22 2008044520

A catalogue record for this book is available from the British Library.

ISBN: 978-0-470-03043-1 (HB)

Typeset in 10/12pt Sabon by Thomson Digital, Noida, India.

To Jolanta, Andrzej and Kristiana
with gratitude and love

Contents

Preface xi

List of Boxed Examples xiii

1 Panta Rei: Everything Flows **1**
- 1.1 Historical Perspective 1
- 1.2 What Lies Ahead? 3
- 1.3 How Nature Uses RD 4
  - 1.3.1 Animate Systems 5
  - 1.3.2 Inanimate Systems 8
- 1.4 RD in Science and Technology 9
- References 12

2 Basic Ingredients: Diffusion **17**
- 2.1 Diffusion Equation 17
- 2.2 Solving Diffusion Equations 20
  - 2.2.1 Separation of Variables 20
  - 2.2.2 Laplace Transforms 26
- 2.3 The Use of Symmetry and Superposition 31
- 2.4 Cylindrical and Spherical Coordinates 34
- 2.5 Advanced Topics 38
- References 43

3 Chemical Reactions **45**
- 3.1 Reactions and Rates 45
- 3.2 Chemical Equilibrium 50
- 3.3 Ionic Reactions and Solubility Products 51
- 3.4 Autocatalysis, Cooperativity and Feedback 52
- 3.5 Oscillating Reactions 55

3.6 Reactions in Gels 57
References 59

4 Putting It All Together: Reaction–Diffusion Equations and the Methods of Solving Them 61
4.1 General Form of Reaction–Diffusion Equations 61
4.2 RD Equations that can be Solved Analytically 62
4.3 Spatial Discretization 66
4.3.1 Finite Difference Methods 66
4.3.2 Finite Element Methods 70
4.4 Temporal Discretization and Integration 80
4.4.1 Case 1: $\tau_{Rxn} \geq \tau_{Diff}$ 81
4.4.1.1 Forward Time Centered Space (FTCS) Differencing 81
4.4.1.2 Backward Time Centered Space (BTCS) Differencing 81
4.4.1.3 Crank–Nicholson Method 82
4.4.1.4 Alternating Direction Implicit Method in Two and Three Dimensions 83
4.4.2 Case 2: $\tau_{Rxn} \ll \tau_{Diff}$ 83
4.4.2.1 Operator Splitting Method 83
4.4.2.2 Method of Lines 84
4.4.3 Dealing with Precipitation Reactions 86
4.5 Heuristic Rules for Selecting a Numerical Method 87
4.6 Mesoscopic Models 87
References 90

5 Spatial Control of Reaction–Diffusion at Small Scales: Wet Stamping (WETS) 93
5.1 Choice of Gels 94
5.2 Fabrication 98
Appendix 5A: Practical Guide to Making Agarose Stamps 101
5A.1 PDMS Molding 101
5A.2 Agarose Molding 101
References 102

6 Fabrication by Reaction–Diffusion: Curvilinear Microstructures for Optics and Fluidics 103
6.1 Microfabrication: The Simple and the Difficult 103
6.2 Fabricating Arrays of Microlenses by RD and WETS 105
6.3 Intermezzo: Some Thoughts on *Rational* Design 109
6.4 Guiding Microlens Fabrication by Lattice Gas Modeling 111

6.5 Disjoint Features and Microfabrication of Multilevel Structures 117
6.6 Microfabrication of Microfluidic Devices 121
6.7 Short Summary 124
References 124

7 Multitasking: Micro- and Nanofabrication with Periodic Precipitation 127
7.1 Periodic Precipitation 127
7.2 Phenomenology of Periodic Precipitation 128
7.3 Governing Equations 130
7.4 Microscopic PP Patterns in Two Dimensions 137
7.4.1 Feature Dimensions and Spacing 139
7.4.2 Gel Thickness 140
7.4.3 Degree of Gel Crosslinking 142
7.4.4 Concentration of the Outer and Inner Electrolytes 142
7.5 Two-Dimensional Patterns for Diffractive Optics 145
7.6 Buckling into the Third Dimension: Periodic 'Nanowrinkles' 152
7.7 Toward the Applications of Buckled Surfaces 155
7.8 Parallel Reactions and the Nanoscale 158
References 160

8 Reaction–Diffusion at Interfaces: Structuring Solid Materials 165
8.1 Deposition of Metal Foils at Gel Interfaces 165
8.1.1 RD in the Plating Solution: Film Topography 167
8.1.2 RD in the Gel Substrates: Film Roughness 172
8.2 Cutting into Hard Solids with Soft Gels 178
8.2.1 Etching Equations 178
8.2.1.1 Gold Etching 180
8.2.1.2 Glass and Silicon Etching 181
8.2.2 Structuring Metal Films 181
8.2.3 Microetching Transparent Conductive Oxides, Semiconductors and Crystals 186
8.2.4 Imprinting Functional Architectures into Glass 189
8.3 The Take-Home Message 192
References 192

9 Micro-chameleons: Reaction–Diffusion for Amplification and Sensing 195
9.1 Amplification of Material Properties by RD Micronetworks 197

9.2 Amplifying Macromolecular Changes using Low-Symmetry Networks 203
9.3 Detecting Molecular Monolayers 205
9.4 Sensing Chemical 'Food' 208
9.4.1 Oscillatory Kinetics 211
9.4.2 Diffusive Coupling 212
9.4.3 Wave Emission and Mode Switching 213
9.5 Extensions: New Chemistries, Applications and Measurements 215
References 222

10 Reaction–Diffusion in Three Dimensions and at the Nanoscale 227
10.1 Fabrication Inside Porous Particles 228
10.1.1 Making Spheres Inside of Cubes 228
10.1.2 Modeling of 3D RD 230
10.1.3 Fabrication Inside of Complex-Shape Particles 235
10.1.4 'Remote' Exchange of the Cores 236
10.1.5 Self-Assembly of Open-Lattice Crystals 238
10.2 Diffusion in Solids: The Kirkendall Effect and Fabrication of Core–Shell Nanoparticles 240
10.3 Galvanic Replacement and De-Alloying Reactions at the Nanoscale: Synthesis of Nanocages 248
References 253

11 Epilogue: Challenges and Opportunities for the Future 257
References 263

Appendix A: Nature's Art 265

Appendix B: Matlab Code for the Minotaur (Example 4.1) 271

Appendix C: C++ Code for the Zebra (Example 4.3) 275

Index 283

Preface

This book is aimed at all those who are interested in chemical processes at small scales, especially physical chemists, chemical engineers and material scientists. The focus of the work is on phenomena in which chemical reactions are coupled with diffusion – hence *Chemistry in Motion*. Although reaction–diffusion (RD) phenomena are essential for the functioning of biological systems, there have been only a few examples of their application in modern micro- and nanotechnology. Part of the problem has been that RD phenomena are hard to bring under experimental control, especially when the system dimensions are small.

As we will see shortly, these limitations can be lifted by surprisingly simple experimental means. The techniques introduced in Chapters 5 to 10 will allow us to control RD at micro- and nanoscales and to fabricate a variety of small-scale structures: microlenes, complex microfluidic architectures, optical elements, chemical sensors and amplifiers, unusual micro- and nanoparticles, and more. Although these systems are still very primitive compared with the sophisticated RD schemes found in biology, they illustrate one general thought that underlies this book – namely that if RD is properly 'programmed' it can be a very unique and practical way of manipulating matter at small scales. The hope is that those who read this monograph will be able to carry the torch further and construct RD micro-/nanosystems that gradually approach the complexity and usefulness of biological RD.

For this to become a reality, however, we must understand RD in quantitative detail. Since RD phenomena are inherently nonlinear, and the participating chemicals evolve into final structures via nontrivial and sometimes counter-intuitive ways, rational design of RD systems requires a fair degree of theoretical treatment. Recognizing this, we devote Chapters 2 through 4 to a thorough discussion of the physical basis of RD and the theoretical tools that can be used to model it.

In these and other chapters, new concepts are derived from the basics and assume only rudimentary knowledge of chemistry and physics and some familiarity with differential equations. This does not mean that the things covered are necessarily easy – not at all! In all cases, however, the material builds up gradually and multiple examples and literature sources are provided to illustrate the key concepts.

The book can be used as a text for a one-semester, graduate elective course in chemical engineering (combining elements of transport and kinetics), and in materials science or chemistry classes on chemical self-organization, self-assembly or micro-/nanotechnology. When taught to chemical engineers, Chapters 2 through 4 should be covered in detail. For more practically minded audiences, it might be reasonable to focus the class on specific self-organization phenomena and their applications (Chapters 5 to 10), consulting the theory from Chapters 2 to 4 as needed.

Finally, several acknowledgements are due. The National Science Foundation generously provided the funding under the CAREER award. I hope the money was well spent! Sincere thanks go to my graduate students – Kyle Bishop, Siowling Soh, Paul Wesson, Rafal Klajn, Chris Wilmer and Chris Campbell – who helped enormously at every stage of the writing process. Last but not least, the book would have never come into being if not for the constant support, love and patience of my family – my most fantastic parents and wife to whom I dedicate this monograph.

Bartosz A. Grzybowski, Evanston, USA

List of Boxed Examples

2.1 Unsteady Diffusion in an Infinite Tube 30
2.2 Unsteady Diffusion in a Finite Tube 31
2.3 Is Diffusion Good for Drug Delivery? 37
2.4 Random Walks and Diffusion 42
3.1 More Than Meets the Eye: Nonapparent Reaction Orders 46
3.2 Sequential Reactions 49
4.1 How Diffusion Betrayed the Minotaur 68
4.2 The Origins of the Galerkin Finite Element Scheme 74
4.3 How Reaction–Diffusion Gives Each Zebra Different Stripes 89
6.1 A Closer Look at Gel Wetting 106
6.2 Is Reaction–Diffusion Time-Reversible? 114
6.3 Optimization of Lens Shape Using a Monte Carlo Method 116
7.1 Periodic Precipitation via Spinodal Decomposition 131
7.2 Wave Optics and Periodic Precipitation 144
7.3 Calculating Diffraction Patterns 149
8.1 Stokes–Einstein Equation 176
8.2 RD Microetching for Cell Biology: Imaging Cytoskeletal Dynamics in 'Designer' Cells 184
9.1 Patterning an Excitable BZ Medium with WETS 210
9.2 Calculating Binding Constants from RD Profiles 219
10.1 Transforming Surface Rates into Apparent Bulk Rates 232

1

Panta Rei: Everything Flows (*Heraclitus, 535–475 BC*)

1.1 HISTORICAL PERSPECTIVE

Change and motion define and constantly reshape the world around us, on scales from molecular to global. Molecules move, collide and react to generate new molecules; components of cells traffic to places where they are needed to participate in and maintain life processes; organisms congregate to perform collective tasks, produce offspring or compete against one another. The subtle interplay between change and motion gives rise to an astounding richness of natural phenomena, and often manifests itself in the emergence of intricate spatial or temporal patterns.

Formal study of such pattern-forming systems began with chemists. Chemistry as a discipline has always been concerned with both molecular change and motion, and by the end of the nineteenth century the basic laws describing the kinetics of chemical reactions as well as the ways in which molecules migrate through different media had been firmly established. At that point, it was probably inevitable that sooner or later some curious chemist would 'mix' (as the profession prescribes) these two ingredients to 'synthesize' a system, in which chemical reactions were coupled in some nontrivial way to the motions of the participating compounds.

This was actually done in 1896 by a German chemist, Raphael Liesegang.[1] In his seminal experiment, Liesegang observed that when certain pairs of inorganic salts move and react in a gel matrix, they produce periodic bands of a precipitate (Figure 1.1).

Chemistry in Motion: Reaction–Diffusion Systems for Micro- and Nanotechnology Bartosz A. Grzybowski

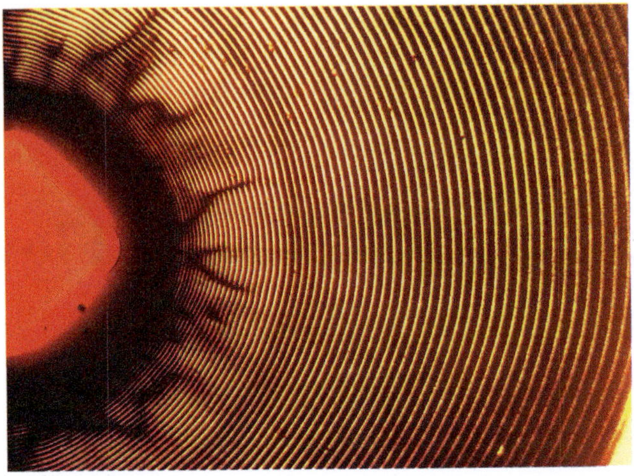

Figure 1.1 Classical Liesegang rings. A small droplet of silver nitrate (red region on the left) is placed on a thin film of gelatin containing potassium dichromate. As $AgNO_3$ diffuses into the gel and outwards from the drop, it reacts with $K_2Cr_2O_7$ to give regular, periodic bands of insoluble $Ag_2Cr_2O_7$. The bands in this picture are all thinner than a human hair

The surprising aspect of this discovery was that there was nothing in the mechanism of a chemical reaction itself that would explain or even hint at the origin of the observed spatial periodicity. Although Liesegang recognized that the patterns had something to do with how the molecules move with respect to one another, he was unable to explain the origin of banding, and the finding remained – at least for the time being – a scientific curiosity. By the early 1900s, however, examples of intriguing spatial patterns resulting from reactions of migrating chemicals in various arrangements had become quite abundant. In 1910, a nearly forgotten French chemist, Stéphane Le Duc, catalogued them in a book titled *Théorie Physico-Chimique de la Vie et Générations Spontanées* (Physical–Chemical Theory of Life and Spontaneous Creations),[2] in which he also alluded to the potential biological significance of such structures. Although his analogies between patterns in salt water and mitotic spindle or polygonal salt precipitates and confluent cells were certainly naive, Le Duc's work was in some sense prophetic. Several decades later, when his static patterns were supplemented by structures varying both in space and time, changing colors and propagating chemical waves, the analogy to living systems became clear. The ability to recreate life-like behavior in a test tube fuelled interest in migration/reaction systems. Chemists teamed up with biologists, physicists, mathematicians and engineers to explore the new universe of reactions in motion. Theory caught up, and several new branches of science – notably, nonlinear chemical kinetics and dynamic system theory – flourished. Mathematical tools and computational resources became available with which to model and explain a wide range of previously puzzling phenomena, including the formation of skin patterns in certain animals, or the functioning of cellular skeleton. By the end of the last century, migration/reaction systems were certainly no longer considered a scientific oddity, but rather a key element of the evolving world.

And yet, despite these undeniable achievements, the nonlinear, pattern-forming chemical systems have not been widely incorporated in modern technology. Historically, the field focused on explaining the underlying physical phenomena, on model experiments in macroscopic arrangements and, more recently, on using the acquired knowledge to understand the existing biological systems. At the same time, we have not been able to apply this knowledge to mimic nature and to design new, artificial constructs that would use migration/reaction to make and control small-scale structures. Nevertheless, we argue in this book that such capability is within our reach and that migration/reaction is perfectly suited for applications in micro- and even nanotechnology. The underlying theme of this monograph is that by setting chemistry in motion in a proper way, it is not only possible to discover a variety of new phenomena, but also to build – importantly, without human intervention – micro-/nano-architectures and systems of practical importance. While we are certainly not attempting to create artificial life – a term that has become somewhat of a scientific cliché – we are motivated by and keen on learning from nature's ability to synthesize systems of chemical reactions programmed in space and time to perform desired tasks. In trying to do so, we limit ourselves to the most common and probably the simplest mode of migration – diffusion – and henceforth focus on the so-called reaction–diffusion (RD) systems.

1.2 WHAT LIES AHEAD?

Our discussion begins with illustrative examples of RD in both animate and inanimate formations chosen to emphasize the universality of RD at different length scales and the creativity with which nature uses it to build and control various types of structures and systems. Inspired by these examples, we then set a stage for the development of our own RD microsystems. In Chapters 2–4, we review the basics of relevant chemical kinetics and diffusion, set up a mathematical framework of RD equations and outline the types of methods that are used to solve them (this part is somewhat mathematically advanced and can probably be skipped on a first reading of the book). With these important preliminaries, we turn our attention to specific classes of micro- and nanoscopic systems, discuss the phenomena that underlie them and the technologically important structures they can produce. By the end of this journey, we will learn how to use RD to make microlenses and diffraction gratings, microfluidic devices and nanostructured supports for cell biology; we will see how RD can be applied in chemical sensing, amplification of molecular events and in biological screening studies. Although the examples we cover span several disciplines, we try to keep the discussion accessible to a general reader and avoid specialized nomenclature wherever possible (after all, this book is not about some specific application, but about the generality of the RD approach to make small things). Since we envision this book to be not only instructive but also thought-provoking, we wish to leave the reader with a set of open-ended questions/problems (Chapter 11) that – in our opinion – will determine the future development of this rapidly evolving field of research. Throughout the text, we include over twenty boxed examples that are intended to highlight specific (and often more mathematical) aspects of the described phenomena. Finally, for those who would like to take a break from equations and strictly scientific arguments, we also provide some artistic respite in Appendix A, which deals with the application of RD to create microscale artwork.

1.3 HOW NATURE USES RD

Some examples of RD in nature are shown in Figure 1.2.

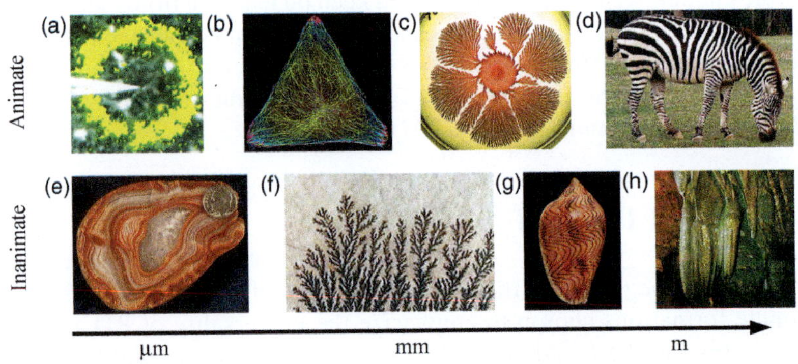

Figure 1.2 Examples of animate (a–d) and inanimate (e–h) reaction–diffusion systems on various length scales. (a) Calcium waves propagating in a retinal cell after mechanical stimulation (scale bar: 50 μm). (b) Fluorescently labeled microtubules in a cell confined to a 40 μm triangle on a SAM-patterned surface of gold (staining scheme: green = microtubules; red = focal adhesions; blue = actin filaments; scale bar: 10 μm). (c) Bacterial colony growth (scale bar: 5 mm). (d) Turing patterns on a zebra. (e) Polished cross-section of a Brazilian agate (scale bar: 200 μm) containing iris banding with a periodicity of 4 μm. (f) Dendritic formations on limestone (scale bar: 5 cm). (g) Patterns formed by reaction–diffusion on the sea shell *Amoria undulate*. (h) Cave stalactites (scale bar: 0.5 m). Image credits: (a) Ref. 6 (1997), *Science*, **275**, 844. Reprinted with permission from AAAS. (b) and (c) reprinted with permission from soft matter, micro- and nanotechnology via reaction diffusion, B. A. Grzybowski *et al.*, copyright (2005), Royal Society of Chemistry (d) Ref. 30, copyright (1995), Nature Publishing Group. (e) Ref. 39, copyright (1995), AAAS. (f) Courtesy of Geoclassics.com. (g) Ref. 48, reproduced by permission of the Association for Computing Machinery. (h) Courtesy of M. Bishop, Niagara Cave, Minnesota.

1.3.1 Animate Systems

The idea that RD phenomena are essential to the functioning of living organisms seems quite intuitive – indeed, it would be rather hard to envision how any organism could operate without moving its constituents around and using them in various (bio)chemical reactions. Surprisingly, however, rigorous evidence that links RD to living systems is relatively fresh and dates back only to the discoveries of Alan Turing[3] in the 1940s and Boris Belousov[4] in the 1950s. Turing recognized that an initially uniform mixture containing diffusing, reactive activator and inhibitor species can spontaneously break symmetry and give rise to stationary concentration variations (i.e., to spatially extended patterns; Figure 1.3(a)). Belousov, on the other hand, discovered a class of systems in which nonlinear coupling between reactions and diffusion gives rise to chemical oscillations in time and/or in space (the latter, in the form of chemical waves; Figure 1.3(b)).

While at first sight these findings might not seem directly relevant to living species, it turns out that Turing's and Belousov's systems contain the essential 'ingredients' – nonlinear coupling and feedback loops – whose various combinations provide a versatile basis for regulatory processes in cells, tissues, organisms and even organism assemblies. For instance, Turing-like, instability-mediated processes can differentiate initially uniform chemical mixtures into regions of distinct composition/function and can thus underlie organism development; chemical oscillations can serve as clocks synchronizing biological events, and the waves can transmit chemical signals. The examples below illustrate how these elements are integrated into biological systems operating at various length scales.

A great variety of regulatory processes inside of cells rely on calcium signals mediated by oscillations or chemical waves. The temporal oscillations in Ca^{2+} concentration are a consequence of a complex RD mechanism (Figure 1.4), in which an external 'signal' first binds to a surface receptor and then triggers the synthesis of inositol-1,4,5-triphosphate (*IP3*) messenger. Subsequently, this

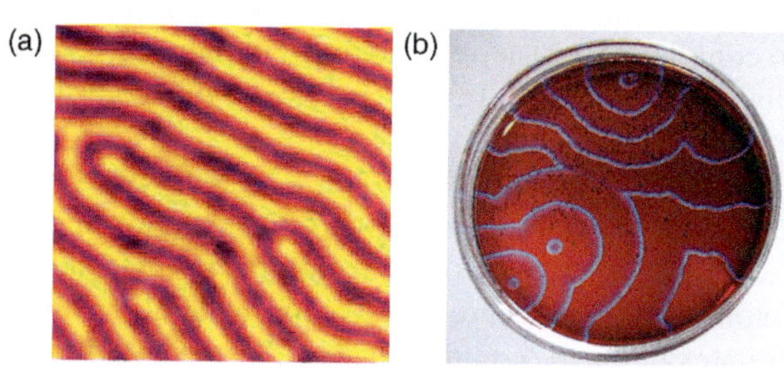

Figure 1.3 (a) Turing pattern formed by CIMA reaction. (b) Traveling waves in the Belousov–Zhabotinsky chemical system. (Image credits: (a) Courtesy of J. Boissonade, CRPP Bordeaux. (b) Courtesy of I. Epstein, Brandeis University. Reproduced by permission of the Royal Society of Chemistry.)

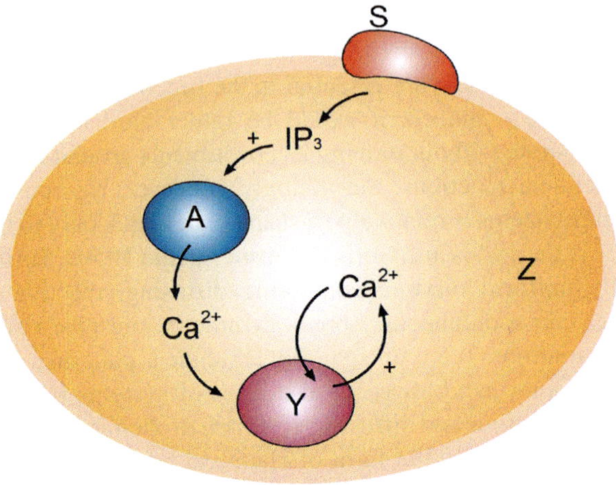

Figure 1.4 Schematic representation of a RD process controlling intracellular oscillations of Ca^{2+}

messenger causes the release of Ca^{2+} from the so-called *IP3* sensitive store, *A*, whose calcium influx into a cytosol (*Z*) activates an insensitive store *Y*. The net diffusion into and out of *Y* is regulated by a positive feedback loop regulated by calcium concentration in the *Z* region. Ultimately, this mechanism causes and controls rhythmic variations in the concentration of Ca^{2+} ions within the cell. These oscillations, for example, increase the efficiency of gene expression, where the oscillating signals enable transcription at Ca^{2+} levels lower than for steady-concentration inputs. In addition, changes in the oscillation frequency allow entrainment and activation of only specific targets on which Ca^{2+} acts, thereby improving the specificity of gene expression.[5] When calcium signals propagate through space (Figure 1.2(a)) in the form of chemical waves,[6,7] the steep transient concentration gradients of Ca^{2+} interact with various types of calcium binding sites (e.g., calcium pumps like ATPase;[8] buffers like calbindin, calsequestrin and calretinin;[9] enzymes like phospholipases[10] and calmodulin[11]) and give rise to complex RD systems synchronizing intracellular and intercellular events as diverse as secretion from pancreatic cells, coordination of ciliary beating in bacteria or wound healing.[12]

Many aspects of cellular metabolism and energetics also rely on RD. For example, glucose-induced oscillations help coordinate the all-important process of glycolysis (i.e., breaking up sugars to make high-energy ATP molecules), induce NADH and proton waves and can regulate other metabolic pathways.[13]

RD also facilitates efficient 'communication' between ATP generation (mitochondria) and ATP consumption sites (e.g., cell nucleus and membrane metabolic 'sensors'), which is essential for normal functioning of a cell.[14,15] In order to ferry ATP timely to ATP-deficient sites, nature has developed a sophisticated RD system of spatially distributed enzymes, collectively known as a 'phosphoryl wires'

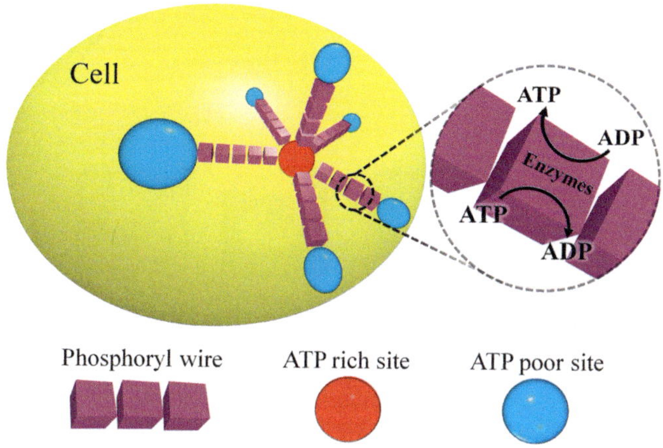

Figure 1.5 Cellular transport of ATP along 'phosphoryl wires' (purple) from an ATP generation site (red) to ATP consumption sites (blue). The panel on the right magnifies one unit of the wire. This unit comprises of a pair of enzymes: the first enzymes hydrolyzes ATP to ADP and generates chemical energy that triggers the reverse, ADP-to-ATP reaction on the second enzyme. The regenerated ATP diffuses to the next unit of the wire and the cycle repeats

(Figure 1.5). These enzymes hydrolyze ATP to ADP at one catalytic site while generating ATP from ADP at a neighboring site. The newly generated ATP then diffuses to another nearby enzyme and the process iterates along the wire. In this way, the ATP is 'pushed' along the wire in a series of domino-like moves called 'flux-waves'. Overall, ATP is delivered to a desired location rapidly, in a time significantly shorter than would be expected for a random, purely diffusive transport through the same distance.[14–16]

Finally, cells use RD to build and dynamically maintain their dynamic 'bones' called microtubules (cf. Figure 1.2(b)), which are constantly growing (at the so called plus-ends pointing toward the cell's periphery) and shrinking (at the minus-ends near centrosome). The balance between these processes depends on the local supply of monomeric tubulin components and a variety of auxiliary microtubule-binding proteins and GTP.[17]

As we have already mentioned in the context of calcium waves, RD can span more than a single cell. In some cases, such long-range processes can have severe consequences to our health. For instance, if RD waves of electrical excitation in the heart's myocardiac tissue propagate as spirals (Figure 1.6),[18] they can lead to life-threatening reentrant cardiac arrhythmias such as ventricular tachycardia and fibrillation.[19] Another prominent example is that of periodically firing neurons synchronized through RD-like coupling,[20] which can extend over whole regions of the brain and propagate in the form of the so-called spreading depressions – that is, waves of potassium efflux followed by sodium influx.[21] These waves temporarily shut down neuronal activity in the affected regions and can cause migraines and peculiar visual disturbances ('fortifications').[21]

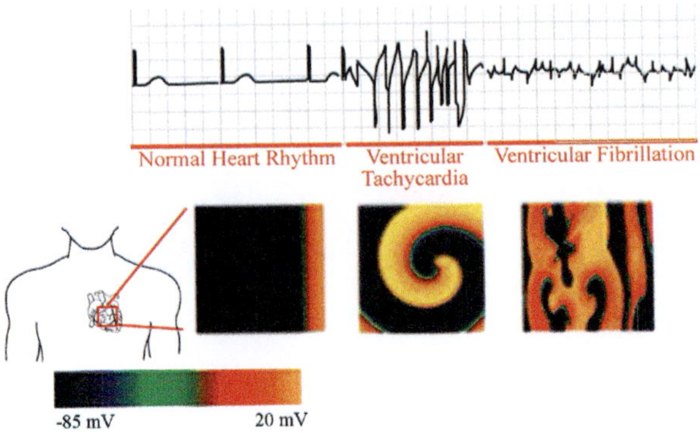

Figure 1.6 The top panel shows a representative ECG recording following the transition from a normal heart rhythm to ventricular fibrillation, an arrhythmia that can lead to sudden cardiac death. The bottom panel shows computer-generated images of RD electrical-activity waves involved in the transition. Left: a single electrical wave produced by the heart's natural pacemaker spreads throughout the heart and induces a contraction. These waves normally occur about once every 0.8 s. Middle: a spiral wave with a period of about 0.2 s can produce fast oscillations characteristic of an arrhythmia called tachycardia, which often directly precedes the onset of fibrillation. Right: multiple spiral waves produced by the breakup of a spiral wave can lead to fast, irregular oscillations characteristic of fibrillation. (Images courtesy of the Center for Arrhythmia Research at Hofstra.)

In organism development, RD is thought to mediate the directed growth of limbs. This process has been postulated[23–25] to involve transforming growth factor (TGFβ), which stimulates production of fibronectin (a 'cell-sticky' protein) and formation of fibronectin prepatterns (nodes) linking cells together into precartilageous nodules. The nodules, in turn, actively recruit more cells from the surrounding area and inhibit the lateral formation of other foci of condensation and potential limb growth.

RD is sometimes used to coordinate collective development or defense/survival strategies of organism populations. For example, starved amoebic slime molds (e.g., *Dictyostelium discoideum*) emit spiral waves of cAMP that cause their aggregation into time-dependent spatial patterns.[26] Similarly, initially homogeneous bacterial cultures grown under insufficient nutrient conditions form stationary, nonequilibrium patterns (Figure 1.2(c)) to minimize the effects of environmental stress.[27–29]

Lastly, some biological RD processes give rise to patterns of amazing aesthetic appeal. Skin patterns emerging through Turing-like mechanisms in marine angelfish *Pomacanthus*,[30] zebras (Figure 1.2(d)), giraffes or tigers[31,32] are but a few examples.

1.3.2 Inanimate Systems

While living systems use complex RD schemes mostly for regulatory/signaling purposes, inanimate creations employ RD based on simple, inorganic chemistries to build spatially extended structures. Many natural minerals have textures characterized by compositional zoning (examples include plagioclase, garnet, augite or zebra spa rock)[33–38] with alternating layers composed of different types of precipitates. An interesting example of two-mineral deposition is the alternation of defect-rich chalcedony and defect-poor quartz observed in iris agates (Figure 1.2(e)).[39] Interestingly, the striking similarity to Liesegang rings[40–43] created in 'artificial' RD systems suggests that banding of mineral textures is governed by similar (Ostwald–Liesegang[44] or two-salt Liesegang[45]) mechanisms. Cave stalactites (Figure 1.2(h)) owe their shapes to RD processes[33] involving (i) hydrodynamics of a thin layer of water carrying Ca^{2+} and H^+ ions and flowing down the stalactite, (ii) calcium carbonate reactions and (iii) diffusive transport of carbon dioxide. Formation of a stalactite is a consequence of the locally varying thickness of the fluid layer controlling the transport of CO_2 and the precipitation rate of $CaCO_3$. RD-driven dendritic structures (Figure 1.2(f)) appear on surfaces of limestone.[46] These dendrites are deposits of hydrous iron or manganese oxides formed when supersaturated solutions of iron or manganese diffuse through the limestone and precipitate at the surface on exposure to air. The structure of these mineral dendrites can be successfully described in terms of simple redox RD equations.[47] Finally, RD has been invoked to explain the formation and pigmentation of intricate seashells,[48] such as those shown in Figure 1.2(g).

1.4 RD IN SCIENCE AND TECHNOLOGY

The range of tasks for which nature uses RD in so many creative ways is really impressive. RD appears to be not only a very flexible but also a 'convenient' way of manipulating matter at small-scales – once RD is set in motion, it builds and controls its creations spontaneously, without any external guidance, and apparently without much effort. From a practical perspective, this sounds very appealing, and one might expect that hosts of smart scientists and engineers all over the world are working on mimicking nature's ability to make technologically important structures in this nature-inspired way. After all, would it not be great if we could just set up a desired micro-/nanofabrication process and have nature do all the tedious work for us (while we pursue one of our multiple hobbies)?

Probably yes, but for the time being it is more of a Huxley-type vision. In reality, until very recently, there have been virtually no applications of RD either in micro- or in nanoscience. Worse still, a significant part of research on RD is focused on how to avoid it! For example, engineers are striving to eliminate oxidation waves emerging via a RD mechanism on catalytic converters in automobiles and in catalytic packed-bed reactors (Figure 1.7). Such waves introduce highly nonlinear – and potentially even chaotic[49] – temperature and concentration variations that are challenging to design around, problematic to control and can drastically affect automobile emissions.[50] In catalytic packed-bed reactors, RD nonlinearities introduce hot zones,[51] concentration waves[52] and unsteady-state temperature

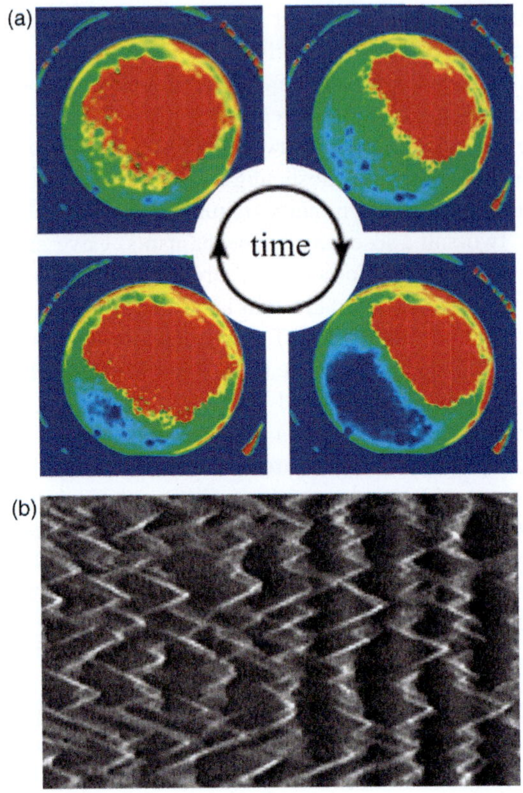

Figure 1.7 Reaction–diffusion in catalytic systems (a) Periodic temperature variations on the top surface of a packed-bed reactor (times are 40 s, 2 min, 4 min and 25 min starting in the upper left-hand corner and moving clockwise). (b) Feedback-induced transition from chemical turbulence to homogeneous oscillations in the catalytic oxidation of CO on Pt(110); this photo was taken between 85 and 125 s with homogeneous oscillations occurring around 425 s. (Image credits: (a) Ref. 51 © 2004 American Chemical Society. (b) Ref. 68 © 2003 American Physical Society.)

profiles[53] that can prevent the system from attaining optimal performance. These phenomena have significant impacts for industry (economic), as well as for the environment (societal).

There are several reasons why RD has not yet found its rightful place in modern technology. First, RD is difficult to bring under experimental control, especially at small scales. As we will see in the chapters to come, RD phenomena can be very sensitive to experimental conditions and to environmental disturbances. In some cases (albeit, rare), changing the dimensions of a RD system by few thousandths of a millimeter (cf. Chapter 9) can change the entire nature of the process this system supports. No wonder that working with such finicky phenomena has not yet become the bread-and-butter of scientists or engineers, who are probably accustomed to more robust systems. Second, even if this and other practical issues were resolved, RD would still present many conceptual challenges since the nonlinearities it involves often make the relationship between a system's ingredients and its final structure/function rather counterintuitive. It takes some skill – and often some serious computing power – to see how and why the various feedback loops and autocatalytic steps involved in RD give rise to a particular structure/function that ultimately emerges. It is even harder to reverse-engineer a problem and choose the ingredients in such a way that a RD process would evolve these ingredients into a desired architecture. Third, there appear to be some ‘sociological’ barriers related to the interdisciplinarity of RD systems. Although RD phenomena are inherently linked to chemistry, relatively few chemists feel comfortable with coupled partial differential equations, Hopf bifurcations or instabilities. These aspects of RD are more familiar and interesting to physicists – few of them, however, are intimate with or interested in such mundane things as solubility products of the participating chemicals, ionic strengths of the solutions used, or the kinetics of the reactions involved. Finally, materials engineers who are often the avant-garde of micro- and nanotechnology, focus – quite understandably – on currently practical and economical techniques, and not on some futuristic schemes requiring basic researches. Given this state of affairs, it becomes apparent that implementing RD means making all these scientists talk to one another and cross the historical boundaries of traditional disciplines. As many readers have probably experienced in their own academic or industrial careers, it is not always a trivial task.

And yet, at least the author of this book believes that RD has a bright future – especially in micro- and nanotechnology. As we will learn shortly, RD can be controlled experimentally with astonishing precision and by sometimes surprisingly simple means. It can be made predictable and it can build structures for which current fabrication methods do not offer viable solutions. It can be made robust, economical and even interesting to students from different backgrounds. The following are some additional arguments.

(i) Micro- and nanoscales are just right for RD-based fabrication. Since the times required for molecules to diffuse through a given distance scale with a square of this distance (cf. Chapter 2), the smaller the dimensions of a RD system, the more rapid the fabrication process. With a typical diffusion coefficient in a (soft) medium supporting an RD process being $\sim 10^{-5}\,\mathrm{cm^2\,s^{-1}}$, RD can build a 10 μm structure in about one-tenth of a second. To build a millimeter-sized structure, RD would have to toil for 1000 seconds, and making an object 1 cm across would keep it busy for 100 000 seconds. For RD, smaller is better.

(ii) RD can be initiated at large, easy-to-control scales and still generate structures with significantly *smaller* dimensions, down to the nanoscale. Liesegang rings can be much thinner that the droplet of the outer electrolyte from which they originate (cf. Figure 1.1), arms of growing dendrites are minuscule in comparison with the dimensions of the whole structure, and the characteristic length of the pattern created through the Turing mechanism is usually much smaller than the dimensions of the system containing the reactants.

(iii) The emergence of small structures can be *programmed* by the initial conditions of an RD process – that is, by the initial concentrations of the chemicals and by their spatial locations. In particular, several chemical reactions can be started simultaneously, each performing an independent task to enable parallel fabrication. No other micro- or nanofabrication method can do this.

(iv) RD can produce patterns encoding spatially continuous concentration variations (*gradients*). This capability is especially important in the context of surface micro-patterning – methods currently in use (photolithography,[54,55] printing[56,57]) modify substrates only at the locations to which a modifying agent (whether a chemical[58,59] or radiation[60,61]) is delivered, and they do so to produce 'binary' patterns. In contrast, RD can evolve chemicals from their initial (patterned) locations in the plane of the substrate and deposit them onto this substrate at quantities proportional to their local concentrations. Gradient-patterned surfaces are of great interest in cell motility assays,[62,63] biomaterials[64,65] and optics.[66,67]

(v) RD processes can be *coupled* to chemical reactions modifying the material properties of the medium in which they occur. In this way, RD can transform initially uniform materials into composite structures, and can selectively modify either their bulk structure or surface topographies. Moreover, the possibility of coupling RD to other processes occurring in the environment can provide a basis for new types of sensing/detection schemes. The inherent nonlinearity of RD equations implies their high sensitivity to parameter changes, and suggests that they can *amplify* small 'signals' (e.g., molecular-scale changes) influencing RD dynamics into, ideally, macroscopic visual patterns. While this idea might sound somewhat fanciful, we point out that chameleons have realized it long time ago and use it routinely to this day to change their skin colors. In Chapter 9 you will see how we, the humans, can learn something from these smart reptiles.

The rest of the book is about realizing at least parts of the ambitious vision outlined above. We will start with the very basics of reaction and diffusion, and then gradually build in new elements and classes of phenomena. Although by the end of our story we will be able to synthesize several types of RD systems rationally and flexibly, we do not forget that even the most advanced of our creations are still no match for the complex RD machinery biology uses. What we do hope for is that this book inspires some creative readers to narrow this gap and ultimately match biological complexity in human-designed RD.

REFERENCES

1. Liesegang, R.E. (1896) *Naturwiss. Wochenschr.*, **11**, 353.
2. Le Duc, S. (1910) *Théorie Physico-Chimique de la Vie et Générations Spontanées* (ed. A. Poinat), Paris.
3. Turing, A.M. (1952) The chemical basis of morphogenesis. *Phil. Trans. R. Soc. B*, **237**, 37.
4. Zaikin, A.N. and Zhabotinsky, A.M. (1970) Concentration wave propagation in 2-dimensional liquid-phase self-oscillating system. *Nature*, **225**, 535.
5. Dolmetsch, R.E., Xu, K.L. and Lewis, R.S. (1998) Calcium oscillations increase the efficiency and specificity of gene expression. *Nature*, **392**, 933.
6. Newman, E.A. and Zahs, K.R. (1997) Calcium waves in retinal glial cells. *Science*, **275**, 844.
7. Straub, S.V., Giovannucci, D.R. and Yule, D.I. (2000) Calcium wave propagation in pancreatic acinar cells: functional interaction of inositol 1,4,5-trisphosphate receptors, ryanodine receptors and mitochondria. *J. Gen. Physiol.*, **116**, 547.

8. Garrahan, P.J. and Rega, A.F. (1990) In *Intracellular Calcium Regulation* (ed. F. Bronner), John Wiley & Sons, Ltd, New York.
9. Baimbridge, K.G., Celio, M.R. and Rogers, J.H. (1992) Calcium-binding proteins in the nervous-system. *Trends Neurosci.*, **15**, 303.
10. Exton, J.H. (1997) Phospholipase D: enzymology, mechanisms of regulation, and function. *Physiol. Rev.*, **77**, 303.
11. Farnsworth, C.L., Freshney, N.W. and Rosen, L.B. *et al.* (1995) Calcium activation of Ras mediated by neuronal exchange factor Ras-GRF. *Nature*, **376**, 524.
12. Falcke, M. (2004) Reading the patterns in living cells: the physics of Ca^{2+} signaling. *Adv. Phys.*, **53**, 255.
13. Mair, T., Warnke, C., Tsujiy, K. and Muller, S.C. (2005) Control of glycolytic oscillations by temperature. *Biophys. J.*, **88**, 639.
14. Dzeja, P.P. and Terzic, A.Phosphotransfer networks and cellular energetics J. Exp. Biol. **206**, 2039.
15. Reich, J.G. and Sel'kov, E.E. (1981) *Energy Metabolism of the Cell: A Theoretical Treatise*, Academic Press, London.
16. Dzeja, P. and Terzic, A. (1998) Phosphotransfer reactions in the regulation of ATP-sensitive K^+ channels. *FASEB J.*, **12**, 523.
17. Tabony, J., Glade, N., Demongeot, J. and Papaseit, C. (2002) Biological self-organization by way of microtubule reaction–diffusion processes. *Langmuir*, **18**, 7196.
18. Hess, B. (2000) Periodic patterns in biology. *Naturwiseenschaften*, **87**, 199.
19. Garfinkel, A., Kim, Y.H., Voroshilovsky, O., Qu, Z.L., Kil, J.R., Lee, M.H., Karaguezian, H.S., Weiss, J.N. and Chen, P.S. (2000) Preventing ventricular fibrillation by flattening cardiac restitution. *Proc. Natl. Acad. Sci. USA*, **97**, 6061.
20. Gray, C.M., Konig, P., Engel, A.K. and Singer, W. (1989) Oscillatory responses in cat visual-cortex exhibit inter-columnar synchronization which reflects global stimulus properties. *Nature*, **338**, 334.
21. Terman, D. and Wang, D.L. (1995) Global competition and local cooperation in a network of neural, oscillators. *Physica D*, **81**, 148.
22. Dahlem, M.A. and Muller, S.C. (2004) Reaction–diffusion waves in neuronal tissue and the window of cortical excitability. *Ann. Phys. (Leipzig)*, **13**, 442.
23. Newman, S.A. (1988) Lineage and pattern in the developing vertebrate limb. *Trends Genet.*, **4**, 329.
24. Newman, S.A. and Frisch, H.L. (1979) Dynamics of skeletal pattern formation in developing chick limb. *Science*, **205**, 662.
25. Leonard, C.M., Fuld, H.M., Frenz, D.A., Downie, S.A., Massague, J. and Newman, S.A. (1991) Role of transforming growth-factor-beta in chondrogenic pattern-formation in the embryonic limb – stimulation of mesenchymal condensation and fibronectin gene-expression by exogenous Tgf-beta and evidence for endogenous Tgf-beta-like activity. *Dev. Biol.*, **145**, 99.
26. Camazine, S. (2001) *Self-Organization in Biological Systems*, Princeton University Press, Princeton, NJ.
27. Budrene, E.O. and Berg, H.C. (1991) Complex patterns formed by motile cells of Escherichia coli. *Nature*, **349**, 630.
28. Budrene, E.O. and Berg, H.C. (1995) Dynamics of formation of symmetrical patterns by chemotactic bacteria. *Nature*, **376**, 49.
29. Ben-Jacob, E., Cohen, I. and Levine, H. (2000) Cooperative self-organization of micro-organisms. *Adv. Phys.*, **49**, 395.

30. Kondo, S. and Asai, R.A reaction–diffusion wave on the skin of the marine angelfish pomacanthus. Nature, **376**, 765.
31. Jiang, T.X., Widelitz, R.B. and Shen, W.M. *et al.* (2004) Integument pattern formation involves genetic and epigenetic controls: feather arrays simulated by digital hormone models. *Int. J. Dev. Biol.*, **48**, 117.
32. Kondo, S. (2002) The reaction–diffusion system: a mechanism for autonomous pattern formation in the animal skin. *Genes Cells*, **7**, 535.
33. Short, M.B., Baygents, J.C. and Beck, J.W. *et al.* (2005) Stalactite growth as a free-boundary problem: a geometric law and its platonic ideal. *Phys. Rev. Lett.*, **94**, 18501.
34. Haase, C.S., Chadam, J., Feinn, D. and Ortoleva, P. (1980) Oscillatory zoning in plagioclase feldspar. *Science*, **209**, 272.
35. Allegre, C.J., Provost, A. and Jaupart, C. (1981) Oscillatory zoning: a pathological case of crystal-growth. *Nature*, **294**, 223.
36. Reeder, R.J., Fagioli, R.O. and Meyers, W.J. (1990) Oscillatory zoning of Mn in solution-grown calcite crystals. *Earth-Sci. Rev.*, **29**, 39.
37. Yardley, B.W.D., Rochelle, C.A., Barnicoat, A.C. and Lloyd, G.E. (1991) Oscillatory zoning in metamorphic minerals: an indicator of infiltration metasomatism. *Mineral. Mag.*, **55**, 357.
38. Krug, H.J., Jacob, K.H. and Dietrich, S. (1994) In *Fractals and Dynamic Systems in Geoscience* (ed. J.H. Kruhl), Springer-Verlag, New York.
39. Heaney, P.J. and Davis, A.M. (1995) Observation and origin of self-organized textures in agates. *Science*, **269**, 1562.
40. Bensemann, I.T., Fialkowski, M. and Grzybowski, B.A. (2005) Wet stamping of microscale periodic precipitation patterns. *J. Phys. Chem. B*, **109**, 2774.
41. Flicker, M. and Ross, J. (1974) Mechanism of chemical instability for periodic precipitation phenomena. *J. Chem. Phys.*, **60**, 3458.
42. Muller, S.C. and Ross, J. (2003) Spatial structure formation in precipitation reactions. *J. Phys. Chem. A*, **107**, 7997.
43. Hantz, P. (2002) Regular microscopic patterns produced by simple reaction–diffusion systems. *Phys. Chem. Chem. Phys.*, **4**, 1262.
44. Sultan, R., Ortoleva, P., Depasquale, F. and Tartaglia, P. (1990) Bifurcatoin of the Ostwald–Liesegang supersaturation nucleation depletion cycle. *Earth-Sci. Rev.*, **29**, 163.
45. Sultan, R.F., Al-Kassem, N.K., Sultan, A.A.H. and Salem, N.M. (2000) Periodic trends in precipitate patterning schemes involving two salts. *Phys. Chem. Chem. Phys.*, **2**, 3155.
46. Bates, R.L. and Jackson, J.A. (eds) (1987) *Glossary of Geology*, 3rd edn, American Geological Institute, Annapolis Junction.
47. Chopard, B., Herrmann, H.J. and Vicsek, T. (1991) Structure and growth-mechanism of mineral dendrites. *Nature*, **343**, 409.
48. Fowler, D.R., Meinhardt, H. and Prusinkiewicz, P. (1992) *Comp. Graph.*, **26**, 379.
49. Ertl, G. (1991) Oscillatory kinetics and spatiotemporal self-organization in reactions at solid surfaces. *Science*, **254**, 1750.
50. Shuai, S.J. and Wang, J.X. (2004) Unsteady temperature fields of monoliths in catalytic converters. *Chem. Eng. J.*, **100**, 95.
51. Marwaha, B., Sundarram, S. and Luss, D. (2004) Dynamics of transversal hot zones in shallow packed-bed reactors. *J. Phys. Chem. B*, **108**, 14470.
52. Jaree, A., Hudgins, R.R. and Budman, H.M. *et al.* (2003) Hysteresis and extinction waves in catalytic CO oxidation caused by reactant concentration perturbations in a packed-bed reactor. *Ind. Eng. Chem. Res.*, **42**, 1662.

53. Jaree, A., Hudgins, R.R. and Budman, H. *et al.* (2003) Amplification of inlet temperature disturbances in a packed-bed reactor for CO oxidation over Pt/Al_2O_3. *Chem. Eng. Sci.*, **58**, 833.
54. Rai-Choudhury, P. (ed). (1997) *Handbook of Microlithography, Micromachining, and Microfabrication*, IET, London.
55. Thompson, L.F. and Kerwin, R.E. (1976) Polymer resist systems for photolithography and electron lithography. *Annu. Rev. Mater. Sci.*, **6**, 267.
56. Xia, Y.N. and Whitesides, G.M. (1998) Soft lithography. *Angew. Chem. Int.*, **37**, 551.
57. Michel, B., Bernard, A. and Bietsch, A. *et al.* (2001) Printing meets lithography: soft approaches to high-resolution printing. *IBM J. Res. Dev.*, **45**, 697.
58. Larsen, N.B., Biebuyck, H., Delamarche, E. and Michel, B. (1997) Order in microcontact printed self-assembled monolayers. *J. Am. Chem. Soc.*, **119**, 3017.
59. Delamarche, E., Donzel, C. and Kamounah, F.S. *et al.* (2003) Microcontact printing using poly(dimethysiloxane) stamps hydrophilized by poly(ethylene oxide) silanes. *Langmuir*, **19**, 8749.
60. Tolfree, D.W.L. (1998) Microfabrication using synchrotron radiation. *Rep. Prog. Phys.*, **61**, 313.
61. Wallraff, G.M. and Hinsberg, W.D. (1999) Lithographic imaging techniques for the formation of nanoscopic features. *Chem. Rev.*, **99**, 1801.
62. Cunningham, C.C., Stossel, T.P. and Kwiatkowski, D.J. (1991) Enhanced motility in NIH-3T3 fibroblasts that overexpress gelsolin. *Science*, **251**, 1233.
63. Parent, C.A. and Devreotes, P.N. (1999) A cell's sense of direction. *Science*, **284**, 765.
64. Suchanek, W. and Yoshimura, M. (1998) Processing and properties of hydroxyapatite-based materials for use as hard tissue replacement implants. *J. Mater. Res.*, **13**, 94.
65. Minuth, W.W., Sittinger, M. and Kloth, S. (1998) Tissue engineering: generation of differentiated artificial tissues for biomedical applications. *Cell Tissue Res.*, **291**, 1.
66. Manhart, P.K. and Blankenbecler, R. (1997) Fundamentals of macro axial gradient index optical design and engineering. *Opt. Eng.*, **36**, 1607.
67. Ren, H.W. and Wu, S.T. (2002) Inhomogeneous nanoscale polymer-dispersed liquid crystals with gradient refractive index. *Appl. Phys. Lett.*, **81**, 3537.
68. Beta, C., Bertram, M. and Mikhailov, A.S. *et al.* (2003) Controlling turbulence in a surface chemical reaction by time-delay autosynchronization. *Phys. Rev. E*, **67**, 046224.

2

Basic Ingredients: Diffusion

2.1 DIFFUSION EQUATION

Recall from your introductory chemistry or physics classes that molecules are very dynamic entities. In a liquid or a gas, they constantly bounce off their neighbors and change the direction of their motion randomly. Whereas for a spatially uniform substance these so-called Brownian motions average out without producing any noticeable macroscopic changes, they do lead to the redistribution of matter in spatially inhomogeneous solutions. Consider a thin test tube of cross-sectional area dA and length L filled in half by a concentrated water solution of sugar, and in half by pure water (Figure 2.1). Obviously, this initial configuration does not last, and with time the sugar redistributes itself evenly throughout the entire test tube. To see how it happens, let us represent the physical system by a series of histograms whose heights are proportional to the numbers of sugar molecules, $N(x, t)$, contained in cylindrical sections between locations x and $x + \Delta x$ at a given time, t. Starting from the initial configuration ($N(x, 0) = N_0$ for $x < 0$ and $N(x, 0) = 0$ for $x > 0$), we then let each molecule perform its Brownian jiggle by taking a step along the x coordinate either to the left (with probability $p = {}^1\!/_2$) or to the right (also with $p = {}^1\!/_2$). While in the homogeneous region $x < 0$ this procedure has initially little effect since the neighboring particles simply exchange their positions, there is a significant net flow of sugar molecules near the boundary between the two regions, $x = 0$. When the procedure is repeated many times, more and more sugar molecules migrate to the right and eventually the mixture becomes homogeneous. Overall, the microscopic motions of individual particles give rise to a macroscale process, which we call diffusion.

To cast diffusion into a functional form, we first note that the number of molecules crossing the $x = 0$ plane per unit time steadily decreases as the heights of

Chemistry in Motion: Reaction–Diffusion Systems for Micro- and Nanotechnology Bartosz A. Grzybowski

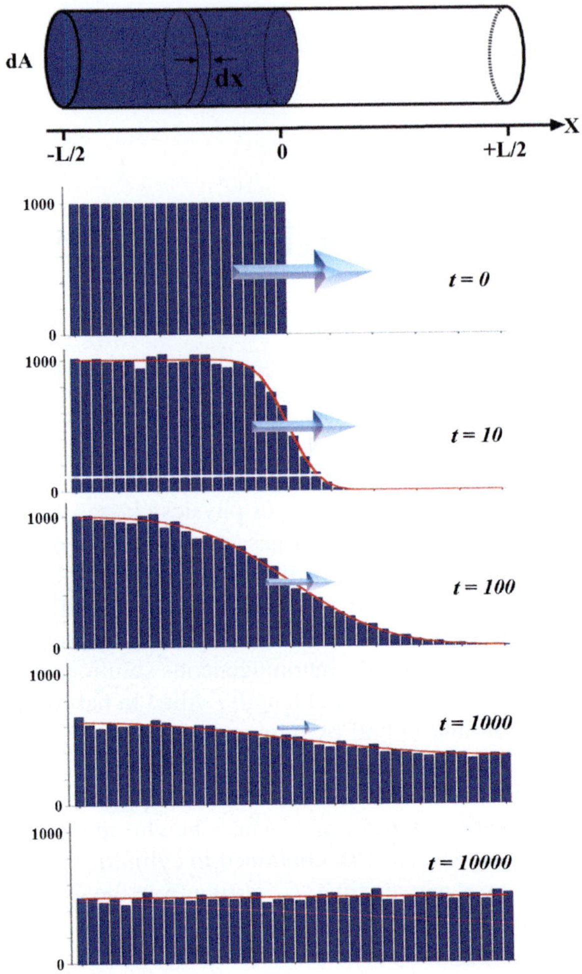

Figure 2.1 Diffusion in a thin tube. Heights of the histograms correspond to the numbers of particles (here, $N(x, 0) = 1000$ for $x < 0$) in discretized slices of thickness dx. Time, t, corresponds to simulation steps, in which each particle performs one Brownian jiggle. Blue arrows illustrate that the diffusive flux across $x = 0$ diminishes with time, when gradients become less steep. Finally, red curves give analytic solutions to the problem

our histograms even out. In other words, the more 'shallow' the concentration variations, the smaller the net flow of matter due to diffusion. This observation was first quantified by Adolf Fick (a German physiologist and, interestingly, inventor of the contact lens) who in 1855 proposed the phenomenological law that today bears his name. According to this law, the number of molecules diffusing through a unit surface area per unit time – the so-called diffusive flux – is linearly proportional to the local slope of the concentration variation. For our discretized test tube system, the flux is along the x-direction and can be written as $j_x(x,t) \propto (N(x+\Delta x,t) - N(x,t))/\Delta x$ or, in terms of experimentally more convenient concentrations (i.e., amounts of substance per unit volume), as $j_x(x,t) \propto (c(x+\Delta x,t) - c(x,t))/\Delta x$. Taking the limit of infinitesimally small Δx and introducing a positive proportionality constant, D, which we will discuss in more detail later, Fick's first law of diffusion for a one-dimensional system becomes $j_x(x,t) = -D[\partial c(x,t)/\partial x]$, where the minus sign sets the directionality of the flow (e.g., in our case, $\partial c(x,t)/\partial x < 0$ and the flux is 'positive', to the right). In three dimensions, matter can flow along all coordinates and the diffusive flux is a vector, which in Cartesian coordinates can be written as

$$\boldsymbol{j}(x,y,z,t) = -D\left(\frac{\partial c}{\partial x}\boldsymbol{e}_x, \frac{\partial c}{\partial y}\boldsymbol{e}_y, \frac{\partial c}{\partial z}\boldsymbol{e}_z\right) = -D\nabla c(x,y,z,t) \tag{2.1}$$

where the second equality simplifies the notation by the use of the gradient operator.

The first law by itself is of rather limited practical value since measuring fluxes is much harder than measuring concentrations. To overcome this problem, we make use of the fact that, during a purely diffusive process, molecules redistribute themselves in space, but their number remains constant. Let us again consider an infinitesimally thin section of our test tube between x and $x + \mathrm{d}x$. As noted above, the number of sugar molecules in this region can only change due to the diffusive transport through the right boundary at $x + \mathrm{d}x$ or the left boundary at x. Specifically, the number of molecules leaving the cylindrical volume element per unit time through the right boundary is $j_x(x+\mathrm{d}x,t)\mathrm{d}A$ and that entering through the left boundary is $j_x(x,t)\mathrm{d}A$. Expanding the value of flux at $x + \mathrm{d}x$ in the Taylor series (to first order), the net rate of outflow is then

$$(j_x(x+\mathrm{d}x,t) - j_x(x,t))\mathrm{d}A = \frac{\partial(j_x(x,t))}{\partial x}\mathrm{d}x\mathrm{d}A \tag{2.2}$$

Now, this outflow has to be equal to the negative of the overall change in the number of molecules in our cylindrical volume element, $-\partial N(x,t)/\partial t$. Equating these two expressions and noting that

$$c(x,t) = \lim_{\mathrm{d}x \to 0} \frac{N(x,t)}{\mathrm{d}A\mathrm{d}x} \tag{2.3}$$

we arrive at the partial differential equation relating spatial variations in the flux to the temporal variation of concentration:

$$\frac{\partial(j_x(x,t))}{\partial x} + \frac{\partial c(x,t)}{\partial t} = 0 \tag{2.4}$$

This derivation can be easily extended to three dimensions, and written concisely with the use of the divergence operator as

$$\nabla \cdot \boldsymbol{j}(x, y, z, t) + \frac{\partial c(x, y, z, t)}{\partial t} = 0 \tag{2.5}$$

The final form of the diffusion equation is now easily obtained by substituting Fick's first law for $\boldsymbol{j}$. In one dimension,

$$\frac{\partial}{\partial x}\left(-D\frac{\partial c(x, t)}{\partial x}\right) + \frac{\partial c(x, t)}{\partial t} = 0 \tag{2.6}$$

and assuming that the diffusion coefficient is constant:

$$\frac{\partial c(x, t)}{\partial t} = D\frac{\partial^2 c(x, t)}{\partial x^2} \tag{2.7}$$

Similarly, in three dimensions and in Cartesian coordinates:

$$\begin{aligned}\frac{\partial c(x, y, z, t)}{\partial t} &= D\left(\frac{\partial^2 c(x, y, z, t)}{\partial x^2} + \frac{\partial^2 c(x, y, z, t)}{\partial y^2} + \frac{\partial^2 c(x, y, z, t)}{\partial z^2}\right) \\ &= D\nabla^2 c(x, y, z, t)\end{aligned} \tag{2.8}$$

where we have simplified the notation by introducing the so-called Laplace operator, ∇^2.

2.2 SOLVING DIFFUSION EQUATIONS

Mathematically, Equation (2.8) is a second-order partial differential equation (PDE), whose general solutions can be found by several methods. As with every PDE, however, the knowledge of a general solution is not automatically equivalent to solving a physical problem of interest. To find the 'particular' solution describing a given system (e.g., our test tube), it is necessary to make the general solution congruent with the boundary and/or initial conditions. By matching the general solution with these 'auxiliary' equations, we then identify the values of parameters specific to our problem. Of course, this procedure is sometimes tedious, but in many physically relevant cases it is possible to use symmetry arguments and/or geometrical simplifications to obtain exact solutions. And even if this fails, one can always seek numerical solutions (Chapter 4), which – though not as intellectually exciting – are virtually always guaranteed to work. In this section, we will focus on some analytically tractable problems and will discuss two general strategies of solving them.

2.2.1 Separation of Variables

Separation of variables is probably the simplest and most widely used method for solving various types of linear PDEs, including diffusion, heat or wave equations.

Although it does not work for all specific problems, it is usually the first option to try.

Let us apply this method to a two-dimensional diffusive process that has reached its steady, equilibrium profile. In that case, the concentration no longer evolves in time, and the diffusion equation simplifies to the so-called Laplace equation:

$$\frac{\partial^2 c(x,y)}{\partial x^2} + \frac{\partial^2 c(x,y)}{\partial y^2} = 0 \tag{2.9}$$

Let the domain of the process be a square of side L (i.e., $0 \le x \le L$ and $0 \le y \le L$), and the boundary conditions be such that the concentration is zero on three sides of this square and equal to some function $f(x)$ along the fourth side: $c(0,y) = 0$, $c(L,y) = 0$, $c(x,0) = 0$, $c(x,L) = f(x)$ (Figure 2.2).

To solve this problem, we *postulate* the solution to be a product of two functions, each dependent on only one variable, $c(x,y) = X(x)Y(y)$. By substituting this 'trial' solution into the diffusion equation, we obtain $X''(x)Y(y) + X(x)Y''(y) = 0$, where single prime denotes first derivative with respect to the function's arguments, and double prime stands for the second derivative. Rearranging, we obtain

$$\frac{Y''(y)}{Y(y)} = -\frac{X''(x)}{X(x)} \tag{2.10}$$

Note that the variables are now *separated* in the sense that the left-hand side of (2.10) depends only on y, while the right-hand side depends only on x. The crucial insight at this point is to recognize that in order for two functions depending on different

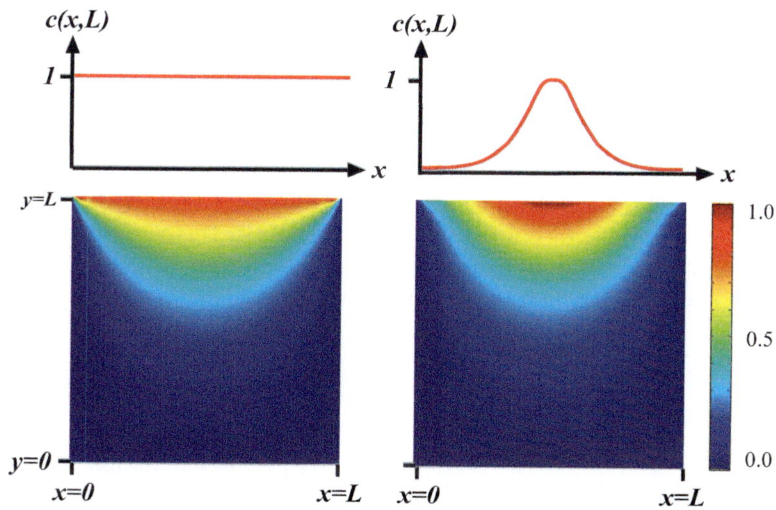

Figure 2.2 Steady-state concentration profiles over a square domain for (left) $c(x,L) = f(x) = 1$ and (right) $c(x,L) = f(x) = \exp(-(x - L/2)^2/b)$

variables to be always equal, both of these functions must be equal to the same, constant value, λ:

$$\frac{Y''(y)}{Y(y)} = \lambda = -\frac{X''(x)}{X(x)} \tag{2.11}$$

This operation allows us to rewrite the original PDE as two ordinary differential equations (ODEs), each of which can be solved separately:

$$X''(x) + \lambda X(x) = 0 \quad \text{and} \quad Y''(y) - \lambda Y(y) = 0 \tag{2.12}$$

The first equation has a general solution $X(x) = A\cos\sqrt{\lambda}x + B\sin\sqrt{\lambda}x$, where A and B are some constants. The particular solution is found by making use of the boundary conditions along the x direction. Specifically, because at $x=0$, $C(0, y)=0$, it follows that $X(0)Y(y)=0$, so that either $X(0)=0$ or $Y(y)=0$. The second condition is trivial and not acceptable since the concentration profile would then loose all dependence on y. Therefore, we use $X(0)=0$ to determine $A=0$. The same argument can be used for the other boundary condition, $c(L, y)=0$, so that $X(L)=0$. This gives $B\sin\sqrt{\lambda}L = 0$, which is satisfied if $\sqrt{\lambda}L = n\pi$ or, equivalently, $\lambda = n^2\pi^2/L^2$ with $n=0, 1, 2 \ldots \infty$. With these simplifications, the solution for a given value of n becomes $X_n(x) = B_n\sin(n\pi x/L)$. The second equation in (2.12) is derived by similar reasoning. Here, the general solution is given by $Y(y) = E\exp(\sqrt{\lambda}y) + F\exp(-\sqrt{\lambda}y)$ and the boundary condition at $y=0$, $c(x,0)=0$, leads to $Y(0)=0$ and $E=-F$. Substituting for λ, we then obtain $Y_n(y) = E_n(\exp(n\pi y/L) - \exp(-n\pi y/L)) = 2E_n\sinh(n\pi y/L)$. Finally, multiplying the particular solutions and noting that the product $X_n(x)Y_n(y)$ satisfies the diffusion equation for all values of n, we can write the expression for the overall concentration profile as the following series:

$$c(x, y) = \sum_{n=1}^{\infty} G_n\sin(n\pi x/L)\sinh(n\pi y/L) \tag{2.13}$$

in which $G_n=2B_nE_n$ and the summation starts from $n=1$ since $c(x, y)$ is equal to zero for $n=0$. Our last task is now to find the values of G_n. This can be done by using the remaining piece of information specifying the problem – namely, boundary condition $c(x, L)=f(x)$. Substituting into (2.13), we obtain

$$\sum_{n=1}^{\infty} G_n\sin(n\pi x/L)\sinh(n\pi) = f(x) \tag{2.14}$$

In order to proceed further, we need to recognize the last expression as a Fourier series and recall that the 'basis' sine functions in this series are mutually orthogonal. Mathematically, this orthogonality condition is expressed as $\int_0^L \sin(n\pi x/L)\sin(m\pi x/L)dx = (L/2)\delta(m, n)$, where the Kronecker delta

$\delta(m, n) = 1$ if $n = m$ and 0 otherwise. With this relation at hand, we can multiply both sides of (2.14) by $\sin(m\pi x/L)$ and integrate over the entire domain $(0 \leq x \leq L)$. Because the only term that survives on the left-hand side is that for $m = n$, we can now find G_m

$$G_m = \frac{2}{\sinh(m\pi)L} \int_0^L \sin(m\pi x/L) f(x) \mathrm{d}x \tag{2.15}$$

and, finally, write the full solution to our diffusion problem as

$$c(x, y) = \sum_{n=1}^{\infty} \left(\frac{2}{\sinh(n\pi)L} \int_0^L \sin(n\pi x/L) f(x) dx \right) \sin(n\pi x/L) \sinh(n\pi y/L) \tag{2.16}$$

The difficulty in evaluating this expression lies mostly in its definite integral part. In some cases, this integral can be found analytically – for example, with $f(x) = 1$, (2.16) simplifies to

$$c(x, y) = \sum_{n=1,3,5\ldots}^{\infty} \frac{4}{n\pi \sinh(n\pi)} \sin(n\pi x/L) \sinh(n\pi y/L) \tag{2.17}$$

and gives the concentration profile shown on the left side of Figure 2.2. For more complex forms of $f(x)$, it is always possible to find $c(x, y)$ from Equation (2.16) by using one of the computational software packages (e.g., Matlab or Mathematica). For instance, the right side of Figure 2.2 shows the $c(x, y)$ contour for a Gaussian function of $f(x) = \exp(-(x - L/2)^2/b)$ with $b = 0.2$ generated using Matlab.

The above example permits some generalizations. First, the solutions of diffusive problems obtained by the separation of variables are always written in the form of Fourier/eigenfunction series. What changes from problem to problem are the details of the expansion and possibly the basis functions. The latter can happen, for example, when $B = 0$ and A is nonzero. In that case, the series expansion would be in the form of cosine functions, and in order to get the expansion coefficients, G_n, one would have to use the orthogonality property

$$\int_0^L \cos(n\pi x/L) \cos(m\pi x/L) \mathrm{d}x = \begin{cases} 0 & \text{if } m \neq n \\ L/2 & \text{if } m = n \neq 0 \\ L & \text{if } m = n = 0 \end{cases} \tag{2.18}$$

(for further discussion and other orthogonality expressions, see Ref. 1 and Ref. 2).

Second, the method is extendable to problems with several nonzero boundary conditions imposed simultaneously (e.g., $c(0, y) = f_1(y)$, $c(L, y) = f_2(y)$,

$c(x,0)=f_3(x)$, $c(x,L)=f_4(x)$). The algorithmic approach here is to solve the 'sub-problem' cases with only one of the nonzero conditions present while setting others to zero (e.g., $c(0,y)=f_1(y)$, $c(L,y)=0$, $c(x,0)=0$, $c(x,L)=0$). The overall solution is then obtained by simply summing up the solutions to these sub-problems.

Third, and probably most important, the method is certainly not limited to steady-state problems. Since extension to time-dependent problems requires some mathematical finesse (or at least some 'tricks'), we will now discuss the issue in more detail.

Let us return to the example of two-dimensional diffusion on a square domain. We will now consider a time-dependent case described by the equation

$$\frac{\partial c(x,y,t)}{\partial t}=D\left(\frac{\partial^2 c(x,y,t)}{\partial x^2}+\frac{\partial^2 c(x,y,t)}{\partial y^2}\right) \tag{2.19}$$

with boundary conditions $c(0,y,t)=0$, $c(L,y,t)=0$, $c(x,0,t)=0$, $c(x,L,t)=f(x)=1$ and an initial condition $c(x,y,0)=0$. First, we note that, in its present form, the problem cannot be solved by the separation of variables. Although we could write $c(x,y,t)=X(x)Y(y)T(t)$, separate the variables into

$$\frac{T'(t)}{T(t)}-\frac{Y''(y)}{Y(y)}=\frac{X''(x)}{X(x)}=-\lambda \tag{2.20}$$

and then solve for $X_n(x)=B_n\sin(n\pi x/L)$ with $\lambda=n^2\pi^2/L^2$, the rest of the procedure would not work. This is so because after substituting the solution for X to obtain

$$\frac{T'(t)}{T(t)}+\frac{n^2\pi^2}{L^2}=\frac{Y''(y)}{Y(y)}=-\beta \tag{2.21}$$

(where β is some constant), and writing out the general solution for $Y(y)=E\sin(\sqrt{\beta}y)+F\cos(-\sqrt{\beta}y)$, we would have to determine β using the relevant boundary conditions. In the previous example, this last step was feasible, because both boundary conditions were homogeneous (i.e., equal to zero; cf. discussion after Equation (2.12)). Now, however, one boundary condition is nonzero, and β cannot be determined independently of the unknown constants E and F. Specifically, although $Y(0)=0$ gives neatly $F=0$, condition $Y(L)=1$ leaves us with one equation for two unknowns, $E\sin(\sqrt{\beta}L)=1$. We have run up against a problem.

The way out of this predicament is to play with the variables and boundary conditions slightly to first make the latter homogeneous and only then solve the diffusion equation. Let us define a new variable $\theta(x,y,t)=c(x,y,t)-c_{ss}(x,y)$, in which $c_{ss}(x,y)$ stands for the steady-state solution we obtained in Equation (2.16). Substituting this into (2.19), we get

$$\frac{\partial\{\theta(x,y,t)+c_{ss}(x,y)\}}{\partial t}=\frac{\partial^2\{\theta(x,y,t)+c_{ss}(x,y)\}}{\partial x^2}+\frac{\partial^2\{\theta(x,y,t)+c_{ss}(x,y)\}}{\partial y^2} \tag{2.22}$$

Because, by definition of the steady-state solution $\partial c_{ss}(x,y)/\partial t=0$, and because $c_{ss}(x,y)$ must obey (2.9), we can simplify the diffusion equation to

$$\frac{\partial\theta(x,y,t)}{\partial t}=\frac{\partial^2\theta(x,y,t)}{\partial x^2}+\frac{\partial^2\theta(x,y,t)}{\partial y^2} \tag{2.23}$$

with the following transformed initial and boundary conditions:

$$\theta(x,y,0)=-c_{ss};\ \ \theta(0,y,t)=0;\ \ \theta(L,y,t)=0;\ \ \theta(x,0,t)=0;\ \ \theta(x,L,t)=0 \tag{2.24}$$

The purpose of the transformation of variables becomes obvious if we note that the problem has now been changed to one in which the initial condition is nonhomogeneous, but all boundary conditions are equal to zero. In this case, we can easily solve (2.21) to obtain

$$\begin{aligned}
X_n &= A_n\sin(n\pi x/L), \quad n=1,2,3\ldots\\
Y_m &= E_m\sin(m\pi y/L), \quad m=1,2,3\ldots\\
T_{nm} &= \Omega_{nm}e^{-(m^2+n^2)\pi^2 t/L^2}
\end{aligned} \tag{2.25}$$

and the combined solution for $\theta(x,y,t)$

$$\theta(x,y,t)=\sum_{n=1}^{\infty}\sum_{m=1}^{\infty}G_{nm}\sin(n\pi x/L)\sin(m\pi y/L)e^{-(m^2+n^2)\pi^2 t/L} \tag{2.26}$$

where coefficients $G_{nm}=A_nE_m\Omega_{nm}$ are determined by applying the initial condition and the orthogonal property of the sine functions (cf. Equations (2.13)–(2.16)). The final, time-dependent solution is found by using the expression for $c_{ss}(x,y)$ from Equation (2.17):

$$\begin{aligned}
c(x,y,t) &= \theta(x,y,t)+c_{ss}(x,t)\\
&= \sum_{n=1,3,5\ldots}^{\infty}\sum_{m=1}^{\infty}\frac{8(-1)^m m}{n\pi^2(n^2+m^2)}\sin(n\pi x/L)\sin(m\pi y/L)e^{-(m^2+n^2)\pi^2 t/L^2}\\
&\quad+\sum_{n=1,3,5..}^{\infty}\frac{4}{n\pi\sinh(n\pi)}\sin(n\pi x/L)\sinh(n\pi y/L)
\end{aligned} \tag{2.27}$$

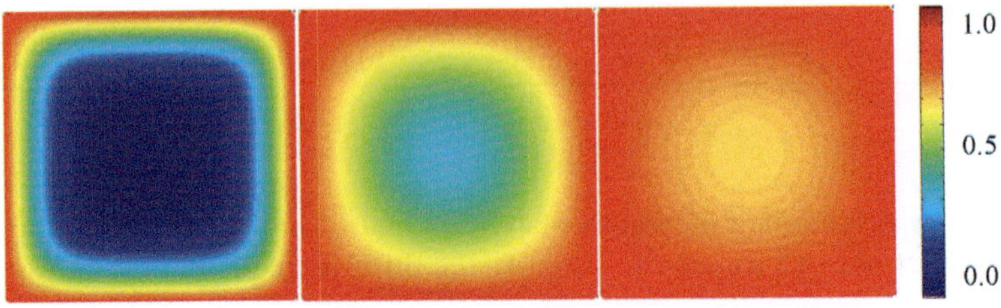

Figure 2.3 Time-dependent concentration profiles for a diffusive process initiated from the boundaries of a square kept at constant concentration equal to unity and with initial concentration in the square set to zero. The snapshots shown here correspond to times $t=0.01$, $t=0.05$, $t=0.1$

To put this exercise into a wider context, the procedure of eliminating nonhomogeneous boundary conditions by the use of an 'auxiliary' steady-state solution is a general approach to solving time-dependent diffusion equations by the separation of variables method. In more complex cases where several boundary conditions are nonzero, the overall strategy is the same, but one has to first 'split' the problem of interest into sub-problems each with only one nonzero boundary condition (as in (2.22)), then solve these sub-problems separately (as in (2.27)), and finally add up their solutions. For example, if the boundary conditions were $c(0,y,t)=1, c(L,y,t)=1, c(x,0,t)=1, c(x,L,t)=1$, one would have to solve four problems of type (2.23), in each of them taking a different solution to the steady-state case $c_{ss}^{i}(x,y)$, $i=1, 2, 3, 4$, with one nonzero boundary condition (i.e., in shorthand, $c(0,y)=\delta(i,1)$, $c(L,y)=\delta(i,2)$, $c(x,0)=\delta(i,3)$, $c(x,L)=\delta(i,4)$). The colorful plots in Figure 2.3 illustrate the end result of this procedure for different values of time.

2.2.2 Laplace Transforms

Methods based on the Laplace transform, L, are less 'intuitive' and popular than the separation of variables technique, but are very handy when solving problems with known initial rather than boundary conditions. One typical case here is when the domain of the diffusive process is infinite (e.g., diffusion in an infinitely long tube), and the solution cannot be expanded into a series of finite-period eigenfunctions (cf. Section 2.2.1).

The Laplace transform is a mathematical operation defined as

$$\hat{f}(s)=L\{f(t)\}=\int_{0}^{\infty}\exp(-st)f(t)\mathrm{d}t \tag{2.28}$$

that converts a function f depending on variable t into a new function $\hat{f}$ depending on variable s. Since evaluation of the transform integrals is not always trivial, it is usually more convenient to look up one of the available tables for the transforms of

Table 2.1 Transforms of common functions pertinent to diffusive problems

Function	Transform
1	$1/s$
$t^{n-1}/(n-1)!$	$1/s^n$
$\exp(-at)$	$1/(s+a)$
$\sin(at)$	$a/(s^2+a^2)$
$\cos(at)$	$s/(s^2+a^2)$
$t^{-1/2}\exp(-a^2/4t)$	$\sqrt{\pi/s}\exp(-a\sqrt{s})$
$\frac{a}{2t^{3/2}}\exp(-a^2/4t)$	$\sqrt{\pi}\exp(-a\sqrt{s})$

common functions and for useful transform properties. Some examples pertinent to diffusive problems are given in Tables 2.1 and 2.2 (for further information, see Ref. 3).

The main idea behind the use of Laplace transforms for solving differential equations is that they often can transform PDEs into easier-to-solve ODEs. In a typical procedure, one applies L to both sides of the original equation, solves the simple ODE in transformed variables and then transforms the solution back – via a unique, inverse transform, $f(t) = L^{-1}\{\hat{f}(s)\}$ – into the original coordinates. While this procedure cannot deal with all types of equations/functions (especially those for which analytical expressions of L are not available), it yields short and often elegant solutions to seemingly difficult problems. Before we tackle such problems, let us first practice the use of L and L^{-1} on a rather trivial ODE.

Consider the differential equation $\mathrm{d}y(t)/\mathrm{d}t + y(t) = 1$ with initial condition $y(0) = 0$. To solve this problem by Laplace transform, we first make use of the differentiation property from Table 2.2 to write $s\hat{y}(s) - y(0) + \hat{y}(s) = s^{-1}$.

Table 2.2 Transforms of common functions pertinent to diffusive problems

Property	Transform
Linearity	$L\{c_1 f(t) + c_2 g(t)\} = c_1\hat{f}(s) + c_2\hat{g}(s)$
Translation	$L\{\exp(at)f(t)\} = \hat{f}(s-a)$
Differentiation	$L\{\mathrm{d}f/\mathrm{d}t\} = s\hat{f}(s) - f(0)$
	$L\{\mathrm{d}^2 f/\mathrm{d}t^2\} = s^2\hat{f}(s) - sf(0) - \mathrm{d}f(s)/\mathrm{d}t\vert_{s=0}$
	$L\{t^n f(t)\} = (-1)^n \mathrm{d}^n\hat{f}(s)/\mathrm{d}s^n$
Integration	$L\left\{\int_0^t f(\xi)\mathrm{d}\xi\right\} = \frac{1}{s}\hat{f}(s)$
Convolution	$L\left\{\int_0^t f(t-\xi)g(\xi)\mathrm{d}\xi\right\} = L\left\{\int_0^t f(\xi)g(t-\xi)\mathrm{d}\xi\right\} = \hat{f}(s)\hat{g}(s)$

The initial condition neatly eliminates $y(0)$ and after rearranging we get $\hat{y}(s) = s^{-1}(s+1)^{-1}$. For reasons that will become obvious shortly, it is convenient to write this equation as $\hat{y}(s) = \hat{f}(s)\hat{g}(s)$ with $\hat{f}(s) = s^{-1}$ and $\hat{g}(s) = (s+1)^{-1}$. Now comes the crucial point of the solution, when we apply the inverse Laplace transform to get back to the original variable, t: $y(t) = L^{-1}\{\hat{y}(s)\} = L^{-1}\{\hat{f}(s)\hat{g}(s)\} = \int_0^t f(t-\xi)g(\xi)d\zeta$, where we used the convolution property (Table 2.2). The last thing to note is that for $\hat{f}(s) = s^{-1}, f(t-\xi) = 1$ and for $\hat{g}(s) = (s+1)^{-1}$, $g(\xi) = \exp(-\xi)$ (cf. Table 2.1) so that the last integral simplifies to $y(t) = \int_0^t \exp(-\xi)d\zeta = 1 - \exp(-t)$, which is the solution to our problem.

As you see, the method works almost automatically: transform, rearrange, inverse transform; all along, use appropriate tables. Let us now see whether the same scheme can prove successful with a less trivial problem of time-dependent diffusion in a thin, infinitely long tube filled with water. At time $t=0$, a thin plug of dye (in total amount M) is introduced at location $x=0$, from which it subsequently diffuses out in both directions (Figure 2.4). Our task is to find the concentration profiles as a function of both t and x. Since the problem is symmetric around $x=0$, it is sufficient to solve the governing, one-dimensional diffusion equation over a semi-infinite domain, $x>0$. The diffusion equation $\partial c/\partial t = D\partial^2 c/\partial x^2$ has the initial condition $c(x,0) = 0$ and boundary conditions $c(\infty, t) = 0$ and $\int_0^\infty c dx = M/2$. The last equation comes from mass conservation, since the total amount of dye at all times must be conserved and, by symmetry, only half of the dye is present in the $x \geq 0$ domain.

We now perform the Laplace transform on both sides of the equation and on the boundary conditions. Importantly, we note that when L acts on a function of x only or on a partial derivatives with respect to x (i.e., derivatives at constant t), the result is greatly simplified. For example, using definition (2.28), we can write

$$L\left\{\frac{\partial^2 c(x,t)}{\partial x^2}\right\} = \int_0^\infty \frac{\partial^2 c(x,t)}{\partial x^2}\exp(-st)dt = \int_0^\infty \frac{\partial^2}{\partial x^2}[c(x,t)\exp(-st)]dt$$

$$= \frac{\partial^2}{\partial x^2}\int_0^\infty [c(x,t)\exp(-st)]dt = \partial^2 L\{c(x,t)\}/\partial x^2 = \partial^2 \hat{c}(x,s)/\partial x^2 \qquad (2.29)$$

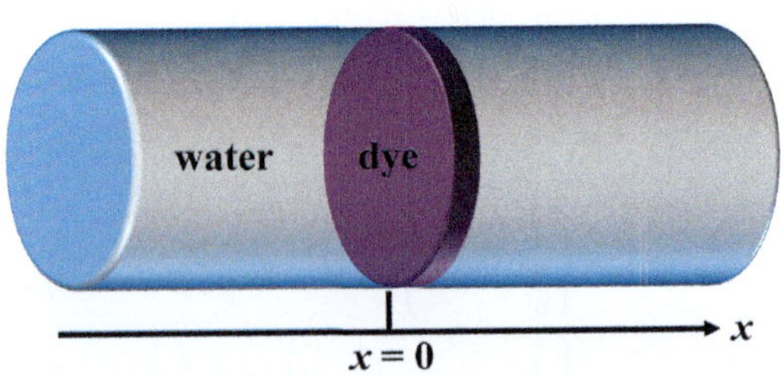

Figure 2.4 Time-dependent diffusion in a thin, infinitely long tube filled with water

By virtue of this property, the transformed diffusion equation becomes an ordinary differential equation $s\hat{c}(x,s) - c(x,0) = Dd^2\hat{c}(x,s)/dx^2$, which using $c(x,0) = 0$ simplifies to $d^2\hat{c}(x,s)/dx^2 - (s/D)\hat{c}(x,s) = 0$. The transformed initial/boundary conditions are $\hat{c}(x,0) = 0$, $\hat{c}(\infty,s) = 0$, $\int_0^\infty \hat{c}\,\mathrm{d}x = M/2s$, where in the last equation we made use of the fact that integration is over variable x only. This ODE problem has a general solution $\hat{c}(x,s) = A\exp(\sqrt{s/D}x) + B\exp(-\sqrt{s/D}x)$. Because $\hat{c}(\infty,s) = 0$, we get $A = 0$. Similarly, using the integral condition for the conservation of mass, it is easy to show that $B = M/2\sqrt{Ds}$ and $\hat{c}(x,s) = (M/2\sqrt{Ds})\exp(-\sqrt{s/D}x)$. Finally, using the penultimate entry in Table 2.1, we inverse-transform into the (x, t) variables to obtain $c(x,t) = (M/2\sqrt{\pi Dt})\exp(-x^2/4Dt)$, which is the solution to our original problem (Figure 2.5).

In summary, the Laplace transform deals swiftly with problems that would be quite hard to solve by other approaches (e.g., seeking self-similar solutions[4]). This method – if applicable to a particular problem – can offer brevity and satisfying mathematical elegance. With some practice in manipulating the forward and backward transforms (see Example 2.1 for some more fun) it can become a powerful tool for solving diffusive problems over infinite domains.

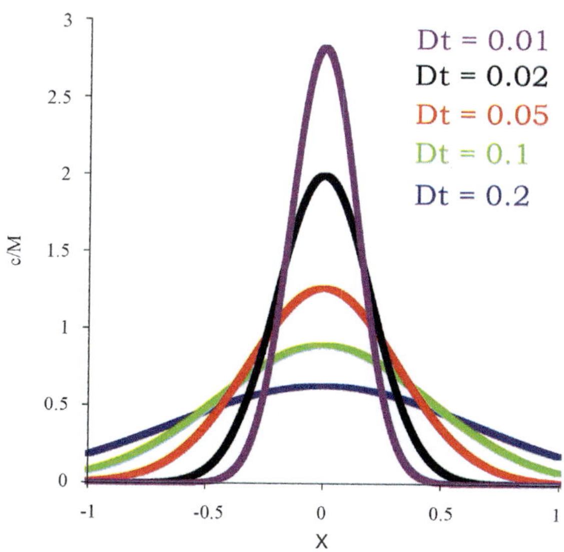

Figure 2.5 Normalized concentration profiles for the diffusion problem with planar source. Concentration slowly levels out with increasing values of *Dt*

Example 2.1 Unsteady Diffusion in an Infinite Tube

Consider a cylindrical tube of infinite length. Initially, the tube is divided into two equal parts by a thin partition located at $x = 0$ (Figure 2.1). The left portion of the tube ($x < 0$) contains an aqueous solution of sugar of concentration C_0; the right portion ($x > 0$) contains pure water. At time $t = 0$, the partition is removed, and sugar molecules start migrating from regions of high concentration to regions of low concentration – here, from left to right. Assuming that transport is purely diffusive (i.e., there are no convective flows) and approximating the tube as one-dimensional (because there is no concentration dependence across tube's cross-section), solve the diffusion equation for the time-dependent concentration profiles within the tube.

Since the problem is one-dimensional, we have $\partial c(x,t)/\partial t = D\partial^2 c(x,t)/\partial x^2$. We will first solve this equation for the pure-water portion of the tube, $x \geq 0$. For the initial/boundary conditions in this region, we have $c(x,0) = 0$ and $c(x \to \infty, t) = 0$, since the concentration far to the right always approaches zero. Because the equation is second order in spatial variables, we need one more boundary condition, which we now find using symmetry arguments. Specifically, we observe that the diffusion problem is physically unchanged, if we rotate the tube by 180° so that diffusion proceeds from right to left. Such process can be described by a new variable $u(x,t) = C_0 - c(x,t)$ – of course, if $c(x,t)$ is a solution of the diffusion equation, so is $u(x,t)$. In particular, the values of both functions at $x = 0$ are equal at all times (by symmetry!), and so $C_0 - c(0,t) = c(0,t)$ or $c(0,t) = C_0/2$, which is our missing condition.

The problem we formulated can now be solved by the Laplace transform method. Taking the Laplace transform of the diffusion equation and boundary/initial conditions, we find

$$\mathrm{d}^2\hat{c}(x,s)/\mathrm{d}x^2 - (s/D)\hat{c}(x,s) = 0 \quad \text{with} \quad \hat{c}(0,s) = C_0/2s, \quad \hat{c}(x \to \infty, s) = 0$$

where $\hat{c}$ denotes the Laplace transform of the concentration, c. Thus, our partial differential equation has been reduced to a second-order ordinary differential equation with general solution $\hat{c}(x,s) = A\exp(\sqrt{s/D}x) + B\exp(-\sqrt{s/D}x)$ where A and B are unknown constants. Applying the boundary conditions, we find $A = 0$ and $B = C_0/2s$ and $\hat{c}(x,s) = (C_0/2s)\exp(-\sqrt{s/D}x)$. Finally, taking the inverse Laplace transform and using the convolution property (Tables 2.1 and 2.2), we find (the reader is encouraged to verify this step) the solution in the original variables:

$$c(x,t) = \int_0^t \frac{C_0 x}{4\sqrt{\pi\tau^3}} \exp\left(-\frac{x^2}{4D\tau}\right) \mathrm{d}\tau$$

Substituting $\eta = x/\sqrt{4D\tau}$ in the integral yields

$$c(x,t) = \frac{C_0}{\sqrt{\pi}} \int_{x/\sqrt{4Dt}}^{\infty} \exp(-\eta^2)\mathrm{d}\eta = \frac{C_0}{\sqrt{\pi}} \left[\int_0^{\infty} \exp(-\eta^2)\mathrm{d}\eta - \int_0^{x/\sqrt{4Dt}} \exp(-\eta^2)\mathrm{d}\eta \right]$$

or

$$c(x,t) = \frac{C_0}{2}\left[1 - \mathrm{erf}\left(\frac{x^2}{\sqrt{4Dt}}\right)\right]$$

Extension of this solution to the $x < 0$ region is left as an exercise for the reader.

Answer. For $x < 0$, $c(x,t) = \frac{C_0}{2}\left[1 - \mathrm{erf}\left(\frac{x^2}{\sqrt{4Dt}}\right)\right]$

2.3 THE USE OF SYMMETRY AND SUPERPOSITION

As illustrated by Examples 2.1 and 2.2, the symmetry of a problem can greatly simplify the procedure of solving diffusion equations. Formally, symmetry is associated with certain operations (e.g., translation, rotation, reflection) under which the modeled system remains unchanged. By identifying such operations, it is often possible to first solve the diffusion equation over a smaller region, and then use it to construct the solution valid for the problem's entire domain.

Example 2.2 Unsteady Diffusion in a Finite Tube

If the tube discussed in Example 2.1 is finite, it is necessary to consider the boundaries at $x = \pm L/2$ explicitly. Of course, the governing equation and the symmetry of the problem remain the same, which enables us to treat each half of the tube independently of the other. Considering the domain $0 \leq x \leq L/2$, the initial and boundary conditions for this case are written $c(x,0) = 0$, $c(0,t) = C_0/2$, $\partial c(x,t)/\partial x|_{L/2} = 0$. Here, only the condition at $x = L/2$ has changed and accounts for the fact that this boundary is impermeable to diffusion – that is, the diffusive flux therein is necessarily zero. Since the domain is finite, we will use the separation of variables for our solution.

First, anticipating further calculations, we rescale the variables according to $\bar{x} = 2x/L$, $\bar{t} = 4Dt/L^2$ and $\bar{c} = (C_0 - 2c)/C_0$. This so-called nondimensionalization procedure allows us to simplify the diffusion equation and make the boundary conditions homogeneous (i.e., equal to zero):

$$\frac{\partial \bar{c}}{\partial \bar{t}} = \frac{\partial^2 \bar{c}}{\partial \bar{x}^2} \quad \mathrm{IC}: \bar{c}(\bar{x},0) = 1; \quad \mathrm{BC}: \bar{c}(0,\bar{t}) = 0, \quad \partial \bar{c}(\bar{x},\bar{t})/\partial \bar{x}|_1 = 0$$

Assuming that the concentration profile can be written as $\bar{c}(\bar{x},\bar{t}) = X(\bar{x})T(\bar{t})$ and substituting into the governing equation, we obtain

$$\frac{1}{T}\frac{\partial T}{\partial \bar{t}} = \frac{1}{X}\frac{\partial^2 X}{\partial \bar{x}^2} = -\kappa$$

Since one side of this equation depends only on x, and the other side only on t, both must be equal to the same constant denoted here as $-\kappa$. Using general solution $X(\bar{x}) = A\sin(\sqrt{\kappa}\bar{x}) + B\cos(\sqrt{\kappa}\bar{x})$ and applying boundary conditions yields $X_n(\bar{x}) = A_n\sin((n+\frac{1}{2})\pi\bar{x})$ where $n=0, 1, 2, \ldots \infty$ and A_n is a set of unknown constants (note that if the boundary conditions were nonhomogeneous, we would not be able to determine κ). Similarly, solving for $T(\bar{t})$ we obtain $T_n(\bar{t}) \propto \exp\left[-(n+\frac{1}{2})^2\pi^2\bar{t}\right]$. The total solution is then given by

$$\bar{c}(\bar{x},\bar{t}) = \sum_{n=0}^{\infty} A_n\sin\left[(n+\tfrac{1}{2})\pi\bar{x}\right]\exp\left[-(n+\tfrac{1}{2})^2\pi^2\bar{t}\right]$$

The unknown coefficients A_n are determined from the initial condition and using the orthogonal property of the sine functions:[1,2]

$$A_n = \left(\int_0^1 \bar{c}(\bar{x},0)\sin\left[(n+\tfrac{1}{2})\pi\bar{x}\right]\mathrm{d}\bar{x}\right) \Big/ \left(\int_0^1 \sin^2\left[(n+\tfrac{1}{2})\pi\bar{x}\right]\mathrm{d}\bar{x}\right) = \frac{2}{\pi(n+\frac{1}{2})}$$

The concentration profile is then written as

$$\bar{c}(\bar{x},\bar{t}) = \frac{2}{\pi}\sum_{n=0}^{\infty}\frac{\sin\left[(n+\frac{1}{2})\pi\bar{x}\right]}{(n+\frac{1}{2})}\exp\left[-(n+\tfrac{1}{2})^2\pi^2\bar{t}\right]$$

or in terms of the 'original' dimensional variables as

$$c(x,t) = \frac{C_0}{2} + \frac{2C_0}{\pi}\sum_{n=0}^{\infty}\frac{\sin[(2n+1)\pi x/L]}{(2n+1)}\exp\left[-(2n+1)^2\pi^2 Dt/L^2\right]$$

Although this solution is derived specifically for the domain $0 \leq x \leq L/2$, it is in fact valid over the entire domain $-L/2 \leq x \leq L/2$ (the proof is left to the reader; hint: the sum in the last equation is an odd function). Also, as might be expected, the solutions for finite and infinite tubes agree quite well for small values of time or for long tube dimension ($\bar{t} \leq 0.1$) but diverge for larger values of $\bar{t}$.

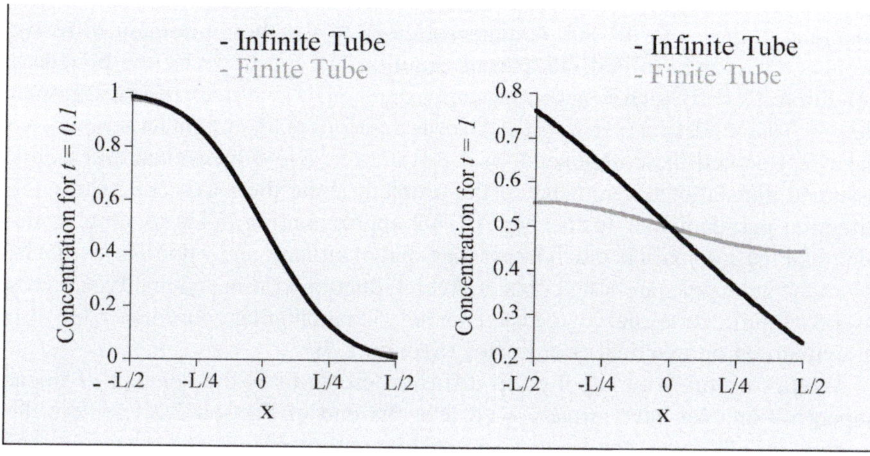

To illustrate this procedure, consider the example shown in Figure 2.6. Here, a small block of agarose gel soaked with substance A is placed onto a larger block of pure gelatin (we will see more of these gels in later chapters). Upon contact, A begins to diffuse into the gelatin along directions indicated by the arrows.[5] By inspection, we immediately see that this problem is invariant with respect to reflection about $x = 0$, and that diffusive profiles in the 'left' and the 'right' subdomains are identical at all times. Therefore, we can first solve the diffusion equation for, say, $x > 0$ and then simply reflect it around the symmetry axis (by

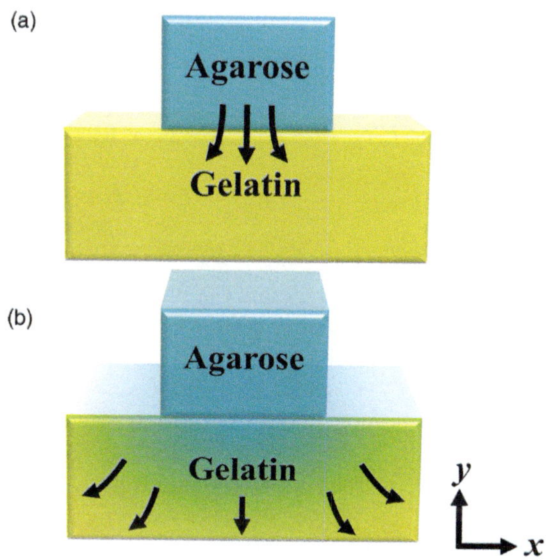

Figure 2.6 Small block of agarose gel soaked with substance A is placed onto a larger block of pure gelatin

changing x to $-x$ in the sub-domain solution) to get the all-domain diffusive profiles. Of course, to find the 'partial' solution, we must specify the boundary condition at $x=0$ which is, not surprisingly, dictated by symmetry (no flux through the $x=0$ plane, $\partial c(x,t)/\partial x|_0 = 0$). Also, as a curious reader might have noted, we have represented the components of the system as two-dimensional and tacitly assumed translational symmetry of the problem along the z axis (i.e., along the direction perpendicular to the page). This approximation holds as long as the length of the z domain is much larger than that of either x and y domains – that is, when the gel blocks are long cuboids. Such reduction of dimensionality is a very useful simplification, and can be used not only in rectangular coordinates, but also in cylindrical or spherical coordinates (Section 2.4).

Another useful 'trick' applied to diffusion problems is the concept of linear superposition. We have already used it at the end of Section 2.2.1, where the original problem was first broken up into four sub-problems, and their solutions were then added up. Such additivity is a general property of linear systems to which the simple diffusion equation belongs. To see superposition at work, consider the following one-dimensional, time-dependent diffusion problem, with constant diffusion coefficient, D:

$$\frac{\partial c(x,t)}{\partial t} = D\frac{\partial^2 c(x,t)}{\partial x^2} \quad \text{BC}: c(-L,t) = C_L,\ c(L,t) = C_R;\ \text{IC}: c(x,0) = C_0 \tag{2.30}$$

We begin by decomposing this problem into two sub-problems, such that the sum of their solutions $c_1(x,t)$ and $c_2(x,t)$ is equal to the solution of (2.30) and satisfies the pertinent boundary and initial conditions. We can write

$$\frac{\partial c_1(x,t)}{\partial t} = D\frac{\partial^2 c_1(x,t)}{\partial x^2} \quad \text{BC}: c_1(-L,t) = C_{L1},\quad c_1(L,t) = C_{R1};\quad \text{IC}: c_1(x,0) = C_{01} \tag{2.31}$$

$$\frac{\partial c_2(x,t)}{\partial t} = D\frac{\partial^2 c_2(x,t)}{\partial x^2} \quad \text{BC}: c_2(-L,t) = C_{L2},\quad c_2(L,t) = C_{R2};\quad \text{IC}: c_2(x,0) = C_{02} \tag{2.32}$$

where $C_{L1} + C_{L2} = C_L$, $C_{R1} + C_{R2} = C_R$ and $C_{01} + C_{02} = C_0$. The key step now is to recognize that the boundary/initial conditions can be set to arbitrary values provided that they sum up to those of the original problem. In particular, it is convenient to choose $C_{L1} = C_L$, $C_{R1} = 0$, $C_{L2} = 0$, $C_{R2} = C_R$ such that each sub-problem now has only one inhomogeneous (nonzero) boundary condition, and can be readily solved by the methods discussed in Section 2.2. For more practice in the use of symmetry and superposition, see Ref. 6 and Ref. 7.

2.4 CYLINDRICAL AND SPHERICAL COORDINATES

So far, we have solved diffusion equation using rectangular coordinates, which are convenient for processes occurring over rectangular or cuboidal domains. In many

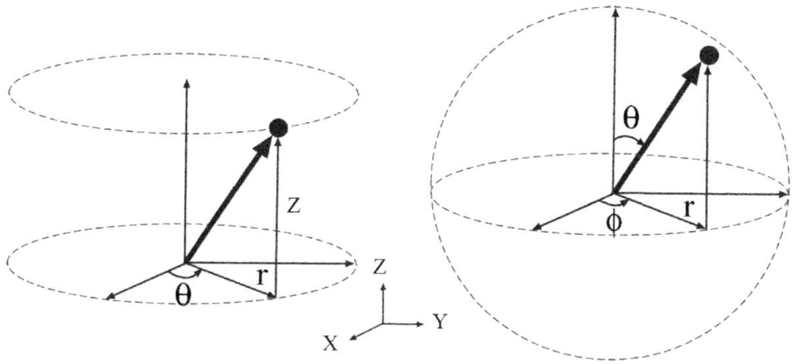

Figure 2.7 Cylindrical (left) and spherical (right) coordinate systems

real-life cases, however, diffusive fields have symmetries such as cylindrical (diffusion of a dye from a long, thin filament) or spherical (diffusion from a spherical drop) for which it is more natural to use (r, θ, z) or (r, θ, ϕ) coordinates, respectively (Figure 2.7). Importantly, a judicious choice of coordinate system can often simplify the problem at hand.

As an example, consider diffusion of a dye initially constrained to a small spherical droplet. Intuitively, we 'feel' that the right coordinate system should reflect the droplet's symmetry – hence, we decide to use spherical coordinates (r, θ, ϕ) in which the diffusion equation is

$$\frac{\partial c(r,\theta,\phi,t)}{\partial t} = D\left(\frac{1}{r^2}\frac{\partial}{\partial r}\left(r^2\frac{\partial c(r,\theta,\phi,t)}{\partial r}\right) + \frac{1}{r^2\sin\theta}\frac{\partial}{\partial\theta}\left(\sin\theta\frac{\partial c(r,\theta,\phi,t)}{\partial\theta}\right) + \frac{1}{r^2\sin^2\theta}\frac{\partial^2 c(r,\theta,\phi,t)}{\partial\phi^2}\right) \tag{2.33}$$

Before we start regretting our choice that has led to such a monstrous equation, let us note that the concentration of the dye at a given distance r does not depend on the angular coordinates (θ, ϕ). Therefore, the partial derivatives of concentration with respect to both of these angles vanish and the equation reduces to a much more manageable form:

$$\frac{\partial c(r,t)}{\partial t} = \frac{D}{r^2}\frac{\partial}{\partial r}\left(r^2\frac{\partial c(r,t)}{\partial r}\right) = \frac{D}{r^2}\left(2r\frac{\partial c(r,t)}{\partial r} + r^2\frac{\partial^2 c(r,t)}{\partial r^2}\right) = D\left(\frac{\partial^2 c(r,t)}{\partial r^2} + \frac{2}{r}\frac{\partial c(r,t)}{\partial r}\right) \tag{2.34}$$

The common and useful 'trick' is now to assume that the solution is of the form $u(r, t) = c(r, t)r$. With this educated guess, the equation transforms to

$$\frac{\partial u(r,t)}{\partial t} = D\frac{\partial^2 u(r,t)}{\partial r^2} \tag{2.35}$$

which is effectively a one-dimensional problem that can be solved as in Example 2.3.

In cylindrical coordinates, the diffusion equation is

$$\frac{\partial c(r,z,\theta,t)}{\partial t} = D\left(\frac{1}{r}\frac{\partial}{\partial r}\left(r\frac{\partial c(r,z,\theta,t)}{\partial r}\right) + \frac{\partial^2 c(r,z,\theta,t)}{\partial z^2} + \frac{1}{r^2}\frac{\partial^2 c(r,z,\theta,t)}{\partial \theta^2}\right) \tag{2.36}$$

and has its own set of useful simplifications. Somewhat surprisingly, however, z and θ are the 'easy' variables and r is the 'difficult' one. For example, if the problem we try to solve has no dependence on r, the first term on the right-hand side vanishes and the equation can be solved by separation of variables (Section 2.2.1). If, however, concentration depends on r, the situation becomes more complicated and the solution often involves Bessel functions. In one illustrative example, we consider the apparently simple problem of finding a steady-state concentration profile around a long gel cylinder with one of its ends immersed in a volatile liquid which diffuses along the cylinder (liquid concentration depends on z) and also evaporates from its surface (r dependence). In this case, the diffusion equation is written in terms of r and z:

$$\frac{1}{r}\frac{\partial}{\partial r}\left(r\frac{\partial c(r,z)}{\partial r}\right) + \frac{\partial^2 c(r,z)}{\partial z^2} = 0 \tag{2.37}$$

We let $c(r, z) = R(r)Z(z)$ and apply separation of variables to obtain

$$\frac{R''}{R} + \frac{R'}{rR} = -\frac{Z''}{Z} = \lambda \tag{2.38}$$

The left-hand side of this expression gives $r^2R'' + rR' + \lambda r^2 R = 0$, whose general solution is $R(r) = AJ_0(\sqrt{\lambda}r) + BY_0(\sqrt{\lambda}r)$, where A and B are some constants found by matching particular boundary/initial conditions, J_0 is the Bessel function of the first kind of order 0 and Y_0 is the Bessel function of the second kind of order 0. The functions look a bit like 'damped' sines and cosines (Figure 2.8) and have

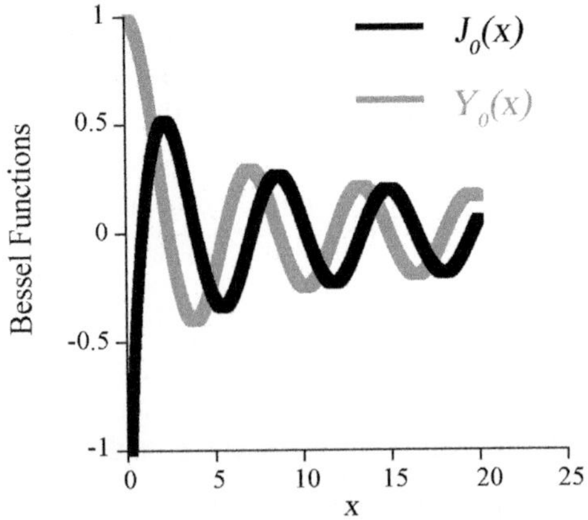

Figure 2.8 Bessel functions

similar orthogonality relationships (e.g., $\int_0^1 J_0(\sqrt{\lambda_n}r)J_0(\sqrt{\lambda_m}r)rdr = 0$ for $m \neq n$). More detailed discussion of their properties and uses in solving diffusion equations in cylindrical coordinates can be found elsewhere.[1,2,7]

To recapitulate, the choice of proper coordinate system reflecting the symmetry of the physical problem is often the key to solving diffusion equations. In many common cases, transformation to spherical/cylindrical coordinates simplifies calculations significantly – and although this does not necessarily mean that these calculations are always trivial, there are usually much less involved than in rectangular coordinates (if you have plenty of time to kill, try solving any of the problems from this section in x, y, z coordinates).

Example 2.3 Is Diffusion Good for Drug Delivery?

Consider a spherical gel capsule of radius a filled uniformly with a concentration C_0 of a new life-saving drug. When ingested, the drug begins to diffuse out of the gel and into the stomach, where it is quickly absorbed into the bloodstream. Therefore, the concentration of the drug outside of the capsule is much less than C_0 and can be safely neglected – i.e., $c(r = a) \ll C_0$ such that $c(r = a) \approx 0$. What is the rate of drug delivery as a function of time?

Because of the spherical symmetry of this problem, there are no concentration gradients in the θ or ϕ directions, and the terms in the diffusion equation involving $\partial c/\partial\phi$ or $\partial c/\partial\theta$ are equal to zero. Initially, the gel capsule ($0 \leq r \leq a$) is filled with a uniform concentration C_0, and the outer boundary is maintained at a constant, zero concentration. At $r = 0$, we require that the concentration be bounded (i.e., it does not diverge to $\pm\infty$). Mathematically, these statements are expressed as follows:

$$\frac{\partial c}{\partial t} = \frac{D}{r^2}\frac{\partial}{\partial r}\left(r^2\frac{\partial c}{\partial r}\right) \quad \text{with} \quad c(r,0) = C_0, \quad c(0,t) \text{ bounded and } c(a,t) = 0$$

Although this problem may look more difficult than those discussed in previous examples, there is a useful trick that can be applied to any spherically symmetric diffusion problem. By simply substituting $u(r, t) = rc(r, t)$, this equation is transformed into the one-dimensional diffusion equation in Cartesian form:

$$\frac{\partial u}{\partial t} = D\frac{\partial^2 u}{\partial r^2} \quad \text{with} \quad u(r,0) = rC_0, \quad u(0,t) = 0 \quad \text{and} \quad u(a,t) = 0$$

Because the boundary conditions are homogeneous, this equation can be solved by separation of variables (Section 2.2.1):

$$c(r,t) = -\frac{2C_0 a}{\pi r}\sum_{n=1}^{\infty}\frac{\cos(n\pi)}{n}\sin\left(\frac{n\pi r}{a}\right)\exp\left(\frac{-(n\pi)^2 Dt}{a^2}\right)$$

From the time-dependent concentration profile $c(r, t)$, we may now find the rate at which the drug is being delivered. This rate – defined as the number of moles, M, of drug molecules leaving the capsule per unit time – is equal to the flux through the interface at $r = a$ multiplied by the area of that interface:

$$\frac{dM}{dt} = \left(-D\frac{\partial c}{\partial r}\bigg|_a \right)(4\pi a^2) = 8\pi a D C_0 \sum_{n=1}^{\infty} \exp\left(\frac{-(n\pi)^2 Dt}{a^2} \right)$$

As illustrated in the rightmost graph below (note the logarithmic y-scale!), this function is infinite at $t = 0$, decreases faster than exponentially for $t < 0.1a^2/D$ and slows to an exponential decay for $t > 0.1a^2/D$, where the higher order terms of the summation become negligible. Ideally, a drug delivery system should release the drug at a constant rate over a prolonged period of time; thus, the system described here, in which the delivery rate decreases exponentially, is clearly not ideal. For modern drug delivery technologies circumventing these diffusional limitations, the reader is referred to Langer.[8]

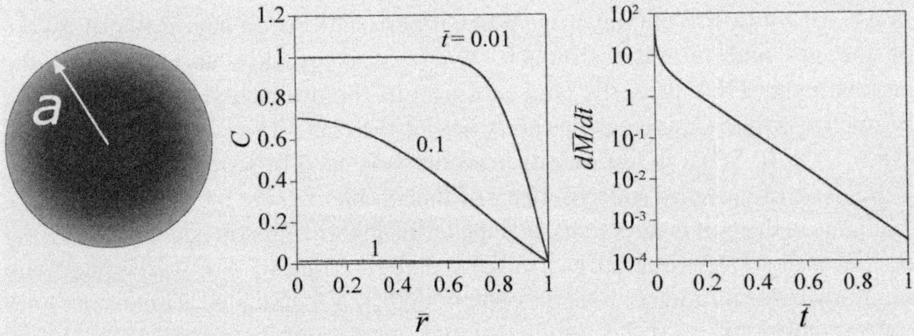

Note. For generality, the graphs use dimensionless variables $\bar{C} = C/C_0$, $\bar{r} = r/a$, $\bar{t} = Dt/a^2$, $\bar{M} = 3M/4\pi a^3 C_0$.

2.5 ADVANCED TOPICS

After some twenty pages of solving diffusion equations in various arrangements, it is probably easy to get somewhat complacent about diffusion. By now we surely know how it works. But do we really? In this section, we will revisit some of the diffusion's basics and also venture into some of its more exotic forms such as subdiffusion and superdiffusion. While this material is not an integral part of the systems we will discuss later in the book, it might come handy in treating problems such as diffusion in nonhomogeneous media or diffusion in biological/cellular systems.

Let us first consider the case of a spatially nonhomogeneous system in which the diffusion coefficient is not constant over the entire domain, but rather varies from one location to another, $D(x, y, z)$. Under these circumstances, the diffusive flux is given by $\vec{j}(x, y, z, t) = -D(x, y, z)\nabla c(x, y, z, t)$. When we substitute this relation into the mass conservation equation

$$\nabla \cdot \vec{j}(x, y, z, t) + \frac{\partial c(x, y, z, t)}{\partial t} = 0, \tag{2.5}$$

we can no longer pull the diffusion coefficient out from the divergence operator. Instead, the diffusion equation becomes

$$\frac{\partial c(x, y, z, t)}{\partial t} = \nabla \cdot (D(x, y, z)\nabla c(x, y, z, t)) \tag{2.39}$$

Even with simple boundary/initial conditions this equation might be hard to solve and has to be treated on a case-by-case basis, often using numerical methods (cf. Chapter 4). At the same time, this formulation of the Fick's law is practically very useful and can be applied to describe diffusion in nonhomogeneously wetted gels, in polymers with spatial gradients of crosslinking, in mixed solids, or in emulsions.

One of the main challenges in modeling these systems is to find the appropriate functional forms of the diffusion coefficient– many of these expressions have been developed heuristically or through simple models (for examples of gel systems, see Amsden[9]). Another important and sometimes confusing point is that the spatial variations in D do not automatically translate into spatial variations in the steady-state concentrations. For instance, consider a one dimensional domain $(0, L)$ over which the diffusion coefficient is a linear function of position, $D(x) = ax$ and the boundary conditions are $c(0) = c(L) = C_O$. When asked about the steady-state concentration profile, the students often answer "intuitively" that the regions of low mobility/small D should "hold" more molecules than the regions of large D. Consequently, the steady-state concentration should increase with decreasing x, possibly dropping to C_0 at $x = 0$ to match the "left" boundary condition. This reasoning is, of course, plain wrong. For the steady-state of this one-dimensional systems, the diffusion equation is

$$0 = \frac{\partial}{\partial x}\left(ax\frac{\partial c(x)}{\partial x}\right) \quad \text{or} \quad x\frac{\partial^2 c(x)}{\partial^2 x} + \frac{\partial c(x)}{\partial x} = 0 \tag{2.40}$$

for which the solution is a uniform concentration profile $c(x) = C_0$. The take home lesson from this example is that while the nonhomogeneity of the diffusion coefficient certainly affects the *kinetics* of the diffusion process, it has nothing to do with the ultimate, steady-state for a closed system at *equilibrium* at which the concentrations are all spatially uniform (more accurately, the chemical potential of the diffusing species are spatially uniform).

We now turn to the subject of anomalous diffusion in which even the use Fick's relation for the diffusive flux ceases to apply. Recall from Section 2.1 that the $\bar{j} = -D\nabla c$ relationship is based on empirical observations rather than derived from first principles. Thus, Fick's law, and the diffusion equation $\partial c/\partial t = D\nabla^2 c$ can only be treated as approximations or limiting cases of more general models. A more fundamental understanding of diffusion can be gained using the concepts from the so called continuous time random walk (CTRW) formalism, which we will outline for a one-dimensional case. In the CTRW approach, a single diffusing particle undergoes a random (stochastic) process whereby it executes a series of "jumps" between which it waits for a prescribed period of time. In general, the length of a given jump and the waiting time between two successive jumps are not fixed quantities but rather random variables drawn from an appropriate joint probability distribution, $\psi(x, t)$, where x is the jump length and t is the waiting time. Furthermore, if the jump length and the waiting time are independent random variables (a common assumption), the joint probability distribution may be decoupled to give $\psi(x, t) = \lambda(x)w(t)$, in which $\lambda(t)$ is the probability distribution for the jump length, and $w(t)$ is the probability distribution for the waiting time. From these functions, the characteristic/expected jump length, Σ, and waiting time, T, are described by the variance and the mean of the respective distributions, $\lambda(t)$ and $w(t)$.

$$\Sigma^2 = \int_{-\infty}^{\infty} x^2\lambda(x)dx \quad \text{and} \quad T = \int_{0}^{\infty} tw(t)dt \tag{2.41}$$

Notice that the characteristic jump length is derived from the variance since the mean jump displacement is zero for an unbiased random walker (i.e., jumps to the right are equally probable as those to the left, see Example 2.4).

Example 2.4 Random Walks and Diffusion

Consider a particle undergoing a one-dimensional random walk, in which it moves a fixed distance, $\pm l$, at each time step, τ. Here, the probability of stepping to the right $(+l)$ is p, and the probability of stepping to the left $(-l)$ is $1-p$. What is the 'root-mean-squared' deviation from the mean displacement after N steps? How is this related to the macroscopic phenomenon of diffusion?

First, let us examine the mean displacement and the variance associated with a single step. The mean displacement per step is simply $\bar{s} = pl + (1-p)(-l) = (2p-1)l$. The variance of s is defined as $\overline{(\Delta s)^2} \equiv \overline{(s-\bar{s})^2}$, and may also be expressed as $\overline{(\Delta s)^2} = \overline{(s^2)} - (\bar{s})^2$. Because $\overline{(s^2)} = pl^2 + (1-p)(-l)^2 = l^2$, it follows that $\overline{(\Delta s)^2} = l^2[1-(2p-1)^2] = 4l^2p(1-p)$. Note that for a symmetric random walker ($p=0.5$), there is no average displacement at each step, $\bar{s} = 0$, and the step size, l, is equal to the square root of the variance, $\Delta s_{\mathrm{rms}} = \sqrt{\overline{(\Delta s)^2}} = l$.

Let us now calculate the mean and the variance of the total displacement after N steps. We first make an important assumption that there are no correlations between consecutive steps – i.e., the direction the particle chooses at time $t = t + \tau$ is *independent* of its choice at an earlier time, t. With this assumption, the total displacement, x, can be treated as the sum of independent random variables s, $x = \sum_{i=1}^{N} s_i$. Taking the mean values of both sides, we obtain $\bar{x} = \sum_{i=1}^{N} \bar{s} = N\bar{s}$; in other words, the mean of N steps is simply N times the mean of each step. To calculate the variance after N steps, $\overline{(\Delta x)^2} \equiv \overline{(x-\bar{x})^2}$, note that $x - \bar{x} = \sum_{i=1}^{N} s_i - \bar{s}$ or in more concise notation $\Delta x = \sum_{i=1}^{N} \Delta s_i$. By squaring both sides, we then obtain

$$(\Delta x)^2 = \left(\sum_{i=1}^{N} \Delta s_i\right)\left(\sum_{j=1}^{N} \Delta s_j\right) = \sum_{i=1}^{N} (\Delta s_i)^2 + \sum_{i=1}^{N}\sum_{j\neq i}(\Delta s_i)(\Delta s_j)$$

and the mean of this equation is

$$\overline{(\Delta x)^2} = \sum_{i=1}^{N} \overline{(\Delta s_i)^2} + \sum_{i=1}^{N}\sum_{j\neq i}\overline{(\Delta s_i)(\Delta s_j)}$$

For the cross terms ($j \neq i$), we make use of the fact that each step is statistically independent, such that $\overline{(\Delta s_i)(\Delta s_j)} = (\overline{\Delta s_i})(\overline{\Delta s_j})$, which is identically zero as $\overline{\Delta s_i} = \bar{s}_i - \bar{s} = 0$. Thus, the variance is given by

$$\overline{(\Delta x)^2} = \sum_{i=1}^{N} \overline{(\Delta s_i)^2} = N\overline{(\Delta s)^2}$$

and

$$\Delta x_{\mathrm{rms}} = \sqrt{\overline{(\Delta x)^2}} = \sqrt{N}\Delta s_{\mathrm{rms}}$$

gives a typical distance from the origin at which a random walker is found after N steps.

We are now in position to relate the physical phenomenon of diffusion to our random walk problem. Recall from the main text that diffusing particles are supposed to 'jiggle randomly' (that is, $p = 0.5$) and that on the microscopic level, the diffusion coefficient, D, can be related to a typical distance traveled by the particle over a given time interval. Since for each step of a symmetric random walker, the step size is $\Delta s_{\mathrm{rms}} = \sqrt{\overline{(\Delta s)^2}} = l$, the diffusion coefficient can be written as $D = (\Delta s_{\mathrm{rms}})^2/2\tau$. Then, using the fact that for N steps, $\Delta x_{\mathrm{rms}} = \sqrt{N}\Delta s_{\mathrm{rms}}$, and identifying $N\tau$ as the total time t of the random walk, we obtain $\Delta x_{\mathrm{rms}} = \sqrt{2Dt}$. This equation tells us that the characteristic distance traveled by a randomly walking particle increases with time as $\sqrt{2Dt}$.

Importantly, an identical result is obtained from the continuous/macroscopic diffusion equation. For a collection of M particles initially located at $x = 0$, this equation yields the time-dependent Gaussian concentration profile, $c(x,t) = (M/\sqrt{4\pi Dt})\exp(-x^2/4Dt)$, whose standard deviation increases with time as $\sigma(t) = \sqrt{2Dt}$. This standard deviation tells us that after time t

from the start of the experiment, a diffusing particle will likely deviate from its mean value of zero by a characteristic displacement $\sqrt{2Dt}$. We note that while the macroscopic description provides only 'average' information, the microscopic, random-walk description can be applied for the rigorous treatment of 'small' systems, in which concentration fluctuations may be significant.

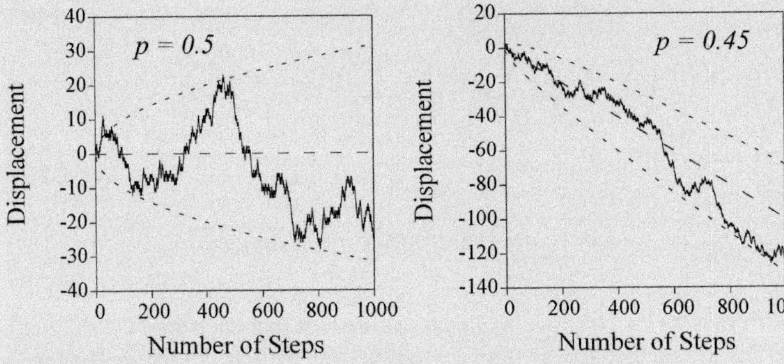

The figure above shows two computer-generated realizations of a symmetric (left) and biased (right) random walks. Dashed lines give the mean values, $\bar{x}$, as a function of the number of steps; dotted lines correspond to $\bar{x} \pm \Delta x_{\mathrm{rms}}$.

Importantly, if both the characteristic jump length, Σ, and waiting time, T, are finite, the long-time limit (i.e., after many jumps), corresponds to the familiar Brownian motion as described by a diffusion equation, $\partial p(x,t)/\partial t = D\partial^2 p(x,t)/\partial x^2$, where $p(x, t)$ is the probability of finding the particle at position x and time t, and the diffusion coefficient is defined as $D = \Sigma^2/2T$. Thus, for many diffusing particles, the concentration $c(x, t)$ is directly analogous to the probability of finding a single particle, $p(x, t)$. A specific example of such a random walk with finite jump lengths and waiting times is discussed in Example 2.4 where we find that the asymptotic form of $p(x, t)$ is a normal distribution, and that the mean-square-displacement of the diffusing particle increases linearly with time as $\langle x^2(t)\rangle = 2Dt$. This linear dependence is the hallmark of "normal" diffusion processes and holds asymptotically regardless of the specific functional forms of $\lambda(t)$ and $w(t)$ provided that Σ and T are finite.

But what happens if these quantities are not finite? For example, when cream is added to a stirred cup of coffee, it disperses much faster than would be predicted by normal diffusion alone. Nevertheless, owing to the chaotic and turbulent flows created by stirring, the motions of the cream "particles" are often well described by the CTRW formalism introduced above. Unlike the case of normal diffusion, however, the characteristic jump length Σ diverges as the particles may execute arbitrarily large displacements. Physically, one may imagine a process whereby the cream "particle" tumbles about locally within a small eddy before taking a large jump from one eddy to another. Such a process known as "superdiffusion" and is characterized by a mean squared displacement that increases faster than linearly, in many cases as a power law $\langle x^2(t)\rangle \sim t^{\alpha}$ with $\alpha > 1$.

Alternatively, if the characteristic waiting time diverges to infinity and Σ is finite, a random walker/particle can actually diffuse slower than expected by normal diffusion. In this case, the mean squared displacement of the particle increases slower than linearly with time as $\langle x^2(t)\rangle \sim t^{\alpha}$ with $\alpha < 1$. Physically, this behavior arises when particles diffuse among "traps" that capture particles and delay them along their otherwise Brownian journey. For example, diffusion of colloidal particles within a polymer network or their self-diffusion near the glass transition both exhibit anomalous subdiffusion due to confinement of the particles in transient "cages" created by the polymer network or neighboring particles.

Finally, if both Σ and T are infinite, the random walk may be super- or subdiffusive depending on the specific forms of the jump probability functions, $\lambda(t)$ and $w(t)$.

In any of the above cases of anomalous diffusion, the diffusion equation (or even the more general Fokker-Planck equation) is no longer appropriate, and more sophisticated mathematical methods based on fractional kinetics are needed to describe the evolution of the probability distribution, $p(x, t)$, of a continuous-time random walker. For example, in the case of subdiffusion, this distribution is governed by the fractional partial differential equation

$$\frac{\partial p(x,t)}{\partial t} = {}_0D_t^{1-\alpha}\kappa_\alpha \nabla^2 p(x,t) \quad \text{for} \quad 0 < \alpha < 1 \quad \text{(subdiffusion)} \tag{2.42}$$

in which κ_α is the generalized diffusion coefficient, and ${}_0D^{1-\alpha}{}_t^{1-\alpha}$ is the Riemann-Liouville operator which accounts for the fractional dependence of the mean squared displacement on time. There exist similar fractional kinetic equations for superdiffusion, for which the mathematics is continuously developing. Details on the derivation of the fractional diffusion equations and their solution can be found in reference[10]. Introductory accounts of super- and subdiffusion can be found in Klafter[11] and Sokolov[12], respectively.

REFERENCES

1. Haberman, R. (2003) *Applied Partial Differential Equations with Fourier Series and Boundary Value Problems,* Prentice Hall, Upper Saddle River, NJ.
2. Asmar, N.H. (2005) *Partial Differential Equations with Fourier Series and Boundary Value Problems,* Prentice Hall, Upper Saddle River, NJ.
3. Abramowitz, M. and Stegun, I.A. (1972) *Handbook of Mathematical Functions with Formulas, Graphs and Mathematical Tables,* Dover Publications, New York.
4. Deen, W.M. (1998) *Analysis of Transport Phenomena,* Oxford University Press, New York.
5. Fialkowski, M., Campbell, C.J., Bensemann, I.T. and Grzybowski, B.A. (2004) Absorption of water by thin, ionic films of gelatin. *Langmuir*, **20**, 3513.
6. Crank, J. (2002) *The Mathematics of Diffusion,* Oxford Science Publications, New York.

7. Farlow, S.J. (1993) *Partial Differential Equations for Scientists and Engineers*, Dover Publications, New York.
8. Langer, R. (1998) Drug delivery and targeting. *Nature*, **392**, 5.
9. Amsden, B. (1998) Solute diffusion within hydrogels. Mechanisms and models. *Macromolecules*, **31**, 8382.
10. Metzler, R. and Klafter, J. (2002) The random walk's guide to anomalous diffusion: a fractional dynamics approach. *Phys. Rep.*, **339**, 1.
11. Klafter, J., Shlesinger, M.F. and Zumofen, G. (1996) Beyond Brownian motion. *Phys. Today*, **49**, 33.
12. Sokolov, I.M., Klafter, J. and Blumen, A. (2002) Fractional kinetics. *Phys. Today*, **55**, 48.

3

Chemical Reactions

3.1 REACTIONS AND RATES

Now that the reader is a seasoned expert on diffusion, it is time to explore the world of reactions. As already discussed in Chapter 2, molecules are very dynamic entities, constantly moving and colliding with their neighbors. This 'aggressive' behavior is the basis for chemical reactions, and if the energy supplied by the colliding molecules is enough to break their bonds, they can combine to give new products. Chemical kinetics links these microscopic collisions to the macroscopically observable changes in reaction/product concentrations.

Consider a chemical reaction in which x molecules of type A react with y molecules of type B to give z molecules of C and w molecules of D. In chemical notation,

$$xA + yB \rightarrow zC + wD \tag{3.1}$$

The rate of this process in a closed and well-mixed system is defined as the change in the concentration of a given species over a given period of time, dt:

$$R = -\frac{1}{x}\frac{d[A]}{dt} = -\frac{1}{y}\frac{d[B]}{dt} = \frac{1}{z}\frac{d[C]}{dt} = \frac{1}{w}\frac{d[D]}{dt} \tag{3.2}$$

where square brackets denote concentrations (typical chemical notation), the minus sign corresponds to the disappearance of substrates, and division by the appropriate stoichiometric coefficients (x, y, z, w) ensures that R characterizes the entire reaction and is the same for all participating chemicals. To relate the reaction rate to the absolute values of concentrations, we first observe that R should be proportional to the frequency of intramolecular collisions necessary for the reaction to occur. For a simple bimolecular reaction, A + B $\rightarrow$ C, the

Chemistry in Motion: Reaction–Diffusion Systems for Micro- and Nanotechnology Bartosz A. Grzybowski

so-called collision theory shows rigorously that the frequency of collisions is proportional to the product of [A] and [B]. Consequently, the rate of reaction becomes $R = k[\mathrm{A}][\mathrm{B}]$, where the proportionality constant k is known as the reaction constant (large for 'fast' reactions; small for 'slow' reactions). Extrapolating this result to other types of reactions, we have

$$R = k[\mathrm{A}]^n[\mathrm{B}]^m \tag{3.3}$$

in which n and m are the so-called reaction orders with respect to the particular species. These orders depend on the atomic details of the reaction mechanism, and are equal to the stoichiometric coefficients (x, y) only for elementary reactions – that is, reactions taking place in one step, in which x molecules of A collide with y molecules of B. When, however, a net reaction is a sum of several elementary steps, the overall reaction orders do not reflect the molecular-scale events, and sometimes can even have fractional values (see Example 3.1).

Example 3.1 More Than Meets the Eye: Nonapparent Reaction Orders

Based on the stoichiometries alone, the formation of HBr from gaseous hydrogen and bromide, $H_2 + Br_2 \xrightarrow{k_x} 2HBr$, might appear to follow a simple rate law such as $R = k_x[H_2][Br_2]$. However, the experimentally measured rate has a much more complex dependence on the concentrations of reactants and even of products:

$$R = k' \frac{[H_2][Br_2]^{1/2}}{1 + k''[HBr][Br_2]^{-1}}$$

where k' and k'' are some constants. This is so because this chemical transformation is not an elementary but rather a radical reaction involving initiation, propagation, and termination steps:

$$\text{Initiation}: \quad Br_2 \xrightarrow{k_1} 2Br^{\bullet}$$

$$\text{Propagation}: \quad Br^{\bullet} + H_2 \xrightarrow{k_2} HBr + H^{\bullet} \quad \text{and} \quad H^{\bullet} + Br_2 \xrightarrow{k_3} HBr + Br^{\bullet}$$

$$\text{Termination}: \quad H^{\bullet} + HBr \xrightarrow{k_4} H_2 + Br^{\bullet} \quad \text{and} \quad 2Br^{\bullet} \xrightarrow{k_{-1}} Br_2$$

To derive the experimental rate expression we apply the steady-state approximation (see Example 3.2) commonly used to describe the kinetics of multistep reactions. This approximation assumes that the concentrations of intermediates – that is, chemical species that do not appear in the overall reaction because they are generated and consumed within the reaction cycle – are constant and do not change in time. Here, the intermediates are the two radicals, $H^\bullet$ and $Br^\bullet$, for which

$$\frac{d[Br^\bullet]}{dt} = 2k_1[Br_2] - k_2[Br^\bullet][H_2] + k_3[H^\bullet][Br_2] + k_4[H^\bullet][HBr] - 2k_{-1}[Br^\bullet]^2 = 0$$

and

$$\frac{d[H^\bullet]}{dt} = k_2[Br^\bullet][H_2] - k_3[H^\bullet][Br_2] - k_4[H^\bullet][HBr] = 0$$

By adding the two equations, we obtain $2k_1[Br_2] = 2k_{-1}[Br^\bullet]^2$ or $[Br^\bullet] = (k_1/k_{-1})[Br_2])^{1/2}$ and then solve for the concentration of $H^\bullet$:

$$[H^\bullet] = \frac{k_2\sqrt{k_1/k_{-1}}[H_2][Br_2]^{1/2}}{k_3[Br_2] + k_4[HBr]}$$

Because the rate of formation of HBr is $d[HBr]/dt = k_2[Br^\bullet][H_2] + k_3[H^\bullet][Br_2] - k_4[H^\bullet][HBr]$ and because $k_2[Br^\bullet][H_2] = k_3[H^\bullet][Br_2] + k_4[H^\bullet][HBr]$ (from $d[H^\bullet]/dt = 0$), we derive

$$\frac{d[HBr]}{dt} = 2k_3[H^\bullet][Br_2] = \frac{2k_2\sqrt{k_1/k_{-1}}[H_2][Br_2]^{1/2}}{1 + (k_4/k_3)[HBr][Br_2]^{-1}}$$

The approximated and experimental equations are equal when $k' = 2k_2\sqrt{k_1/k_{-1}}$ and $k'' = k_4/k_3$.

In practice, reaction orders are determined by changing the initial concentrations of the reactants and monitoring how these changes affect the rate. As an example, consider a reaction between A and B for which some hypothetical values of the reaction rates are listed in Table 3.1. To isolate the effect of each substance, we compare experiments that differ in the concentration of only one substance at a time. For A, experiments number 1 and 3 give a quadratic dependence of the rate on [A]; for B, experiments 1 and 2 indicate that R scales linearly with [B]. Based on these observations, the reaction is second order in A and first order in B: $R = k[A]^2[B]$.

Table 3.1 Hypothetical values of reaction rates for a reaction between A and B

Experiment	Concentration of A (M)	Concentration of B (M)	Initial rate ($M\,s^{-1}$)
(1)	0.100	0.200	9.0×10^{-4}
(2)	0.100	0.400	1.8×10^{-3}
(3)	0.050	0.200	2.25×10^{-4}

Combination of the reaction rate expression (3.3) and definition (3.2) leads to an ordinary differential equation (ODE), whose integration with respect to time gives concentrations of the reacting chemicals at different times. For example, the decomposition of N_2O over a Pt surface, $2N_2O(g) \rightarrow 2N_2(g) + O_2(g)$, where the substrate N_2O is present in excess and saturates the catalytic Pt, is zero order, $R = d[N_2O]/dt = k$. Separation of variables and subsequent integration of this equation gives $[N_2O] = [N_2O]_0 - kt$. For a first-order reaction, such as the S_N1-type solvolysis reaction of *tert*-butyl chloride, $(CH_3)_3CCl \xrightarrow{H_2O} (CH_3)_3COH + HCl$, the reaction rate depends only on the concentration of $(CH_3)_3CCl$. Therefore, the rate of disappearance of *tert*-butyl chloride is given by $d[(CH_3)_3CCl]/dt = -k[(CH_3)_3CCl]$. Separation of variables and integration of this equation gives the final rate law: $[(CH_3)_3CCl] = [(CH_3)_3CCl]_0 e^{-kt}$. Finally, second-order reaction rates can depend on the concentrations of either one or two reactants. The former case corresponds to a situation where two molecules of the same reactant combine to give one molecule of product, as in the dimerization of aluminum chloride: $2AlCl_3 \rightarrow Al_2Cl_6$. The rate of this reaction can be written as $R = -1/2d[AlCl_3]/dt = k[AlCl_3]^2$ which after integration gives

$$\frac{1}{[AlCl_3]} = \frac{1}{[AlCl_3]_0} + 2kt \qquad (3.4)$$

In the latter case, the two reactant molecules are different, as in a typical S_N2 reaction between ethanol (C_2H_5OH) and *n*-propyl iodide (C_3H_7I) under basic conditions:

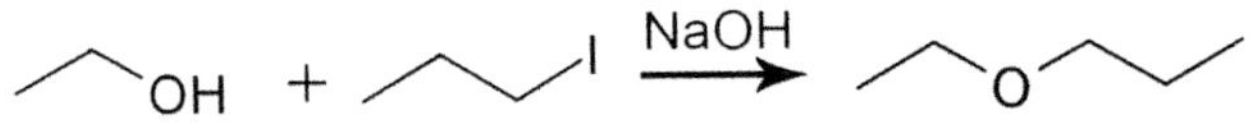

Here, the ethanolate anion 'attacks' the carbon atom adjacent to iodide and displaces the iodide while forming a new ether bond. The rate relationship describing this process is $R = -d[C_2H_5OH]/dt = k[C_2H_5OH][C_3H_7I]$ which is simplified by introducing an auxiliary variable $p = ([C_2H_5OH]_0 - [C_2H_5OH]) = ([C_3H_7I]_0 - [C_3H_7I])$. Substituting this expression into the rate equation, we have $dp/dt = k([C_2H_5OH]_0 - p)([C_3H_7I]_0 - p)$, which is integrated relatively easily to give

$$\frac{1}{[C_3H_7I]_0 - [C_2H_5OH]_0} \ln\left(\frac{[C_2H_5OH]_0([C_3H_7I]_0 - p)}{[C_3H_7I]_0([C_2H_5OH]_0 - p)}\right) = kt \qquad (3.5)$$

For more practice in the integration of rate expressions, see Example 3.2.

Example 3.2 Sequential Reactions

Consider a sequence of two elementary reactions: $A \xrightarrow{k_1} B \xrightarrow{k_2} C$. We wish to express [C] in terms of the rate constants and the initial concentration $[A]_0$. First, we write out the equations for the concentration changes of all species:

$$\begin{aligned} d[A]/dt &= -k_1[A] \\ d[B]/dt &= k_1[A] - k_2[B] \\ d[C]/dt &= k_2[B] \end{aligned}$$

To find [A], we integrate the first expression to get $[A] = [A]_0 e^{-k_1 t}$ and substitute this result into the equation describing the rate of change in [B]: $d[B]/dt = k_1[A]_0 e^{-k_1 t} - k_2[B]$. Solving for [B], we then obtain

$$[B] = [B]_0 e^{-k_2 t} + \frac{k_1[A]_0}{k_2 - k_1}(e^{-k_1 t} - e^{-k_2 t})$$

which, with the initial condition $[B]_0 = 0$, simplifies to

$$[B] = \frac{k_1[A]_0}{k_2 - k_1}(e^{-k_1 t} - e^{-k_2 t})$$

To find [C], we make use of the conservation of mass, $[A]_0 = [A] + [B] + [C]$, and use the integrated expressions for [A] and [B] to arrive at

$$\begin{aligned} [C] &= [A]_0\left[1 - e^{-k_1 t} - \frac{k_1}{k_2 - k_1}(e^{-k_1 t} - e^{-k_2 t})\right] \\ &= [A]_0\left[1 - \frac{k_2}{k_2 - k_1}e^{-k_1 t} + \frac{k_1}{k_2 - k_1}e^{-k_2 t}\right] \end{aligned}$$

The problem can also be tackled with the steady-state approximation applied to [B], $d[B]/dt = -k_2[B] + k_1[A] = 0$, such that $[B] = (k_1/k_2)[A]$. Using the integrated expression for [A], we then have $[B] = (k_1/k_2)[A]_0 e^{-k_1 t}$, and by mass conservation

$$[C] = [A]_0\left[1 - \frac{k_1 + k_2}{k_2}e^{-k_1 t}\right]$$

Notice that the steady-state solution is a valid approximation to the exact solution only when $k_2 \gg k_1$; in that case, both solutions converge to $[C] = [A]_0(1 - e^{-k_1 t})$. Physically, the limit $k_2 \gg k_1$ implies that any B generated by reaction 1 is very quickly consumed to form C. In other words, reaction 1 is the rate-limiting step because it alone controls the rate at which C is produced.

3.2 CHEMICAL EQUILIBRIUM

Until now, we have described reactions as one-way, irreversible processes in which colliding reactants transform into products. This is not the full picture – in reality, the product molecules also collide with one another and can revert back into reactants. Therefore, every reaction should be represented as a two-way process, with rate constant k describing the 'forward' conversion and k_{-1} describing the 'backward' conversion. The interplay between these reactions determines the concentrations of the reactants and of the products. Let us consider a simple example of an elementary reaction $x\mathrm{A} \leftrightarrow z\mathrm{C}$, for which the forward rate is $R = k[\mathrm{A}]^x$ and the backward rate is $R_{-1} = k_{-1}[\mathrm{C}]^z$. Importantly, when the magnitudes of the forward and backward rates are equal, the concentrations of A and C do not change in time, and the reaction is said to have come to equilibrium. The equilibrium condition can be expressed succinctly as $K_{\mathrm{eq}} = k/k_{-1} = [\mathrm{C}]^z/[\mathrm{A}]^x$, where K_{eq} is the so-called equilibrium constant. Similar reasoning can be extended to other reactions, with the equilibrium constant being expressed as the ratio of forward and backward rate constants – for example, for reaction (3.1):

$$K_{\mathrm{eq}} = \frac{k}{k_{-1}} = \frac{[\mathrm{C}]^z[\mathrm{D}]^w}{[\mathrm{A}]^x[\mathrm{B}]^y} \tag{3.6}$$

At this point a curious reader might have noticed that the last extension is somewhat suspicious, and should hold only if the $x\mathrm{A} + y\mathrm{B} \rightarrow z\mathrm{C} + w\mathrm{D}$ and $z\mathrm{C} + w\mathrm{D} \rightarrow x\mathrm{A} + y\mathrm{B}$ reactions are elementary and their orders are equal to stoichiometric coefficients, $R = k[\mathrm{A}]^x[\mathrm{B}]^y$ and $R_{-1} = k_{-1}[\mathrm{C}]^z[\mathrm{D}]^w$. While this criticism is certainly correct, a rigorous thermodynamic analysis that is somewhat beyond the scope of this introductory chapter shows that our definition of the equilibrium constant holds for all reactions irrespective of whether they occur in one elementary step or not. In other words, using the 'intuitive' argument based on the reaction rates is an example of getting the right answer using the wrong method. Readers who are still curious as to how this can be are referred to Normand[1] and Hammes.[2]

When the value of K_{eq} is small, the forward reaction is slow and the overall process is shifted toward the reactants. When K_{eq} is large, reactants are rapidly cleared to make products. In the limiting case when K_{eq} is very large and the reverse rate is negligible, the reaction is said to be irreversible. Examples of such reactions include the formation of HgS from mercury and sulfide ions, and also complexation of Fe^{3+} cations with CN^- anions, which yields hexacyanoferrate, $[\mathrm{Fe(CN)}_6]^{3-}$, complex anions.

An important property of equilibrium constants is that although they depend on the ambient parameters (temperature, pressure), they do not change when the concentrations of chemicals are varied. This property allows one to find the compositions of reaction mixtures when some products or reactants are added/removed. For example, if we keep removing the products (C, D) from the reaction 'soup', the system will speed up their production to keep K_{eq} constant. Conversely, if we keep removing substrates (A, B), the system will respond by converting some products into these reactants. Either way, K_{eq} will not change. Also, we note that when one of the reactants/products is a pure solid or liquid, its molar concentration is constant (moles per volume of a pure substance scale with density, which is an intensive quantity) and can be removed from the expression of K_{eq}. For example, the equilibrium constant of the reaction $Ca(OH)_2(s) \leftrightarrow Ca^{2+}(aq) + 2OH^-(aq)$ is simply $K = [Ca^{2+}][OH^-]^2$ and incorporates the constant concentration of solid calcium hydroxide into the value of K.

3.3 IONIC REACTIONS AND SOLUBILITY PRODUCTS

Since the majority of microscopic reaction–diffusion systems discussed later in this book rely on reactions of inorganic ions, it is important to understand factors influencing ionic solubilities and the means of quantifying them. Some ions can coexist in one aqueous solution (e.g., Li^+ and Cl^-) whereas others form insoluble precipitates (e.g., Li^+ and F^-). To a first approximation, these qualitatively different behaviors can be rationalized by comparing the enthalpies of dissociated and hydrated ions in solution with the enthalpies of ions distributed on a crystalline lattice of the precipitate. As a rule of thumb, small, compact ions have highly negative lattice enthalpies indicating that they are energetically 'happy' in the crystalline/precipitate state. To dissolve such substances, there must be a driving force in the form of high enthalpic gain of ion hydration. As an example, consider the solubilities of LiF and LiCl. Because F^- is a smaller anion than Cl^-, the lattice enthalpy of LiF ($\Delta H_{crys} = -618.3\,kJ\,mol^{-1}$) is more negative than that of LiCl ($\Delta H_{crys} = -408.5\,kJ\,mol^{-1}$), and more energy is needed to break up the former crystal. On the other hand, the small fluoride ions are hydrated better (favorable enthalpy of hydration $\Delta H_{hydr} = -335.4\,kJ\,mol^{-1}$) than the larger chloride anions ($\Delta H_{hydr} = -167.1\,kJ\,mol^{-1}$). It follows that the net enthalpic cost of dissolution, $\Delta H_{sol} = \Delta H_{hydr} - \Delta H_{crys}$, is $41.5\,kJ\,mol^{-1}$ lower for LiCl than for LiF (282.9 versus $241.4\,kJ\,mol^{-1}$), and lithium chloride is more soluble in water.

Similar reasoning can be used to explain the very poor water solubility of $Ca_3(PO_4)_2$ ($1.3\,\mu g\,L^{-1}$) versus the relatively good solubility of $CaSO_4$ ($2.0\,g\,L^{-1}$). Although phosphate ions bearing three negative charges are hydrated better ($\Delta H_{hydr} = -1284.1\,kJ\,mol^{-1}$) than sulfate ions ($\Delta H_{hydr} = -909.3\,kJ\,mol^{-1}$), the lattice enthalpy of $Ca_3(PO_4)_2$ crystal ($\Delta H_{crys} = -4138\,kJ\,mol^{-1}$) is also higher than that of $Ca_3(PO_4)_2$ ($\Delta H_{crys} = -1434.5\,kJ\,mol^{-1}$). Consequently, the enthalpic cost of dissolution is much higher (by $2382.7\,kJ\,mol^{-1}$) for calcium phosphate than for calcium sulfate, explaining the large difference in solubilities.

The degree of solubility of an ionic compound can be quantified by the equilibrium constant of a dissolution reaction $A_nB_m(s) \leftrightarrow nA^{m+}(aq) + mB^{n-}(aq)$. By incorporating the concentration of the pure solid precipitate, A_nB_m, into the value of K, we can write the so-called solubility product $K_{sp} = [A]^n[B]^m$.

Importantly, K_{sp} sets the upper limit on the concentration of free ions in solution. For example, for $CaCO_3$ chalk, $K_{sp} = [Ca^{2+}][CO_3^{2-}] = 3.31 \times 10^{-9}$, which means that we can dissolve at most $x = \sqrt{K_{sp}} = 57.53$ μmol or 5.75 mg of chalk in 1 L of water. For $Co(OH)_2$, $K_{sp} = [Co^{2+}][OH^-]^2 = 6.3 \times 10^{-15}$. Denoting $x = [Cr(OH)_2] = [Cr^{2+}] = 0.5[OH^-]$, we have $K_{sp} = 4x^3$, from which the maximum amount of salt soluble in 1 L of water is $x = \sqrt[3]{K_{sp}/4} = 11.6$ μmol or 1.08 mg. For very poorly soluble HgS, $K_{sp} = 6.31 \times 10^{-52}$, indicating that its solubility corresponds to only 3 molecules of HgS in 1 L of water!

An interesting twist to this story is that while the solubility product sets a limit on the concentration of *free* ions in solution, the actual solubility of a compound can be increased by 'masking' the ions by a side reaction with some auxiliary ligand L. For instance, in a complexation reaction $A_nB_m \rightleftarrows nA^{m+} + mB^{n-} \overset{+xL}{\rightleftarrows} n[AL_x]^{m+} + mB^{n-}$, coordination of metal ions A by ligands L shifts the equilibrium to the right and effectively increases the solubility of the A_nB_m reactant. One manifestation of this effect is the solubility of iron contained in the blood. Based on the solubility product describing the $Fe^{3+} + 3OH^- \leftrightarrow Fe(OH)_3$ reaction at physiological pH = 7.4, $K_{sp} = [Fe^{3+}][OH^-]^3 \approx 10^{-39}$, and the concentration of free Fe^{3+} should not exceed $10^{-39}/(2.51 \times 10^{-7})^3 = 6.3 \times 10^{-16}$ mol L^{-1}. In reality, it is more than 13 orders of magnitude greater (7.5×10^{-3} mol L^{-1}) because hemoglobin – an oxygen-transfer protein found in red bloods cells – binds iron with a strong affinity.

3.4 AUTOCATALYSIS, COOPERATIVITY AND FEEDBACK

Not all chemical reactions progress linearly from substrates to products – some can literally loop onto themselves and either speed up or slow down their own progress. In autocatalytic reactions, products accelerate the reaction. As an example, consider a simple reaction A + B → 2B, with rate law $d[A]/dt = -k[A][B]$.[3,4] Noting that $[A]_0 - [A] = [B] - [B]_0$ and defining $x = [A]$, we obtain $[B] = [A]_0 + [B]_0 - [A] = [A]_0 + [B]_0 - x$. After substitution, the rate law becomes $dx/dt = -kx([A]_0 + [B]_0 - x)$, which can be integrated to give

$$\frac{1}{[A]_0 - [B]_0} \ln\left(\frac{[A]_0[B]}{[B]_0[A]}\right) = -kt \tag{3.7}$$

or

$$[B] = \frac{[A]_0 + [B]_0}{1 + ([A]_0/[B]_0)e^{-kt([A]_0 + [B]_0)}} \tag{3.8}$$

which is a sigmoidal function of time (Figure 3.1). Here, the reaction is initially slow due to the scarcity of the B 'catalyst', but it then accelerates as the product

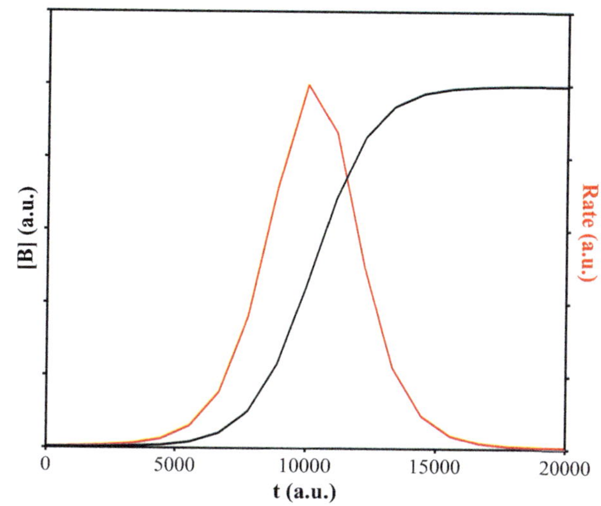

Figure 3.1 Concentration of B (black curve) and the corresponding reaction rate (red) plotted as a function of time for an autocatalytic reaction A + B → 2B

feeds back into the process; ultimately, the reaction slows down and comes to a halt when A is completely consumed.

Examples of autocatalysis include the reaction of permanganate with oxalic acid, spontaneous degradation of aspirin into salicylic and acetic acids (causing old aspirin containers to smell mildly of vinegar) or the acid-catalyzed hydrolysis of ethyl acetate into acetic acid and ethanol (Figure 3.2).

Closely related to autocatalysis is the phenomenon of cooperativity, in which one reaction event (e.g., binding of a ligand to a protein receptor) facilitates further events of the same type. The best known and arguably most important example of cooperativity is the binding of oxygen to hemoglobin. Hemoglobin[5] (Hb) is a large tetrameric metalloprotein found in red blood cells that binds, transports, and

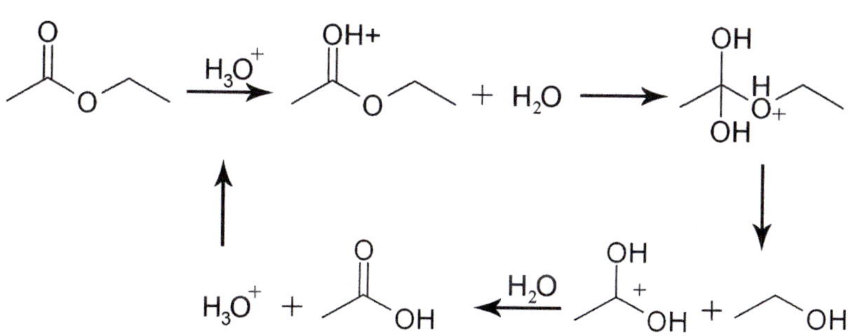

Figure 3.2 Acid-catalyzed hydrolysis of ethyl acetate is an example of an autocatalytic process

releases molecular oxygen. Each Hb protein contains four iron-containing heme groups that can bind oxygen according to a four-step mechanism:[6]

1. $Hb + O_2 \rightleftarrows HbO_2$
2. $HbO_2 + O_2 \rightleftarrows Hb(O_2)_2$
3. $Hb(O_2)_2 + O_2 \rightleftarrows Hb(O_2)_3$
4. $Hb(O_2)_3 + O_2 \rightleftarrows Hb(O_2)_4$

Binding of the first O_2 to the heme's iron causes strain in the protein and induces a structural change from the so-called 'deoxy' to 'oxy' conformation, which is more prone to bind additional O_2 molecules.[6–8] This cooperative effect is reflected in the values of the equilibrium constants – only 5–60 for the first reaction and as much[1] as 3000–6000 for the fourth[5,6] – and allows hemoglobin to act as an efficient oxygen transporter. In oxygen-rich environments (lungs or gills), Hb saturates with oxygen rapidly; when, however, it is transported to oxygen-deficient locations (capillaries), it discards part of its O_2 cargo and remains in a state of low oxygen affinity before being re-circulated to the lungs.

Let us now consider an opposite situation, in which a reaction product slows the reaction down by inhibiting one of the substrates. Such behavior is observed, for instance, during oxidation of biphenyls **I** by molecular oxygen (Figure 3.3).[9] This reaction proceeds via a biphenyl radical **II**, which is ultimately converted into biphenol **III**. This biphenol, however, can readily donate its hydrogen atom back to the substrate radical **II**, converting it to a nonreactive biphenyl **V**. In other words, **III** 'deactivates' **II** and thus inhibits its own production. It is interesting to note that nature invented a similar mechanism to protect some of its key molecules against oxidative degradation by air. One of the most notable examples of such protection is vitamin E, which is chemically a phenol (i.e., counterpart of **III**) and donates its phenolic hydrogen radical to reactive radicals analogous to **II**. In this way, vitamin

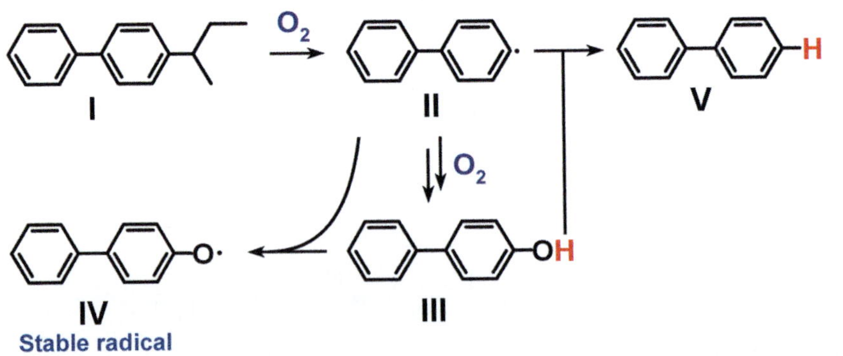

Figure 3.3 Oxidation of 4-alkylbiphenyls as an example of autoinhibition

E deactivates these harmful radicals while converting itself to a relatively safe radical analog of **IV**.[10]

We have now seen examples of reactions that are either accelerated or decelerated by their own products. In many biological systems, *both* of these effects operate simultaneously effectively serving as 'switches' to drive or inhibit specific biotransformations. One important example is glycolysis – that is, a system of concerted biochemical reactions that convert glucose into pyruvate while producing two molecules of energy-rich ATP (adenosine triphosphate).

The rate-limiting step in glycolysis is the catalysis of fructose 6-phosphate (F-6-P) to fructose 1,6-bisphosphate (F-1,6-P) by the enzyme phosphofructokinase (PFK-1).[11] In mammalian liver cells, PFK-1 is regulated by the concentrations of ATP and AMP (adenosine monophosphate, a low-energy product of ATP hydrolysis). High levels of ATP signal the cell that energy is being produced faster than it is consumed, and ATP binds to a noncatalytic site of PFK-1. The presence of ATP effectively inhibits the enzyme's activity towards F-6-P and 'switches' glycolysis off. On the other hand, during times of high energy consumption, when the levels of ATP are low and those of AMP are high, AMP 'unblocks' PFK-1 to accelerate the glycolytic cycle.

Rate-limiting step in glycolysis : (1) $\mathrm{ATP} + \mathrm{F\text{-}6\text{-}}P \xrightarrow{\mathrm{PFK\text{-}1}} \mathrm{F\text{-}1,6\text{-}P} + \mathrm{ADP}$

Regulation of PFK-1 : (2) $\mathrm{PFK\text{-}1} + \mathrm{ATP} \Leftrightarrow \mathrm{PFK\text{-}1\text{-}ATP}$

Glycolosis 'switches off'

(3) $\mathrm{PFK\text{-}1\text{-}ATP} + \mathrm{AMP} \rightarrow \mathrm{PKK\text{-}1} + \mathrm{ATP} + \mathrm{AMP}$

Glycolosis 'switches on'

3.5 OSCILLATING REACTIONS

The autocatalytic and/or inhibitory reactions are also the key components of nonbiological reaction–diffusion systems producing chemical waves and Turing patterns (cf. Chapter 1). Before we see in Chapter 9 how these fascinating phenomena emerge from the coupling between 'looped' chemical transformations and diffusion, we will first discuss their chemical/kinetic component, which gives rise to temporal concentration oscillations in the absence of diffusive flows.

A closed chemical system (i.e., no flow in or out) evolves in time towards a stable equilibrium state, at which the free energy is minimal. Interestingly, while free energy always decreases *monotonically* towards this equilibrium value, the concentrations of the participating chemical species may evolve nonmonotonically, exhibiting oscillatory or even chaotic behaviors. Necessary (but not sufficient) conditions for such chemical oscillations include autocatalysis and nonlinear rate laws.

To get a better feel for oscillatory chemical dynamics, consider a model system described by the following set of kinetic equations and commonly known as the Brusselator (after its creators from the Université Libre in Brussels; other model systems have similar names such as the Oregonator, Palo Altonator, etc.):

$$\left.\begin{array}{ll} A \rightarrow X & (1) \\ 2X + Y \rightarrow 3X & (2) \\ B + X \rightarrow Y + C & (3) \\ X \rightarrow D & (4) \end{array}\right\} \quad (3.9)$$

Here, the overall reactions are simply A → D (adding reactions (1) and (4)) and B → C (adding reactions (2) and (3)); however, the rates of these two reactions are controlled by the production/consumption of reactive intermediates X and Y. Notice that we have neglected the reverse reactions, which are assumed to be negligibly slow compared to the forward reactions. Although with this simplification we may no longer describe the system's approach to equilibrium, we can still describe its dynamics (e.g., oscillations in the concentrations of reactive intermediates X and Y) far from equilibrium, where the reverse reactions are unimportant. Furthermore, we assume that the concentrations of the reactants, A and B, are much greater than those of the intermediates, X and Y; therefore, we may neglect the time dependence of [A] and [B] over short time scales. With these assumptions, the evolution of [X] and [Y] is governed by two coupled, ordinary differential equations:

$$\begin{aligned} \mathrm{d}[X]/\mathrm{d}t &= k_1[\mathrm{A}] + k_2[\mathrm{X}]^2[\mathrm{Y}] - k_3[\mathrm{B}][\mathrm{X}] - k_4[\mathrm{X}] \\ \mathrm{d}[Y]/\mathrm{d}t &= -k_2[\mathrm{X}]^2[\mathrm{Y}] + k_3[\mathrm{B}][\mathrm{X}] \end{aligned} \quad (3.10)$$

which are both nonlinear (specifically, due to the term $k_2[\mathrm{X}]^2[\mathrm{Y}]$) and autocatalytic. Introducing 'rescaled' variables $u = [\mathrm{X}]/[\mathrm{A}]$, $v = [\mathrm{Y}]/[\mathrm{A}]$, $\tau = k_1 t$, these equations can be nondimensionalized into

$$\begin{aligned} \mathrm{d}u/\mathrm{d}\tau &= 1 + \alpha u^2 v - (\beta + \gamma)u \\ \mathrm{d}v/\mathrm{d}t &= -\alpha u^2 v + \beta u \end{aligned} \quad (3.11)$$

where $\alpha = k_2[\mathrm{A}]^2/k_1$, $\beta = k_3[\mathrm{B}]/k_1$, and $\gamma = k_4/k_1$ are dimensionless parameters governing the system's dynamics. Note that by scaling the equations, we have reduced the number of 'control' parameters from six to three (specifically, [X], [Y], k_1, k_2, k_3, $k_4 \rightarrow \alpha$, β, γ), greatly simplifying the problem. Although these equations cannot be solved analytically, they can be integrated numerically (see Chapter 4) using, for example, the popular fourth-order Runge–Kutta method with adaptive time stepping (available through Numerical Recipes[12] or Matlab). For a certain range of values of the parameters α, β, γ, this system exhibits temporal oscillations in the concentrations of X and Y; an example of such oscillatory behavior is shown in Figure 3.4(a).

In some oscillating systems, the periodic variations of concentrations manifest themselves in the form of color changes. For example, the so-called Briggs–Rauscher reaction[13] (Figure 3.4(b)) exhibits temporal oscillations in the concentrations of iodide (I^-) and iodine (I_2), which are reactive intermediates in the overall iodination of malonic acid to 2-iodomalonic acid by iodate (IO_3^-) and hydrogen peroxide. For a more detailed account of this and other oscillating

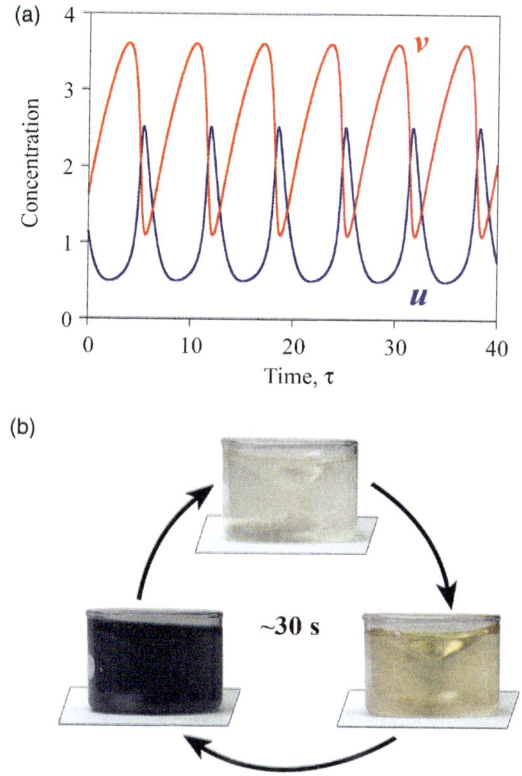

Figure 3.4 (a) Numerical integration of the Brusselator model for $\alpha = 1$, $\beta = 2.5$ and $\gamma = 1$. All variables are presented in their dimensionless forms. (b) The Briggs–Rauscher chemical system exhibits temporal oscillations in the concentrations of iodine (I_2) and iodide (I^-), which are made clearly visible using a thyodene indicator

reactions, the reader is referred to two excellent books on the subject by Field and Burger[14] and Epstein and Pojman.[15]

3.6 REACTIONS IN GELS

Although the coupling between reaction and diffusion can produce spatial patterns in liquid media, these patterns are easily distorted and disappear altogether when the liquid is mixed. In order for reaction–diffusion to make more permanent spatial designs, a more rugged yet permeable support medium is needed. Gels are obvious candidates for the task, and we will encounter them in many systems discussed in subsequent chapters.

Gels are materials in which a porous network of molecules (monomers, oligomers or polymers) or particles spans the volume of a liquid medium.

Depending on the particular system, this network can form in response to various stimuli including temperature decrease, initiation of chemical reaction, or even sound waves.[16] Gels are classified according to the type of solvent they contain and the type of intermolecular interactions supporting the gel structure. Hydrogels absorb aqueous solvents, while organogels absorb organic solvents. Physical gels such as agarose, gelatin or syndiotactic poly(methyl methacrylate) form due to noncovalent, intermolecular interactions between the gel-forming molecules. In contrast, chemical gels such as poly[acrylamide-*co*-(ethylene glycol dimethacrylate)] form as a result of covalent crosslinking of the component monomers/polymers. For both physical and chemical gels, the onset of gelation results in a large increase in viscosity of the solution followed by solidification.

One of the major reasons gels are so commonly used in separation science (e.g., in chromatography and electrophoresis) is that their porous network suppresses convective and/or turbulent flows while allowing transport by diffusion. This 'smoothness' of mass transport makes gels ideal candidates for our reaction–diffusion systems and ensures that the patterns and structures that form will not be washed away by accidental swirls of the underlying fluid.

In addition, gels provide unique supports for various types of chemical reactions including crystallization. Crystallization in gels is dominated by diffusion, and the absence of hydrodynamic flows and reduced diffusion coefficients (due to the gel network 'obstacles') decrease the flux of the molecules at the growing crystal faces.[17] Consequently, the process of crystal growth is slower and more equilibrated, and even large crystals do not sediment from the growing medium. Gel-supported crystallization of inorganic substances can produce excellent quality crystals of centimeter dimensions[18] and can improve the quality of macromolecular/biological crystals.[17] In the context of some of the examples discussed in later

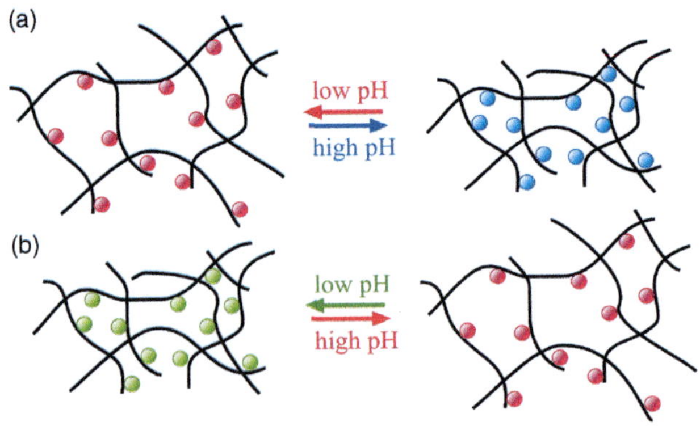

Figure 3.5 Volume changes in pH-sensitive gels. (a) Basic groups such as amines (e.g., NH_2, blue circles) are protonated/positively charged (red circles) at low pH. As a result, the gel swells to relieve unfavorable electrostatic repulsions between these groups. (b) In contrast, acidic groups (e.g., COOH, green circles) become deprotonated/negatively charged at high pH causing gel swelling

chapters, it is worth mentioning that 'hard' crystalline precipitates can buckle the 'soft' gels in which they grow suggesting that this phenomenon can be used to transduce a chemical process into physical deformation.

Deformation of gels can be also induced by other means, both reversible and irreversible. When some of the functionalities on a gel network develop electric charges and repel one another, the gel absorbs more solvent and swells to minimize these repulsions.[19] For example, poly(*N*,*N*-dimethylaminoethyl methacrylate) gel swells under low pH conditions when the free amino groups are protonated; when pH is increased, these groups revert to uncharged amines and the gel shrinks.[20] Qualitatively opposite behavior is seen in gels containing acidic groups[21] that are deprotonated and charged under high pH (Figure 3.5).

Irreversible volumetric response usually requires crosslinking of a gel's backbone. This can be achieved by chemical means, by heating or by using UV light. Light-induced crosslinking has the added advantage that it can easily be confined to selected locations of the gel. For example, placing a photomask (or a patterned transparency) over a gelatin film doped with potassium dichromate, $K_2Cr_2O_7$, and irradiating with 365 nm light causes photoreduction of Cr(VI) to Cr(III) over the exposed regions. The Cr^{3+} cations generated then coordinate and crosslink electron-donating groups of the gelatin's amino acids and cause gel swelling.[22] This method is interesting in the context of reaction–diffusion, since localized crosslinking changes pore sizes and thus the diffusion coefficients of the migrating chemicals; it also locally depletes Cr(VI) which, as we will see in Chapter 7, is an important ingredient of the so-called periodic precipitation reactions.

REFERENCES

1. Normand, L. (2000) *Statistical Thermodynamics*, Cambridge University Press, Cambridge.
2. Hammes, G.G. (2000) *Thermodynamics and Kinetics for the Biological Sciences*, John Wiley & Sons, Ltd, New York.
3. Steinfeld, J.I., Francisco, J.S. and Hase, W.L. (1999) *Chemical Kinetics and Dynamics*, Prentice Hall, Upper Saddle River, NJ.
4. Frost, A. and Pearson, R.G. (1953) *Kinetics and Mechanism*, John Wiley & Sons, Ltd, New York.
5. Dickerson, R.E. and Geis, I. (1983) *Hemoglobin*, Benjamin/Cummings, Menlo Park, CA.
6. Miessler, G. and Tarr, D.A. (2004) *Inorganic Chemistry*, Pearson Prentice Hall, Upper Saddle River, NJ.
7. Baldwin, J. and Chothia, C. (1979) Hemoglobin: structural changes related to ligand binding and its allosteric mechanism. *J. Mol. Biol.*, **129**, 175.
8. Cotton, F.A., Wilkinson, G., Murillo, C.A. and Bochmann, M. (1999) *Advanced Inorganic Chemistry*, John Wiley & Sons, Ltd, New York.
9. Mertwoy, H.E. and Gisser, H. (1967) Substituted biphenyls and terphenyls as oxidatively autoinhibitive compounds. *Ind. Eng. Chem. Proc. Des. Dev.*, **6**, 108.

10. Ames, B.N., Shigenaga, M.K. and Hagen, T.M. (1993) Oxidants, antioxidants, and the degenerative diseases of aging. *Proc. Natl Acad. Sci. USA*, **90**, 7915.
11. Stryer, L. (1995) *Biochemistry*, W.H. Freeman and Company, New York.
12. Press, W.H. (1992) *Numerical Recipes in C: The Art of Scientific Computing*, Cambridge University Press, New York.
13. Briggs, T.S. and Rauscher, W.C. (1973) Oscillating iodine clock. *J. Chem. Educ.*, **50**, 496.
14. Field, R.J. and Burger, M. (1985) *Oscillations and Traveling Waves in Chemical Systems*, John Wiley & Sons, Ltd, New York.
15. Epstein, I.R. and Pojman, J.A. (1998) *An Introduction to Nonlinear Chemical Dynamics: Oscillations, Waves, Patterns, and Chaos*, Oxford University Press, New York.
16. Naota, T. and Koori, H. (2005) Molecules that assemble by sound: an application to the instant gelation of stable organic fluids. *J. Am. Chem. Soc.*, **127**, 9324.
17. Cudney, S., Patel, A. and McPherson, A. (1994) Crystallization of macromolecules in silica-gels. *Acta Crystallogr.*, **D50**, 479.
18. Henisch, H. (1988) *Crystals in Gels and Liesegang Rings*, Cambridge University Press, Cambridge.
19. Peppas, N.A., Hilt, J.Z., Khademhosseini, A. and Langer, R. (2006) Hydrogels in biology and medicine: from molecular principles to bionanotechnology. *Adv. Mater.*, **18**, 1345.
20. Horkay, F., Han, M.-H. and Han, I.S. *et al.* (2006) Separation of the effects of pH and polymer concentration on the swelling pressure and elastic modulus of a pH-responsive hydrogel. *Polymer*, **47**, 7335.
21. Schmaljohann, D. (2006) Thermo- and pH-responsive polymers in drug delivery. *Adv. Drug Del. Rev.*, **58**, 1655.
22. Paszewski, M., Smoukov, S.K., Klajn, R. and Grzybowski, B.A. (2007) Multilevel surface nano- and microstructuring via sequential photoswelling of dichromated gelatin. *Langmuir*, **23**, 5419.

4

Putting It All Together: Reaction–Diffusion Equations and the Methods of Solving Them

4.1 GENERAL FORM OF REACTION–DIFFUSION EQUATIONS

Combining the diffusion and rate equations developed in Chapters 2 and 3, respectively, the general equation describing a reaction–diffusion (RD) process can be written as

$$\frac{\partial C_s}{\partial t} = \nabla \cdot (D_s \nabla C_s) + R_s(C_1 \ldots C_n) \quad \text{for} \quad s = 1, 2, \ldots, n \tag{4.1}$$

Here, $C_s(x, y, z, t)$ is the concentration of species s, $D_s(x, y, z, t)$ is its diffusion coefficient that can, in principle, depend on the spatial location and on time, and $R_s(C_1 \ldots C_n)$ is the reaction term, which is typically (although not necessarily) of the form $R_s(C_1 \ldots C_n) = \sum_{m=1}^{r} (k_m C_1^{\alpha_{m1}} C_2^{\alpha_{m2}} \ldots C_n^{\alpha_{mn}})$, where m is the reaction index, α_{mn} are the reaction orders, and r is the total number of reactions consuming/producing species s. For example, for an autocatalytic reaction $A + B \xrightarrow{k} 2A$ obeying the rate law $R = k[A][B]$ and with both chemicals migrating with constant diffusion coefficients D_A and D_B, the $n = 2$ governing RD equations are $\partial[A]/\partial t = D_A \nabla^2 [A] + k[A][B]$ and $\partial[B]/\partial t = D_B \nabla^2 [B] - k[A][B]$. Unfortunately, even for this simple example, the RD equations cannot be solved analytically and one

Chemistry in Motion: Reaction–Diffusion Systems for Micro- and Nanotechnology Bartosz A. Grzybowski

must rely on numerical methods to compute approximate solutions. The basic approach of most numerical schemes is to divide space and time into discrete 'parcels' and create an 'equivalent' representation of the relevant continuous system. While such discrete systems are only approximately 'equivalent' to their continuous counterparts, they are amenable to direct computation and therefore quite useful.

In the following, we will first discuss the special class of RD systems that permit analytical solutions, and will then survey numerical methods applicable to more general forms of RD equations. We will begin by describing the discretization of the spatial component via (i) *finite difference* and (ii) *finite element* methods and will then address the discrete integration of the time component, where choosing the optimal method depends on the times scales of reaction and diffusion relevant to the problem at hand.

4.2 RD EQUATIONS THAT CAN BE SOLVED ANALYTICALLY

Diffusion coupled with first-order reaction is relatively common in processes such as decomposition,[1] oxidation[2] and etching.[3,4] Importantly, this is the rare subclass of RD systems that can be solved analytically using familiar techniques from Chapter 2. As an illustrative example, let us consider a two-dimensional domain ($0 \leq x \leq L$; $0 \leq y \leq L$), in which the concentrations are specified by boundary conditions $c(0,y,t) = 0$, $c(L,y,t) = 0$, $c(x,0,t) = 0$, $c(x,L,t) = f(x)$, and the diffusing chemical is consumed by a first-order chemical reaction (e.g., etching of a silicon surface by HCl gas[3]). We first solve the steady-state RD equation for this process:

$$D\left(\frac{\partial^2 c(x,y)}{\partial x^2} + \frac{\partial^2 c(x,y)}{\partial y^2}\right) - kc(x,y) = 0 \tag{4.2}$$

where D is the diffusion constant and k is the rate constant of the reaction. Rescaling the variables $\bar{c} = c/C_0$, $\bar{x} = x/L$, $\bar{y} = y/L$ (where C_0 is some characteristic concentration) yields a nondimensional equation

$$\frac{\partial^2 \bar{c}(\bar{x},\bar{y})}{\partial \bar{x}^2} + \frac{\partial^2 \bar{c}(\bar{x},\bar{y})}{\partial \bar{y}^2} - \text{Da}\bar{c}(\bar{x},\bar{y}) = 0 \tag{4.3}$$

with boundary conditions $\bar{c}(0,\bar{y}) = 0, \bar{c}(1,\bar{y}) = 0, \bar{c}(\bar{x},0) = 0, \bar{c}(\bar{x},1) = f(\bar{x})$ and where $\text{Da} = L^2k/D$ is the so-called Damköhler dimensionless number, which quantifies the relative rates of reaction and diffusion. Separating the variables, $\bar{c}(\bar{x},\bar{y}) = X(\bar{x})Y(\bar{y})$, and substituting into the RD equation, we obtain

$$\frac{Y''}{Y} - \text{Da} = -\frac{X''}{X} = \lambda^2 \tag{4.4}$$

where λ^2 is an unknown constant and double primes denote second derivatives with respect to the functions' arguments. Because the boundary conditions prescribe $X(0)=0$ and $X(1)=0$, the solution to the x-component of the equation is $X(\bar{x}) = B\sin(\lambda\bar{x})$, where $\lambda = n\pi$ for $n = 1, 2, 3, \ldots, \infty$. The y-component then becomes

$$\frac{Y''}{Y} - \mathrm{Da} - n^2\pi^2 = 0 \tag{4.5}$$

which has the solution $Y(\bar{y}) = A\sinh\left(\sqrt{\mathrm{Da} + n^2\pi^2}\bar{y}\right)$, subject to the boundary condition $Y(0)=0$. Combining solutions for X and Y and making use of the orthogonality of sine functions, the overall solution is

$$\begin{aligned} \bar{c}(\bar{x},\bar{y}) &= \sum_{n=1}^{\infty} \tilde{A}\sin(n\pi\bar{x})\sinh\left(\sqrt{\mathrm{Da} + n^2\pi^2}\bar{y}\right) \\ \text{where} \quad \tilde{A} &= \frac{2\int_0^1 f(\bar{x})\sin(n\pi\bar{x})dx}{\sinh\left(\sqrt{\mathrm{Da} + n^2\pi^2}\right)} \end{aligned} \tag{4.6}$$

For the simple case of $f(\bar{x}) = 1$, this expression in the original, dimensional variables becomes

$$c(x,y) = C_o \sum_{n=1,3,5\ldots}^{\infty} \frac{4}{n\pi\sinh\left(\sqrt{\mathrm{Da} + n^2\pi^2}\right)} \sin\left(n\pi\frac{x}{L}\right)\sinh\left(\sqrt{\mathrm{Da} + n^2\pi^2}\frac{y}{L}\right) \tag{4.7}$$

The left part of Figure 4.1 shows the concentration profile for $\mathrm{Da} = 30$ and $C_0 = 1$. Comparing this profile with the solution of a corresponding problem without the reaction term (Section 2.2.1; Figure 2.2), we see that the reaction limits propagation of c, as one would intuitively expect. With another boundary condition we considered in Section 2.2.1, $f(x) = \exp(-(x - L/2)^2/b)$, we again find that the concentration profile is more 'flattened' than that in the right part of Figure 2.2.

Having solved the steady-state case, we can now tackle the time-dependent problem with nonzero concentrations on one boundary of the square domain:

$$\frac{\partial c(x,y,t)}{\partial t} = D\left(\frac{\partial^2 c(x,y,t)}{\partial x^2} + \frac{\partial^2 c(x,y,t)}{\partial y^2}\right) - kc(x,y,t) \tag{4.8}$$

with boundary conditions $c(0,y,t) = 0$, $c(x,0,t) = 0$, $c(L,y,t) = 0$, $c(x,L,t) = 1$, $f(x) = 1$ and the initial condition $c(x, y, 0) = 0$. Rescaling the

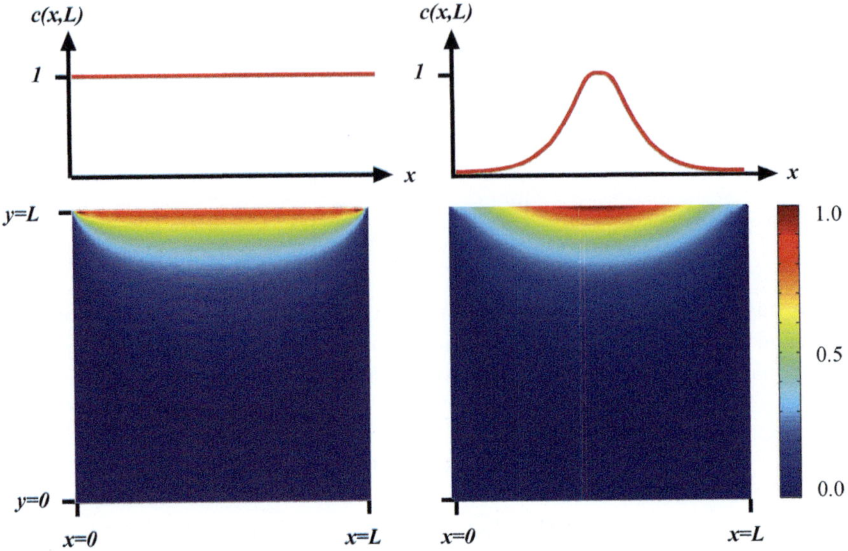

Figure 4.1 Steady-state concentration profiles for a reaction–diffusion process over a square domain for (left) $c(x,L) = f(x) = 1$ and (right) $c(x,L) = f(x)$ where $f(x) = \exp(-(x - L/2)^2/b)$ with $b = 0.2$

variables as in the steady-state case and with dimensionless time $\bar{t} = Dt/L^2$, the RD equation can be rewritten as

$$\frac{\partial \bar{c}(\bar{x},\bar{y},\bar{t})}{\partial \bar{t}} = \frac{\partial^2 \bar{c}(\bar{x},\bar{y},\bar{t})}{\partial \bar{x}^2} + \frac{\partial^2 \bar{c}(\bar{x},\bar{y},\bar{t})}{\partial \bar{y}^2} - \mathrm{Da}\bar{c}(\bar{x},\bar{y},\bar{t}) \tag{4.9}$$

Although nonzero boundary conditions preclude simple solution by the separation of variables, we can circumvent this complication by a shrewd change of variables (Section 2.2.1): $\theta(x, y, t) = c(x, y, t) - c_{SS}(x, y)$, where $c_{SS}(x, y)$ stands for the steady-state solution (4.7). With this change, the RD equation becomes

$$\frac{\partial \theta(x,y,t)}{\partial t} = \frac{\partial^2 \theta(x,y,t)}{\partial x^2} + \frac{\partial^2 \theta(x,y,t)}{\partial y^2} - \mathrm{Da}\theta(x,y,t) \tag{4.10}$$

and the new initial/boundary conditions are $\theta(x, y, 0) = -c_{ss}, \theta(L, y, t) = 0$, $\theta(0, y, t) = 0$, $\theta(x, 0, t) = 0$, $\theta(x, L, t) = 0$. Separation of variables can now be performed to give a solution that is very similar to (2.26) except for the additional Damköhler number:

$$\theta(x,y,t) = \sum_{n=1}^{\infty}\sum_{m=1}^{\infty} G_{nm}\sin(n\pi x/L)\sin(m\pi y/L)e^{-(Da+\pi^2m^2+\pi^2n^2)t/L} \tag{4.11}$$

where G_{nm} is a constant. Using the initial condition and the orthogonal property of the sine functions, the final solution is

$$\begin{aligned} c(x,y,t) &= \theta(x,y,t) + c_{SS}(x,t) \\ &= C_0 \sum_{n=1,3,5\ldots}^{\infty} \sum_{m=1}^{\infty} \frac{8(-1)^m m}{n(\mathrm{Da} + \pi^2 n^2 + \pi^2 m^2)} \\ &\quad \times \sin\left(n\pi \frac{x}{L}\right) \sin\left(m\pi \frac{y}{L}\right) e^{-(\mathrm{Da} + \pi^2 m^2 + \pi^2 n^2)t} \\ &\quad + C_0 \sum_{n=1,3,5\ldots}^{\infty} \frac{4}{n\pi \sinh(\sqrt{Da + n^2\pi^2})} \sin\left(n\pi \frac{x}{L}\right) \sinh\left(\sqrt{\mathrm{Da} + n^2\pi^2}\frac{y}{L}\right) \end{aligned} \tag{4.12}$$

Following the reasoning outlined in Section 2.2.1, the problem can now be extended to nonzero boundary conditions on all boundaries, e.g., $c(x,y,0) = 0$, $c(L,y,t) = 1$, $c(0,y,t) = 1$, $c(x,0,t) = 1$, $c(x,L,t) = 1$. This can be done by first calculating four component solutions (each with only one nonzero boundary condition) and then adding them up. In Figure 4.2, we use this procedure to calculate concentration profiles at different times, t. Comparing these profiles with those of Figure 2.3, we note that the presence of the reaction term slows down the inward propagation of the diffusive front. It is also interesting to note that the steady-state solution is no longer the trivial $c(x,y) = 1$ over the whole domain, but rather resembles the rightmost profile in Figure 4.2 ($t = 0.05$). We will see more profiles of this type in Chapter 9 when we discuss microfabrication protocols based on RD etching.

To conclude the discussion of analytically tractable RD equations, let us have a quick look at processes involving higher order reaction terms. As an example, consider a diffusive process coupled with a second-order reaction for which the non-dimensionalized equation is

$$\frac{\partial^2 \bar{c}(\bar{x},\bar{y})}{\partial \bar{x}^2} + \frac{\partial^2 \bar{c}(\bar{x},\bar{y})}{\partial \bar{y}^2} - \mathrm{Da}\bar{c}(\bar{x},\bar{y})^2 = 0 \tag{4.13}$$

with $\mathrm{Da} = C_0 L^2 k/D$. Attempting to solve this problem by the separation of variables, we write $\bar{c}(\bar{x},\bar{y}) = X(\bar{x})Y(\bar{y})$ and rearrange the governing equation into

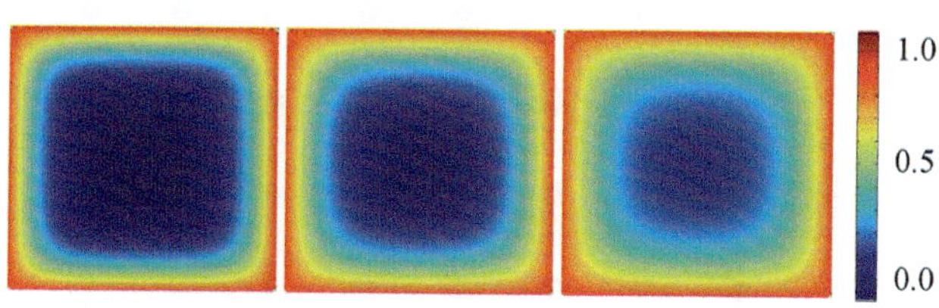

Figure 4.2 Time-dependent concentration profiles for a reaction–diffusion process initiated from the boundaries of a square kept at constant concentration equal to unity and with initial concentration in the square set to zero. The snapshots shown here correspond to times $t = 0.01$, $t = 0.02$, $t = 0.05$. The $t = 0.05$ profile is very close to a steady-state solution

$X''/X + Y''/Y - \mathrm{Da}XY = 0$. Unfortunately, the last term contains *both* X and Y, indicating that it is impossible to transform the equation into a form in which the left- and right-hand sides are functions of different variables (i.e., $f(\bar{x}) = f(\bar{y}) =$ constant). This complication extends to all other reaction orders, n, in which there will always be mixed X^nY^n terms.

4.3 SPATIAL DISCRETIZATION

Faced with RD problems that evade analytical treatment, one mast turn to numerical solutions, in which otherwise continuous coordinates are divided into discrete elements. In this section, we discuss methods for the discretization of the spatial coordinates and of their derivatives.

4.3.1 Finite Difference Methods

In the finite difference (FD) method, one finds an approximate, discrete form of the governing equation, which may then be solved to give an approximate solution. For example, consider the Laplace equation in one dimension (identical to the steady-state diffusion equation), $\partial^2 c(x)/\partial x^2 = 0$, on the domain $0 \leq x \leq L$. This domain may be divided into N points located at $x_1 \ldots x_i \ldots x_N$ and separated by $N-1$ intervals of equal length, $\Delta x = L/(N-1)$, such that $x_i = (i-1)\Delta x$ (Figure 4.3).

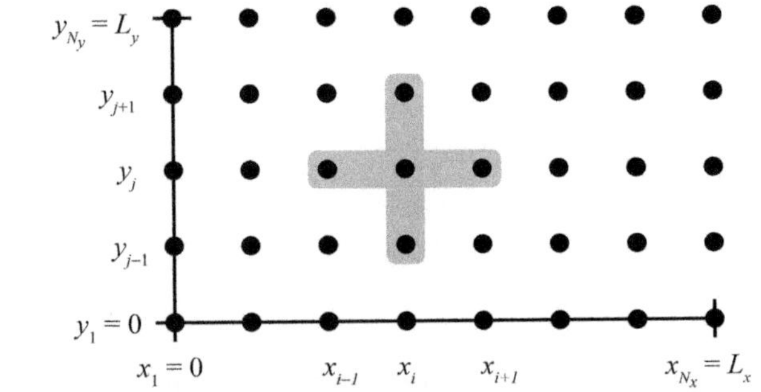

Figure 4.3 Spatial discretization and indexing in the finite difference (FD) method in (a) one dimension and (b) two dimensions. The highlighted areas centered on i in (a) and i, j in (b) illustrate the grid points required to compute the discretized Laplacian operator

To find the FD approximation to this equation, consider the Taylor series for the function $c(x_i + \Delta x)$ about x_i:

$$c(x_i + \Delta x) = c(x) + (\partial c/\partial x)_{x_i}\Delta x + (\partial^2 c/\partial x^2)_{x_i}(\Delta x)^2/2 + \mathrm{O}[(\Delta x)^3] \quad (4.14)$$

Dividing both sides by Δx, we arrive at the forward difference approximation for the first derivative, which is accurate to *first order*:

$$(\partial c/\partial x)_{x_i} = [c(x_i + \Delta x) - c(x)]/\Delta x + \mathrm{O}(\Delta x) \quad (4.15)$$

or, in index notation, $(\partial c/\partial x)_i = [c_{i+1} - c_i]/\Delta x + \mathrm{O}(\Delta x)$. Likewise, from the expansion of $c(x_i - \Delta x)$, we find the *backward difference* approximation for the first derivative

$$c(x_i - \Delta x) = c(x_i) - (\partial c/\partial x)_{x_i}\Delta x + (\partial^2 c/\partial x^2)_{x_i}(\Delta x)^2/2 - \mathrm{O}[(\Delta x)^3] \quad (4.16)$$

or succinctly $(\partial c/\partial x)_i = [c_i - c_{i-1}]/\Delta x + \mathrm{O}(\Delta x)$. Although the simple forward and backward differencing schemes are only accurate to first order, we may derive a *second-order* approximation by subtracting $c(x_i - \Delta x)$ from $c(x_i + \Delta x)$:

$$\begin{aligned} &c(x_i + \Delta x) - c(x_i - \Delta x) = 2(\partial c/\partial x)_{x_i}\Delta x + \mathrm{O}[(\Delta x)^3] \\ &(\partial c/\partial x)_i = [c_{i+1} - c_{i-1}]/2\Delta x + O[(\Delta x)^2] \end{aligned} \quad (4.17)$$

Finally, and most importantly in the context of diffusion, the second-order finite difference approximation to the second derivative $(\partial^2 c/\partial x^2)_{x_i}$ can be derived by adding $c(x_i - \Delta x)$ and $c(x_i + \Delta x)$:

$$\begin{aligned} &c(x_i + \Delta x) + c(x_i - \Delta x) = 2c(x_i) + (\partial^2 c/\partial x^2)_{x_i}(\Delta x)^2 + \mathrm{O}[(\Delta x)^4] \\ &(\partial^2 c/\partial x^2)_i = [c_{i-1} - 2c_i + c_{i+1}]/(\Delta x)^2 + \mathrm{O}[(\Delta x)^2] \end{aligned} \quad (4.18)$$

By analogous reasoning, the Laplacian operator in two and three dimensions may be approximated as, respectively

$$\begin{aligned} \nabla^2 c\big|_{i,j} &= [c_{i+1,j} + c_{i-1,j} + c_{i,j-1} + c_{i,j+1} - 4c_{i,j}]/(\Delta x)^2 + \mathrm{O}[(\Delta x)^2] \\ \nabla^2 c\big|_{i,j,k} &= [c_{i+1,j,k} + c_{i-1,j,k} + c_{i,j-1,k} + c_{i,j+1,k} + c_{i,j,k-1} + c_{i,j,k+1} \\ &\quad - 6c_{i,j,k}]/(\Delta x)^2 + \mathrm{O}[(\Delta x)^2] \end{aligned} \quad (4.19)$$

A variety of boundary conditions (e.g., Dirichlet, von Neumann and Robin types) may be easily incorporated into this scheme using appropriate FD approximations. For example, consider again the Laplace equation, this time in a two-dimensional rectangular domain, $\nabla^2 c(x,y) = 0$ for $0 \le x \le L_x$ and $0 \le y \le L_y$. For Dirichlet boundary conditions of the form $c(0,y) = f(y)$, one may simply specify the desired value at the boundary, $c_{1,j} = f_j$. For von Neumann boundary conditions of the form $\partial c/\partial x|_{0,y} = g(y)$, one may write the central difference form as $(c_{i+1,j} - c_{i-1,j})/2\Delta x = g_j$, such that $c_{i-1,j} = c_{i+1,j} - 2g_j\Delta x$. Substituting the last expression into Equation (4.19) gives the following expression for the boundary at $i = 1$: $[2(-g_j\Delta x - c_{i,j} + c_{i+1,j}) + (c_{i,j-1} - 2c_{i,j} + c_{i,j+1})] = 0$. Note that in the special case of no flux boundaries, $c_{i-1,j} = c_{i+1,j}$, such that there is a 'mirror point' on the other side of the boundary at $i = 1$. Finally, for mixed boundary conditions of the form $\partial c/\partial x|_{0,y} + f(y)c(0,y) = g(y)$, one may again take the central difference approximation and express $c_{i-1,j} = c_{i+1,j} + 2c_{i,j}f_j\Delta x - 2g_i\Delta x$, such that the discrete governing equation at the boundary is given by $[2(-g_j\Delta x - (1 - f_j\Delta x)c_{i,j} + c_{i+1,j}) + (c_{i,j-1} - 2c_{i,j} + c_{i,j+1})] = 0$.

Despite all the ugly-looking subscripts, the FD approach is very easy to implement in various programming languages and is thus a method of choice for standard RD problems. In Example 4.1 we illustrate how it can help the ancient Greek hero Theseus to find the Minotaur beast in the Cretan labyrinth. For more modern examples, the reader is directed elsewhere.[5–8]

Example 4.1 How Diffusion Betrayed the Minotaur

According to Greek mythology, the Minotaur was a fearsome beast, part bull and part man, that dwelled in a labyrinth of winding passages on the island of Crete. When Theseus, the greatest of Greek heroes, arrived to slay the monster, he came armed with a sword and a ball of thread, with which to guide himself out of the maze. However, in a lesser-known version of the myth, Theseus forgets his ball of thread and is forced to go on without it. How does he manage to find the beast and escape an eternity of wandering within the labyrinth? By applying his knowledge of diffusion, of course.

The Minotaur is a foul beast, and its odor permeates every inch of the maze, although in different concentrations. At the center of the maze, in the Minotaur's lair, the concentration of odor molecules is effectively constant with a value C_0. Meanwhile, at the entrances/exits of the maze, which are open to fresh air, the odor concentration is negligible. Furthermore, as the Minotaur has been lurking within the labyrinth for ages, only the 'steady-state' concentration profiles are relevant to Theseus' present challenge (i.e., $\partial c/\partial t \to 0$). Mathematically, this diffusion problem may be stated as follows:

$$0 = D\left(\frac{\partial^2 c(x,y)}{\partial x^2} + \frac{\partial^2 c(x,y)}{\partial y^2}\right) \quad \text{BC}: \; c(\text{center}) = C_0,\; c(\text{exits}) = 0$$

$$\text{and} \quad \partial c/\partial x|_{\text{walls}} = 0$$

Due to complex geometry of the labyrinth, an analytical solution to this problem would be most difficult and impractical; therefore, we adopt a simple numerical approach using the finite difference scheme.

In this computational approach, a two-dimensional grid is created in which each grid point is designated as one of four possible types: 1. source, 2. sink, 3. wall and 4. interior. At the source and sink points, the concentrations are fixed to C_0 and zero, respectively, corresponding to the Minotaur's lair and the exits. Since there is no diffusion through the walls of the labyrinth, these points are not considered in the calculation. To find the steady-state solution, we start with the diffusion equation at time $t=0$ ($c=C_0$ at the Minotaur's lair, zero otherwise) and integrate it with respect to time according to the following finite difference scheme (FTCS method; Section 4.4.1):

$$c_{i,j}^{t+1} - c_{i,j}^{t} = \alpha(c_{i-1,j}^{t} + c_{i+1,j}^{t} + c_{i,j-1}^{t} + c_{i,j+1}^{t} - 4c_{i,j}^{t})$$

where $\alpha = D\Delta t/(\Delta x)^2$. At the boundaries of the labyrinth (i.e., where interior points border wall points), we must alter the above relation to account for the no flux through the walls. For example, for an interior point bordered by two wall points in the positive x-direction and the positive y-direction, the discrete diffusion equation is written as $c_{i,j}^{t+1} - c_{i,j}^{t} = \alpha(2c_{i-1,j}^{t} + 2c_{i,j-1}^{t} - 4c_{i,j}^{t})$, since $c_{i+1,j}^{t} = c_{i-1,j}^{t}$ and $c_{i,j-1}^{t} = c_{i,j+1}^{t}$ ('no flux' boundary conditions). This approach may be generalized to all possible combinations of no flux boundaries (cf. sample code at the end of the chapter).

Upon integrating the numerical scheme described above, we find the steady-state concentration profile of the Minotaur's odor, which may be used to guide our hero in and out of the maze in a most efficient manner. Specifically, Theseus first finds the entrance from which the odor is diffusing at the greatest rate (i.e., largest flux); this is the entrance closest to the Minotaur's lair. To see why, consider steady-state diffusion in one dimension, in which the concentration decreases linearly as $c(x) = C_0(1 - x/L)$ from source ($x=0$) to sink ($x=L$). Thus, the flux, $j(x) = DC_0/L$, depends inversely on the distance between source and sink – i.e., the greater the flux, the shorter the distance. From this entrance, Theseus has to follow the path along which the concentration is increasing most rapidly; this guides him to the Minotaur along the shortest route. After slaying the beast, Theseus returns along the same route by following the path along which the concentration is decreasing most rapidly.

The left-hand picture below shows the labyrinth, locations of the entrances (red) and the Minotaur's location (green). The right-hand picture gives the steady-state concentration of the Minotaur's foul odor (shades of orange).

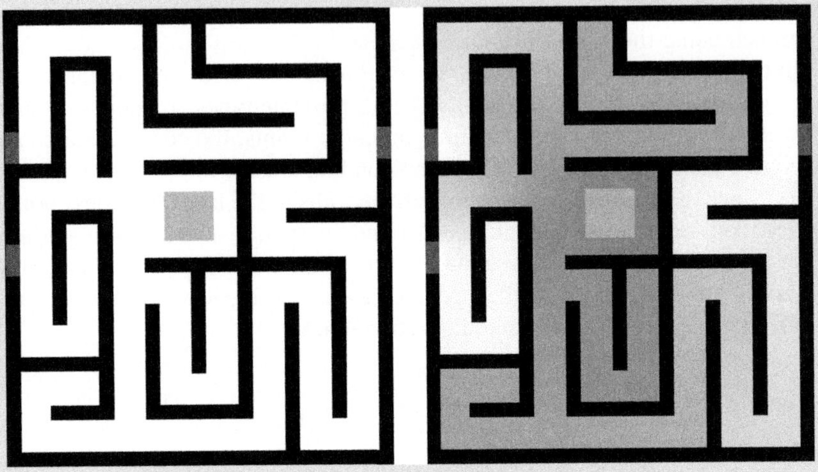

Note. A commented Matlab code for solving this problem is included in Appendix B at the end of the book.

4.3.2 Finite Element Methods

In contrast to the FD approach, in which one approximates the governing equation, the finite element method (FEM) seeks to approximate the solution of the problem. In doing so, it transforms the 'global' differential equations into matrix operations on individual points of a discretized grid. This reduces the complexity of the problem immensely and allows the FEM to find solutions over arbitrary domains, including those of convoluted shapes for which implementing boundary/initial conditions by other methods would be virtually impossible. If one adds to this that the FEM deals efficiently with problems incorporating sources, sinks and steep concentration gradients, it becomes evident why so many engineering software packages (COMSOL, ANSYS, etc.) are based on this method.

Unfortunately, the prowess of the FEM comes at a price. Finite elements are much less intuitive than finite differences, and they are much harder to code. Since voluminous literature[9–12] on the subject is usually too mathematically involved for a FEM novice, we provide this section as a primer that focuses on the conceptual aspects of the method. Nevertheless, some mathematics is inevitable, and readers averse to lengthy equations might choose to skip this section, at least on the first reading of the book.

FEM begins with dividing the problem's spatial domain into discrete (*finite*) elements (Figure 4.4). In one-dimensional problems, these elements are simple

Figure 4.4 Triangular finite elements generated by COMSOL to approximate a circular domain. Colors correspond to the 'quality' of the elements. More isotropic/equilateral triangles have better scores since they usually give better computational accuracy

segments; in two dimensions they are polygons (usually triangles or rectangles); in three, they are polyhedra (e.g., tetrahedrons or parallelepipeds). The shapes and sizes of these elements may be identical over the entire domain or may vary from place to place, with smaller elements usually used to discretize regions near the boundaries. As a rule of thumb, the more elements are used, the more accurate the FEM solution will be.

The solution we seek is then approximated over each element using so-called interpolating (or 'shape') functions (typically linear or polynomial). These component functions are then 'stitched' together in such a way that their values and/or derivatives on common nodes (i.e., vertices common to neighboring elements) match up.

The key step of the FEM procedure is to evaluate and minimize the degree of error in the approximate solution constructed from the interpolating functions. Since we do not know the exact solution of the problem *a priori*, evaluation of this difference is done by a neat mathematical 'trick' in which the approximate solution is substituted back into the governing equation and the discrepancy between the approximate and the true solutions (the 'residual') is minimized using calculus of variations. This approach yields an integral equation defining the most accurate of the approximate solutions. Although the integral equation is rather complicated, it can be translated into simple operations on matrices, which, despite their large sizes, can be easily computed and/or manipulated on a desktop computer to give the final solution to the problem.

To illustrate how this machinery functions in practice, let us solve a simple, one-dimensional problem using the FEM. Consider one-dimensional diffusion with a zeroth-order reaction term

$$\frac{\partial c}{\partial t} = D\frac{\partial^2 c}{\partial x^2} - k \tag{4.20}$$

where L is the size of the domain, and the initial ($c(x,0) = 0.5$) and boundary ($c(0,t>0) = 1$, $c(L,t>0) = 0$) conditions are applied. We proceed as follows.

Step 1. *Discretize solution domain into elements.* Since the problem is one-dimensional, we divide the $[0, L]$ domain into linear 'elements'. For simplicity, we use only three elements, E1, E2, E3, as illustrated in Figure 4.5.

Step 2. *Select appropriate interpolation functions.* Consider now the first element E1 and its nodes 1 and 2. We introduce a local coordinate system with $l = -1$ corresponding to the position of node 1, x_1, and $l = 1$ corresponding to the position of node 2, x_2 (Figure 4.5(b)). For simplicity, we choose linear interpolating functions $S_1 = (1 - l)/2$ and $S_2 = (1 + l)/2$. Using concentrations at nodes 1 and 2 (c_1 and c_2, respectively), the approximate concentration profile over element E1 can be written in terms of the interpolating functions as $c = \frac{1}{2}(1 - l)\, c_1 + \frac{1}{2}(1 + l)c_2 = S_1c_1 + S_2c_2$.

Before proceeding further, we derive some auxiliary identities that will be useful in the next step. For E1, the relationship between the local and 'global' x coordinates is given as

$$l = \frac{2}{x_2 - x_1}(x - x_1) - 1 \tag{4.21}$$

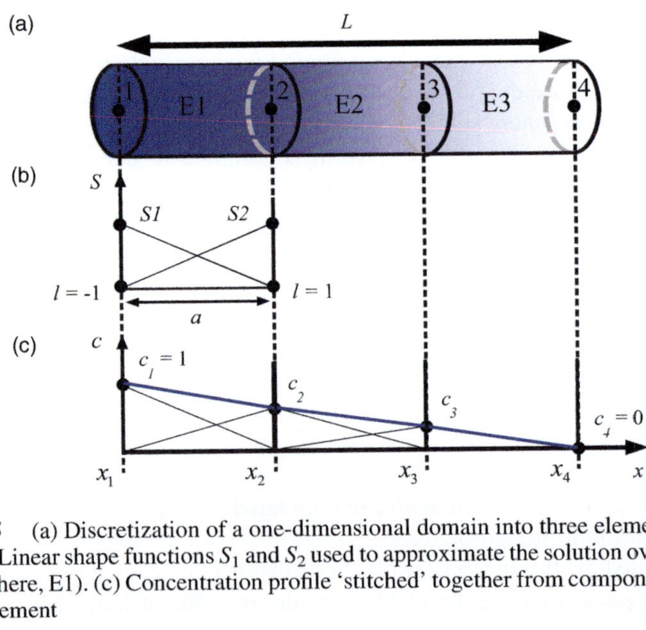

Figure 4.5 (a) Discretization of a one-dimensional domain into three elements and four nodes. (b) Linear shape functions S_1 and S_2 used to approximate the solution over one of the elements (here, E1). (c) Concentration profile 'stitched' together from component solutions on each element

and its differentiation gives

$$\frac{\mathrm{d}l}{\mathrm{d}x} = \frac{2}{x_2 - x_1} = \frac{2}{a} \tag{4.22}$$

where a is the length of E1. This allows for the linearization of concentration gradients:

$$\frac{\partial c}{\partial x} = \frac{\partial c}{\partial l}\frac{\partial l}{\partial x} = \frac{2}{a}\frac{\partial}{\partial l}(S_1 c_1 + S_2 c_2) = \frac{2}{a}\left(c_1 \frac{\partial S_1}{\partial l} + c_2 \frac{\partial S_2}{\partial l}\right) \tag{4.23}$$

The formulas for concentrations and gradients can then be rewritten in matrix notation as, respectively (i) $c = \mathbf{S}\mathbf{c}^T$, where $\mathbf{S} = [S_1 S_2]$, $\mathbf{c} = [c_1 c_2]$ is the vector of node concentrations and 'T' denotes the transpose of matrix $\mathbf{c}$, and (ii)

$$\frac{\partial \mathbf{S}}{\partial x} = \frac{2}{a}\frac{\partial \mathbf{S}}{\partial l} \quad \text{and} \quad \frac{\partial c}{\partial x} = \frac{2}{a}\frac{\partial \mathbf{S}}{\partial l}\mathbf{c}^{\mathrm{T}} \tag{4.24}$$

Finally, introducing the local coordinates for E2 and E3 and using the same form of interpolating functions, we can write similar approximate solutions for these elements. Subsequently, we can 'stitch' these 'local' solutions into a 'global' one, over the entire [0, L] domain (Figure 4.5(c)) as $\tilde{c} = \tilde{S}_1 c_1 + \tilde{S}_2 c_2 + \tilde{S}_3 c_3 + \tilde{S}_4 c_4$, where the interpolating functions are now slightly redefined to avoid double-counting the values of concentrations at the interior nodes 2 and 3:

$$\tilde{S}_1 = S_1 \quad \text{for} \quad x_1 \le x \le x_2; \quad \tilde{S}_2 = \begin{cases} S_2 & x_1 \le x \le x_2 \\ S_1 & x_2 < x \le x_3 \end{cases};$$
$$\tilde{S}_3 = \begin{cases} S_2 & x_2 \le x \le x_3 \\ S_1 & x_3 < x \le x_4 \end{cases}; \quad \text{and} \quad \tilde{S}_4 = S_2 x_3 \le x \le x_4 \tag{4.25}$$

***Step 3**. Derive and simplify an integral equation for the 'best' approximate solution.* Conceptually, this is probably the hardest part of the method, which we illustrate using the popular Galerkin method. First, we substitute the approximate solution $\tilde{c}$ into the original diffusion equation, and define the residual as the value of

$$\frac{\partial \tilde{c}}{\partial t} - D\frac{\partial^2 \tilde{c}}{\partial x^2} + k \tag{4.26}$$

Note that the residual approaches zero as the approximate solution approaches the exact one – in other words, the residual measures how much off we are from the exact solution. Next, we use the variational method outlined in Example 4.2 to show that the residual is minimized if the following integral

Example 4.2 The Origins of the Galerkin Finite Element Scheme

Galerkin's method is arguably the most popular and accurate finite element technique. The cornerstone of this approach is Equation (4.27) in the main text that provides a condition for the minimal difference between the approximate and the exact solutions to a problem of interest. In this example, we outline – at least qualitatively – where this equation comes from.

Let ϕ denote the exact solution to a given problem, and $\tilde{\phi}$ be its FEM approximation expressed in terms of the interpolating functions $\tilde{F}_m$ as $\tilde{\phi} = \sum_{m=1}^{M} a_m \tilde{F}_m$, where the coefficients a_m are the values of $\tilde{\phi}$ at the nodes. The optimal approximate solution should correspond to a minimal sum (integral) of squared deviations between ϕ and $\tilde{\phi}$ over the entire domain of the problem, Ω: $\int_\Omega (\phi - \tilde{\phi})^2 \mathrm{d}\Omega$. To minimize this integral, we seek the zeros of its first derivative with respect to all coefficients a_m:

$$\frac{\partial}{\partial a_m} \int_\Omega (\phi - \tilde{\phi})^2 \mathrm{d}\Omega = \int_\Omega 2(\phi - \tilde{\phi}) \frac{\partial \tilde{\phi}}{\partial a_m} \mathrm{d}\Omega = 0$$

Because $\partial\tilde{\phi}/\partial a_m = \tilde{F}_m$, we have $\int_\Omega (\phi - \tilde{\phi}) \tilde{F}_m \mathrm{d}\Omega = 0$ for all values of m, which is the condition underlying the Galerkin method. This extremum condition corresponds to a minimum (not a maximum or a saddle point) since the second derivative of the integral is positive:

$$\frac{\partial^2}{\partial {a_m}^2} \int_\Omega (\phi - \tilde{\phi})^2 \mathrm{d}\Omega = \int_\Omega 2\tilde{F}_m^2 \mathrm{d}\Omega > 0$$

Note that the symbols for the coefficients a_m and shape functions $\tilde{F}_m$ are different from those used in the RD example described in Section 4.3.2. There, the shape functions $\tilde{S}_m$ and coefficients c_m approximated concentration profile $\tilde{c}$ which entered as *argument* into the functional ϕ ($\tilde{\phi} = (\partial\tilde{c}/\partial t) - D(\partial^2\tilde{c}/\partial x^2) + k$). While it can be shown that the Galerkin method applies to such 'convoluted' cases, the full mathematical derivation is beyond the scope of this book and the reader is directed to Gockenbach[13] for further details.

condition holds:

$$\int_0^L \left\{ \frac{\partial\tilde{c}}{\partial t} - D\frac{\partial^2\tilde{c}}{\partial x^2} + k \right\} \tilde{\mathbf{S}}^T \mathrm{d}x = 0 \tag{4.27}$$

Integrating the diffusive term by parts we obtain

$$\int_0^L \left\{ D\frac{\partial^2\tilde{c}}{\partial x^2} \right\} \tilde{\mathbf{S}}^T \mathrm{d}x = D\tilde{\mathbf{S}}^T \frac{\partial\tilde{c}}{\partial x}\bigg|_0^L - \int_0^L D\frac{\partial\tilde{\mathbf{S}}^T}{\partial x}\frac{\partial\tilde{c}}{\partial x}\mathrm{d}x \tag{4.28}$$

Substituting this equation into (4.27) and using the linearization of concentration gradients (Step 2):

$$\int_0^L D\frac{\partial\tilde{\mathbf{S}}^T}{\partial x}\frac{\partial\tilde{\mathbf{S}}}{\partial x}\mathbf{c}^{\mathrm{T}}\mathrm{d}x + \int_0^L \tilde{\mathbf{S}}^T\tilde{\mathbf{S}}\frac{\partial\mathbf{c}^{\mathrm{T}}}{\partial t}dx = -\int_0^L k\tilde{\mathbf{S}}^{\mathrm{T}}\mathrm{d}x + D\tilde{\mathbf{S}}^{\mathrm{T}}\frac{\partial\tilde{c}}{\partial x}\bigg|_0^L \tag{4.29}$$

At this point it is important to note that owing to the use of shape functions, we have made matrix $\mathbf{c}$ independent of x (since $\mathbf{c}$ stores only the values at the nodes). This property allows one to take $\mathbf{c}$ out of the integrals so that

$$\left(\int_0^L D\frac{\partial\tilde{\mathbf{S}}^{\mathrm{T}}}{\partial x}\frac{\partial\tilde{\mathbf{S}}}{\partial x}\mathrm{d}x\right)\mathbf{c}^{\mathrm{T}} + \left(\int_0^L \tilde{\mathbf{S}}^{\mathrm{T}}\tilde{\mathbf{S}}\mathrm{d}x\right)\frac{\partial\mathbf{c}^{\mathrm{T}}}{\partial t} = \left(-\int_0^L k\tilde{\mathbf{S}}^{\mathrm{T}}\mathrm{d}x + D\tilde{\mathbf{S}}^{\mathrm{T}}\frac{\partial\tilde{c}}{\partial x}\bigg|_0^L\right) \tag{4.30}$$

and all the terms in the parentheses do not depend on time.

Step 4. *Transform the integral equation into a set of algebraic equations for individual elements.* When the domain of integration is subdivided into domains corresponding to the elements E1, E2, E3, Equation (4.30) becomes

$$\begin{aligned}
&\left(\int_{x_1}^{x_2} D\frac{\partial\tilde{\mathbf{S}}^{\mathrm{T}}}{\partial x}\frac{\partial\tilde{\mathbf{S}}}{\partial x}\mathrm{d}x + \int_{x_2}^{x_3} D\frac{\partial\tilde{\mathbf{S}}^{\mathrm{T}}}{\partial x}\frac{\partial\tilde{\mathbf{S}}}{\partial x}\mathrm{d}x + \int_{x_3}^{x_4} D\frac{\partial\tilde{\mathbf{S}}^{\mathrm{T}}}{\partial x}\frac{\partial\tilde{\mathbf{S}}}{\partial x}\mathrm{d}x\right)\mathbf{c}^{\mathrm{T}} \\
&\quad + \left(\int_{x_1}^{x_2} \tilde{\mathbf{S}}^{\mathrm{T}}\tilde{\mathbf{S}}\mathrm{d}x + \int_{x_2}^{x_3} \tilde{\mathbf{S}}^{\mathrm{T}}\tilde{\mathbf{S}}\mathrm{d}x + \int_{x_3}^{x_4} \tilde{\mathbf{S}}^{\mathrm{T}}\tilde{\mathbf{S}}\mathrm{d}x\right)\frac{\partial\mathbf{c}^{\mathrm{T}}}{\partial t} \\
&\quad = \left(-\int_{x_1}^{x_2} k\tilde{\mathbf{S}}^{\mathrm{T}}\mathrm{d}x - \int_{x_2}^{x_3} k\tilde{\mathbf{S}}^{\mathrm{T}}\mathrm{d}x - \int_{x_3}^{x_4} k\tilde{\mathbf{S}}^{\mathrm{T}}\mathrm{d}x\right. \\
&\quad \left. + D\tilde{\mathbf{S}}^{\mathrm{T}}(L)\frac{\partial\tilde{c}(L)}{\partial x} - D\tilde{\mathbf{S}}^{\mathrm{T}}(0)\frac{\partial\tilde{c}(0)}{\partial x}\right)
\end{aligned} \tag{4.31}$$

where the last two terms on the right-hand side of the equation indicate the flux boundary conditions at the ends of the domain. Because the solution has to hold for each finite element, we can consider them separately. For example, over the domain of E1 we have

$$\left(\int_{x_1}^{x_2} D\frac{\partial\mathbf{S}^{\mathrm{T}}}{\partial x}\frac{\partial\mathbf{S}}{\partial x}\mathrm{d}x\right)\mathbf{c}^{\mathrm{T}} + \left(\int_{x_1}^{x_2} \mathbf{S}^{\mathrm{T}}\mathbf{S}\mathrm{d}x\right)\frac{\partial\mathbf{c}^{\mathrm{T}}}{\partial t} = \left(-\int_{x_1}^{x_2} k\mathbf{S}^{\mathrm{T}}\mathrm{d}x - D\mathbf{S}^{\mathrm{T}}(0)\frac{\partial\tilde{c}(0)}{\partial x}\right) \tag{4.32}$$

where the tilde signs have been dropped from the shape functions, $\mathbf{S}$, since $\tilde{\mathbf{S}} = \mathbf{S} = [S_1 \quad S_2]$ in the range $x_1 \le x \le x_2$ for the first element. Now, because $\partial\mathbf{S}/\partial x = [-1\ 1]/2$ and $\partial\mathbf{S}/\partial x = (2/a)(\partial\mathbf{S}/\partial l)$ (4.23) we have $\partial\mathbf{S}/\partial x = [-1\ 1]/a$ and the first term in parentheses on the left-hand side of (4.32) can be rewritten as

$$\int_{x_1}^{x_2} D\frac{\partial\mathbf{S}^{\mathrm{T}}}{\partial x}\frac{\partial\mathbf{S}}{\partial x}\mathrm{d}x = \frac{D}{a^2}\int_{x_1}^{x_2}\begin{bmatrix}-1\\1\end{bmatrix}[-1 \quad 1]\mathrm{d}x = \frac{D}{a}\begin{bmatrix}1 & -1\\-1 & 1\end{bmatrix} \tag{4.33}$$

Similarly, the second term in parentheses on the left-hand side becomes

$$\begin{aligned}\int_{x_1}^{x_2}\mathbf{S}^{\mathrm{T}}\mathbf{S}\mathrm{d}x &= \frac{a}{2}\int_{-1}^{1}\mathbf{S}^{\mathrm{T}}\mathbf{S}\mathrm{d}l = \frac{a}{2}\int_{-1}^{1}\frac{1}{2}\begin{bmatrix}1-l\\1+l\end{bmatrix}\frac{1}{2}[1-l \quad 1+l]\mathrm{d}l\\ &= \frac{a}{8}\int_{-1}^{1}\begin{bmatrix}(1-l)^2 & 1-l^2\\1-l^2 & (1+l)^2\end{bmatrix}\mathrm{d}l = \frac{a}{6}\begin{bmatrix}2 & 1\\1 & 2\end{bmatrix}\end{aligned} \tag{4.34}$$

and the term on the right-hand side is

$$\begin{aligned}-\int_{x_1}^{x_2}k\mathbf{S}^{\mathrm{T}}\mathrm{d}x - D\mathbf{S}^{\mathrm{T}}(0)\frac{\partial\tilde{c}(0)}{\partial x} &= -\frac{ka}{2}\int_{-1}^{1}\frac{1}{2}\begin{bmatrix}1-l\\1+l\end{bmatrix}\mathrm{d}l + \frac{D}{a}\begin{bmatrix}1\\0\end{bmatrix}[-1 \quad 1]\begin{bmatrix}c_1\\c_2\end{bmatrix}\\ &= -\frac{ka}{2}\begin{bmatrix}1\\1\end{bmatrix} + \frac{D}{a}\begin{bmatrix}-1 & 1\\0 & 0\end{bmatrix}\begin{bmatrix}c_1\\c_2\end{bmatrix}\end{aligned} \tag{4.35}$$

Substituting (4.33)–(4.35) into (4.32) gives the algebraic equation for E1:

$$\begin{aligned}&\frac{D}{a}\begin{bmatrix}1 & -1\\-1 & 1\end{bmatrix}\begin{bmatrix}c_1\\c_2\end{bmatrix} + \frac{a}{6}\begin{bmatrix}2 & 1\\1 & 2\end{bmatrix}\begin{bmatrix}\partial c_1/\partial t\\\partial c_2/\partial t\end{bmatrix}\\ &= -\frac{ka}{2}\begin{bmatrix}1\\1\end{bmatrix} + \frac{D}{a}\begin{bmatrix}-1 & 1\\0 & 0\end{bmatrix}\begin{bmatrix}c_1\\c_2\end{bmatrix}\end{aligned} \tag{4.36}$$

Analogous reasoning (i.e., repetition of Equations (4.32) to (4.36)) gives algebraic equations for E2 and E3:

$$\frac{D}{a}\begin{bmatrix}1 & -1\\-1 & 1\end{bmatrix}\begin{bmatrix}c_2\\c_3\end{bmatrix} + \frac{a}{6}\begin{bmatrix}2 & 1\\1 & 2\end{bmatrix}\begin{bmatrix}\partial c_2/\partial t\\\partial c_3/\partial t\end{bmatrix} = -\frac{ka}{2}\begin{bmatrix}1\\1\end{bmatrix} \tag{4.37}$$

$$\frac{D}{a}\begin{bmatrix}1 & -1\\-1 & 1\end{bmatrix}\begin{bmatrix}c_3\\c_4\end{bmatrix} + \frac{a}{6}\begin{bmatrix}2 & 1\\1 & 2\end{bmatrix}\begin{bmatrix}\partial c_3/\partial t\\\partial c_4/\partial t\end{bmatrix} = -\frac{ka}{2}\begin{bmatrix}1\\1\end{bmatrix} + \frac{D}{a}\begin{bmatrix}0 & 0\\-1 & 1\end{bmatrix}\begin{bmatrix}c_3\\c_4\end{bmatrix} \tag{4.38}$$

Since each of the equalities (4.36)–(4.38) represents two equations, each with two unknown concentrations, we have the total of six equations and four unknowns, c_1, c_2, c_3 and c_4. Before such a system can be solved by linear algebra, it needs to be reduced to a system of four equations with four unknowns. This is what we are going to do in the next step.

Step 5. *Assemble element equations for the entire domain.* We now construct the matrix equation that holds for the entire domain of the solution. To start with, let us consider the first term of the E1 solution (4.36); this term can be written in an equivalent but expanded matrix form:

$$\frac{D}{a}\begin{bmatrix} 1 & -1 \\ -1 & 1 \end{bmatrix}\begin{bmatrix} c_1 \\ c_2 \end{bmatrix} \Leftrightarrow \frac{D}{a}\begin{bmatrix} 1 & -1 & 0 & 0 \\ -1 & 1 & 0 & 0 \\ 0 & 0 & 0 & 0 \\ 0 & 0 & 0 & 0 \end{bmatrix}\begin{bmatrix} c_1 \\ c_2 \\ c_3 \\ c_4 \end{bmatrix} \tag{4.39}$$

The first terms of (4.37) and (4.38) can be 'expanded' in an analogous manner and then summed:

$$\begin{aligned} &\frac{D}{a}\left(\begin{bmatrix} 1 & -1 & 0 & 0 \\ -1 & 1 & 0 & 0 \\ 0 & 0 & 0 & 0 \\ 0 & 0 & 0 & 0 \end{bmatrix} + \begin{bmatrix} 0 & 0 & 0 & 0 \\ 0 & 1 & -1 & 0 \\ 0 & -1 & 1 & 0 \\ 0 & 0 & 0 & 0 \end{bmatrix} + \begin{bmatrix} 0 & 0 & 0 & 0 \\ 0 & 0 & 0 & 0 \\ 0 & 0 & 1 & -1 \\ 0 & 0 & -1 & 1 \end{bmatrix}\right)\begin{bmatrix} c_1 \\ c_2 \\ c_3 \\ c_4 \end{bmatrix} \\ &= \frac{D}{a}\begin{bmatrix} 1 & -1 & 0 & 0 \\ -1 & 2 & -1 & 0 \\ 0 & -1 & 2 & -1 \\ 0 & 0 & -1 & 1 \end{bmatrix}\begin{bmatrix} c_1 \\ c_2 \\ c_3 \\ c_4 \end{bmatrix} \end{aligned} \tag{4.40}$$

(Note that this matrix formulation gives a result identical to the direct summation of the first terms of (4.36)–(4.38). Repeating the same procedure for other corresponding terms in (4.36) to (4.38) leads us to the following matrix formulation of the problem:

$$\begin{aligned} &\frac{D}{a}\begin{bmatrix} 1 & -1 & 0 & 0 \\ -1 & 2 & -1 & 0 \\ 0 & -1 & 2 & -1 \\ 0 & 0 & -1 & 1 \end{bmatrix}\begin{bmatrix} c_1 \\ c_2 \\ c_3 \\ c_4 \end{bmatrix} + \frac{1}{6}\begin{bmatrix} 2 & 1 & 0 & 0 \\ 1 & 4 & 1 & 0 \\ 0 & 1 & 4 & 1 \\ 0 & 0 & 1 & 2 \end{bmatrix}\begin{bmatrix} \partial c_1/\partial t \\ \partial c_2/\partial t \\ \partial c_3/\partial t \\ \partial c_4/\partial t \end{bmatrix} \\ &= -\frac{ka}{2}\begin{bmatrix} 1 \\ 2 \\ 2 \\ 1 \end{bmatrix} + \frac{D}{a}\begin{bmatrix} -1 & 1 & 0 & 0 \\ 0 & 0 & 0 & 0 \\ 0 & 0 & 0 & 0 \\ 0 & 0 & -1 & 1 \end{bmatrix}\begin{bmatrix} c_1 \\ c_2 \\ c_3 \\ c_4 \end{bmatrix} \end{aligned} \tag{4.41}$$

After subtraction of the last term on the right-hand side from the first term on the left-hand side we get

$$\underbrace{\frac{D}{a}\begin{bmatrix} 2 & -2 & 0 & 0 \\ -1 & 2 & -1 & 0 \\ 0 & -1 & 2 & -1 \\ 0 & 0 & 0 & 0 \end{bmatrix}\begin{bmatrix} c_1 \\ c_2 \\ c_3 \\ c_4 \end{bmatrix}}_{\mathbf{P}} + \underbrace{\frac{1}{6}\begin{bmatrix} 2 & 1 & 0 & 0 \\ 1 & 4 & 1 & 0 \\ 0 & 1 & 4 & 1 \\ 0 & 0 & 1 & 2 \end{bmatrix}}_{\mathbf{Q}}\begin{bmatrix} \partial c_1/\partial t \\ \partial c_2/\partial t \\ \partial c_3/\partial t \\ \partial c_4/\partial t \end{bmatrix}$$
$$= \underbrace{-\frac{ka}{2}\begin{bmatrix} 1 \\ 2 \\ 2 \\ 1 \end{bmatrix}}_{\mathbf{R}} \tag{4.42}$$

In matrix notation, this is succinctly written as $\mathbf{P}\mathbf{c}^{\mathrm{T}} + \mathbf{Q}(\partial\mathbf{c}^{\mathrm{T}}/\partial t) = \mathbf{R}$.

Step 6. *Solve the given system of equations taking into account specific initial and boundary conditions.* The last thing left to do before solving the problem is to discretize the time derivatives at each node and to approximate the concentrations and concentration gradients at times t and $t + \Delta t$. Using finite difference form, Equation (4.42) can be discretized as

$$\mathbf{P}(\alpha\mathbf{c}_{t+\Delta t}^{\mathrm{T}} + (1-\alpha)\mathbf{c}_t^{\mathrm{T}}) + \mathbf{Q}\left(\alpha\left[\frac{\partial\mathbf{c}^{\mathrm{T}}}{\partial t}\right]_{t+\Delta t} + (1-\alpha)\left[\frac{\partial\mathbf{c}^{\mathrm{T}}}{\partial t}\right]_t\right) = \mathbf{R} \tag{4.43}$$

where α is a scalar defining a specific numerical method one wishes to use (e.g., $\alpha = 0$ corresponds to the simple, fully explicit method and $\alpha = ½$ to Crank–Nicholson averaging over two time steps; for details, see Section 4.7). For example, choosing $\alpha = \frac{1}{2}$ we have

$$\mathbf{P}\left(\frac{1}{2}\mathbf{c}_{t+\Delta t}^{\mathrm{T}} + \frac{1}{2}\mathbf{c}_t^{\mathrm{T}}\right) + \mathbf{Q}\left(\frac{1}{2}\left[\frac{\partial\mathbf{c}^{\mathrm{T}}}{\partial t}\right]_{t+\Delta t} + \frac{1}{2}\left[\frac{\partial\mathbf{c}^{\mathrm{T}}}{\partial t}\right]_t\right) = \mathbf{R} \tag{4.44}$$

To approximate the concentration, we use the simple forward differencing scheme

$$\frac{1}{2}\left(\left[\frac{\partial\mathbf{c}^{\mathrm{T}}}{\partial t}\right]_{t+\Delta t} + \left[\frac{\partial\mathbf{c}^{\mathrm{T}}}{\partial t}\right]_t\right) \approx \frac{\mathbf{c}_{t+\Delta t}^{\mathrm{T}} - \mathbf{c}_t^{\mathrm{T}}}{\Delta t} \tag{4.45}$$

which allows us to write the governing matrix equation as

$$\left(\mathbf{P} + \frac{2}{\Delta t}\mathbf{Q}\right)\mathbf{c}_{t+\Delta t}^{\mathrm{T}} \approx 2\mathbf{R} - \left(P - \frac{2}{\Delta t}\mathbf{Q}\right)\mathbf{c}_t^{\mathrm{T}} \tag{4.46}$$

This final expression can now be solved readily using linear algebra. Since the initial values are known, $\mathbf{c}_t^T$ is known at the first time step and the values of $\mathbf{c}_{t+\Delta t}^T$

can be easily obtained through an inversion of the matrix in parentheses on the left-hand side of (4.46). This process is iterated to get concentrations at different times. A relatively minor but technically important detail is that, at each iteration, one must correct the evolving matrix to satisfy the boundary conditions (otherwise the method gives unreasonable results). As an illustrative example, consider our diffusion equations for which the boundary conditions are $c(0,t) = c_1 = 1$ and $c(L,t) = c_4 = 0$. Taking numerical values, say, $D = a = 1$, $k = 2$, $\Delta t = 1/6$ and assuming that at some time t the concentrations at the nodes are $c_t = [1\ \ 1\ \ 1\ 0]$, Equation (4.46) can be written as

$$\begin{bmatrix} 6 & 0 & 0 & 0 \\ 1 & 10 & 1 & 0 \\ 0 & 1 & 10 & 1 \\ 0 & 0 & 2 & 4 \end{bmatrix} \begin{bmatrix} c_1 \\ c_2 \\ c_3 \\ c_4 \end{bmatrix}_{t+\Delta t} = \begin{bmatrix} 5 \\ 10 \\ 7 \\ 1 \end{bmatrix} \tag{4.47}$$

in which the improper boundary conditions at c_1 have to be corrected in the top (6 and 5 $\rightarrow$ 1 and 1) and bottom (2, 4 and 1 $\rightarrow$ 0, 1 and 0) rows to give

$$\begin{bmatrix} 1 & 0 & 0 & 0 \\ 1 & 10 & 1 & 0 \\ 0 & 1 & 10 & 1 \\ 0 & 0 & 0 & 1 \end{bmatrix} \begin{bmatrix} c_1 \\ c_2 \\ c_3 \\ c_4 \end{bmatrix}_{t+\Delta t} = \begin{bmatrix} 1 \\ 10 \\ 7 \\ 0 \end{bmatrix} \tag{4.48}$$

After these corrections, the inversion of matrix can be performed and the concentration profile can be solved for all time steps.

Now that we have gone through these infernal derivations, it is time to relax and analyze the actual solution we generated. Figure 4.6 compares the FEM results (for $L = 1$ mm, $D = 1 \times 10^{-10}\,\mathrm{m^2\,s^{-1}}$, $k = 2 \times 10^{-4}\,\mathrm{mol\,m^3\,s^{-1}}$) with the analytical solution of the problem (left to the reader to re-derive):

$$\begin{aligned} c &= c_{ss} - \sum_{n=1}^{\infty} A_n \exp[-D(n\pi/L)^2 t]\sin(n\pi x/L) \\ c_{ss} &= kx^2/(2D) - [kL/(2D) + 1/L]x + 1, \\ A_n &= 2[((-1)^n - 1)/2 + 1]/(n\pi) - 2kL^2(1 - (-1)^n)/(Dn^3\pi^3) \end{aligned} \tag{4.49}$$

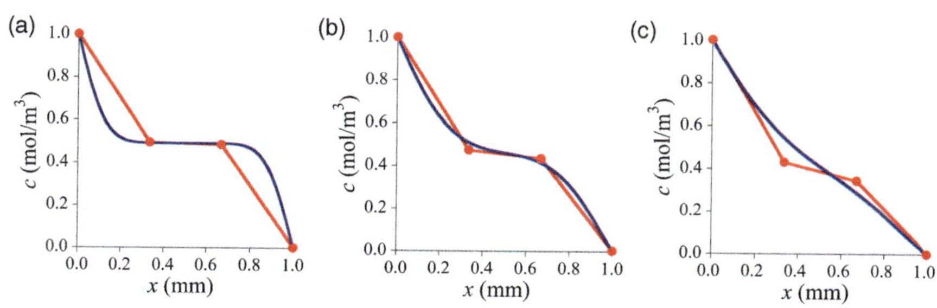

Figure 4.6 Time-dependent concentration profiles from the FEM with three elements (red lines) and from analytical solution (blue lines). The times are (a) 50 s, (b) 200 s and (c) 500 s

where c_{ss} is the steady-state solution. While the FEM captures the general trend, it obviously finds it difficult to approximate the curved regions of the exact solution. With only three finite elements this is hardly surprising. Of course, the quality of solution would increase dramatically if more elements were used – this, however, would increase the size of the matrices, and would make a 'paper-and-pencil' solution (like the one we described in Steps 1–6 above) very tedious. Clearly, the FEM is for computers to calculate, and the most we should do is to either code the algorithm or simply buy commercial FEM software. At the same time, by knowing the nuts-and-bolts of the method, we should be able to judge when the FEM goes astray, and whether the solutions it generates make physical sense.

4.4 TEMPORAL DISCRETIZATION AND INTEGRATION

In this section, we discuss numerical solutions to RD equations, in which both the time and space coordinates are discretized (Figure 4.7). In doing so, we will limit

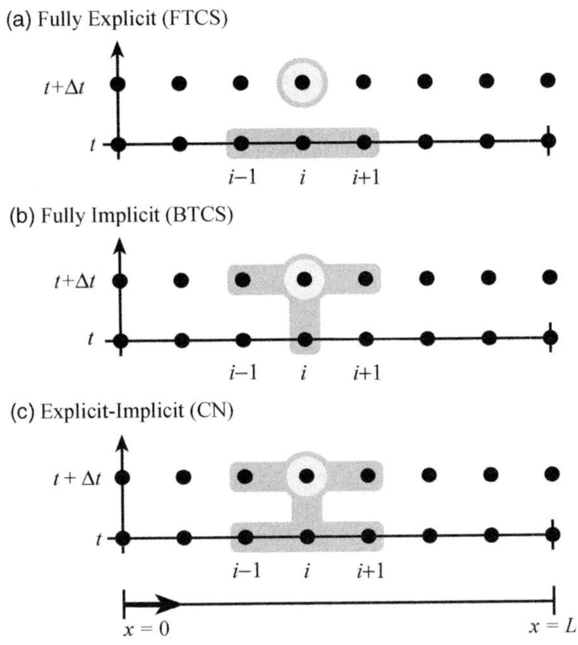

Figure 4.7 (a) Explicit methods like the FTCS scheme calculate the concentration at location x_i and time $t + \Delta t$ (targeted node) using the values of c_{i-1}^t, c_i^t and c_{i+1}^t (nodes highlighted in gray). (b) Implicit methods, like the BTCS scheme, require the simultaneous solution of a set of linear equations for 'future' concentrations at time $t + \Delta t$. Here, the diffusive coupling (i.e., the Laplacian operator) is between 'future' concentrations $c_{i-1}^{t+\Delta t}$, $c_i^{t+\Delta t}$ and $c_{i+1}^{t+\Delta t}$. (c) The Crank–Nicholson method mixes the explicit and implicit approaches to enhance both accuracy and stability. Here, the discretized governing equation centered on location x_i combines all grid points highlighted in gray

ourselves to the most straightforward yet comprehensive difference methods (Section 4.3.1) and distinguish two cases depending on whether the characteristic times of reaction are (i) longer than or commensurate with diffusion times, and (ii) much shorter than diffusion times.

4.4.1 Case 1: $\tau_{\mathrm{Rxn}} \geq \tau_{\mathrm{Diff}}$

When reaction and diffusion take place on similar time scales or when diffusion is faster (i.e., its characteristic time, τ_{Diff}, is smaller than that of reaction, τ_{Rxn}), one may treat the processes simultaneously, using time discretization and integration methods appropriate for diffusion problems.

4.4.1.1 Forward time centered space (FTCS) differencing

In the simplest numerical approach, the RD equations may be solved by replacing the time derivative by its forward difference approximation and the Laplacian operator by its central difference approximation. For example, consider a single chemical species diffusing over a one-dimensional domain while being consumed by a first-order reaction. The FTCS approximation for this problem assuming constant diffusion coefficient is given as

$$\frac{c_i^{t+\Delta t} - c_i^t}{\Delta t} = D\frac{(c_{i-1}^t - 2c_i^t + c_{i+1}^t)}{(\Delta x)^2} - kc_i^t \tag{4.50}$$

From this, the concentrations at time $t+\Delta t$ may be calculated directly as $c_i^{t+\Delta t} = c_i^t + \alpha(c_{i-1}^t - 2c_i^t + c_{i+1}^t) + \Delta t k c_i^t$, where $\alpha = D\Delta t/(\Delta x)^2$. From von Neumann stability analysis,[14] it can be shown that this approach is numerically stable for $[2D/(\Delta x)^2 + k/2]\Delta t < 1$. In other words, the maximum stable time step is inversely related to the sum of the characteristic rates of diffusion and reaction within a cell of width Δx. Thus, for $\tau_{\mathrm{Diff}} \leq \tau_{\mathrm{Rxn}}$, this relation can be approximated as $D\Delta t/(\Delta x)^2 < 0.5$ – i.e., the diffusive time scale determines the stability. Furthermore, as the diffusive time scale over a length L is given by $\tau = L^2/D$, modeling reaction and diffusion via this method requires of the order of $L^2/(\Delta x)^2$ time steps, which may be prohibitively large if $L \gg \Delta x$. The benefit of the FTCS scheme is its simplicity. It is an *explicit* scheme, which means that $c_i^{t+\Delta t}$ for each i can be calculated explicitly from the quantities that are already known at a given time t.

4.4.1.2 Backward time centered space (BTCS) differencing

Similar to the FTCS scheme described above, the BTCS method is also first-order accurate in time and second-order accurate in space. BTCS equations can be succinctly written as

$$c_i^{t+\Delta t} - c_i^t = \alpha(c_{i-1}^{t+\Delta t} - 2c_i^{t+\Delta t} + c_{i+1}^{t+\Delta t}) - \Delta t k c_i^{t+\Delta t} \quad (4.51)$$

and, in contrast to FTCS, are fully *implicit*, meaning that one has to solve a set of simultaneous linear equations at each time step to obtain $c_i^{t+\Delta t}$. Specifically, one must solve a set of equations of the form $-\alpha c_{i-1}^{t+\Delta t} + (1+2\alpha+\Delta t k)c_i^{t+\Delta t} - \alpha c_{i+1}^{t+\Delta t} = c_i^t$ for which, fortunately, there exist very efficient methods of solution.[15] The benefit of the BTCS method is that it is unconditionally stable and evolves toward the correct steady-state solution (i.e., as $\partial c/\partial t \to 0$) even for large time steps.

It should be noted that this implicit scheme is efficient only for *first-order* reactions. For nonlinear reaction terms, $R(c_i^t)$, it is possible to circumvent this limitation by treating the reaction terms explicitly and the diffusion term implicitly as follows:

$$c_i^{t+\Delta t} - c_i^t = \alpha(c_{i-1}^{t+\Delta t} - 2c_i^{t+\Delta t} + c_{i+1}^{t+\Delta t}) + \Delta t R(c_i^t) \quad (4.52)$$

With this modification, however, numerical stability is no longer guaranteed, nor is the unconditional convergence to the correct steady-state solution.

4.4.1.3 Crank–Nicholson method

To achieve second-order accuracy in time, one may combine the fully explicit and the fully implicit approaches described above into the so-called *Crank–Nicholson* scheme.[14] In the case of one-dimensional diffusion with first-order reaction, this combination is simply a linear average of the explicit FTCS and implicit BTCS schemes:

$$\begin{aligned} c_i^{t+\Delta t} - c_i^t = {} & \tfrac{1}{2}\alpha\left[(c_{i-1}^{t+\Delta t} - 2c_i^{t+\Delta t} + c_{i+1}^{t+\Delta t}) + (c_{i-1}^t - 2c_i^t + c_{i+1}^t)\right] \\ & - \tfrac{1}{2}\Delta t k(c_i^{t+1} + c_i^t) \end{aligned} \quad (4.53)$$

Like the fully explicit BTCS scheme, this method is stable for any time step, Δt; however, it provides improved accuracy. These attributes make this the recommended method for most 'traditional' RD problems, in which the time scales of reaction and diffusion are similar (or when $\tau_{\mathrm{Rxn}} \geq \tau_{\mathrm{Diff}}$).

In the case of nonlinear reaction terms, $R(c_i^t)$, one may use the Crank–Nicholson scheme for the diffusion terms and treat the reaction term explicitly:

$$c_i^{t+\Delta t} - c_i^t = \tfrac{1}{2}\alpha\left[(c_{i-1}^{t+\Delta t} - 2c_i^{t+\Delta t} + c_{i+1}^{t+\Delta t}) + (c_{i-1}^t - 2c_i^t + c_{i+1}^t)\right] + \Delta t R(c_i^t) \quad (4.54)$$

This scheme results in second-order accuracy in time for the diffusion term but only first-order accuracy in time for the reaction terms. Thus, for faster reactions or more complex (i.e., nonlinear) chemical kinetics, it is often advisable to use one of the higher order integration schemes described in Section 4.4.2.

4.4.1.4 Alternating direction implicit method in two and three dimensions

Although the above version of the Crank–Nicholson scheme may be extended to two or three dimensions, solving the coupled linear equations becomes increasingly difficult and requires more complex techniques for sparse matrices.[15] The alternating direction implicit (ADI) method avoids these difficulties without sacrificing the stability or accuracy of the Crank–Nicholson approach. Briefly, this scheme divides each time step into two substeps of size $\Delta t/2$, each treated differently:

$$\begin{aligned} c_{i,j}^{t+\Delta t/2} - c_{i,j}^{t} &= \tfrac{1}{2}\alpha\left[\left(c_{i-1,j}^{t+\Delta t/2} - 2c_{i,j}^{t+\Delta t/2} + c_{i+1,j}^{t+\Delta t/2}\right) + \left(c_{i,j-1}^{t} - 2c_{i,j}^{t} + c_{i,j+1}^{t}\right)\right] \\ &\quad + (\Delta t/2)R(c_{i,j}^{t}) \\ c_{i,j}^{t+\Delta t} - c_{i,j}^{t+\Delta t/2} &= \tfrac{1}{2}\alpha\left[\left(c_{i-1,j}^{t+\Delta t/2} - 2c_{i,j}^{t+\Delta t/2} + c_{i+1,j}^{t+\Delta t/2}\right)\right. \\ &\quad \left. + \left(c_{i,j-1}^{t+\Delta t} - 2c_{i,j}^{t+\Delta t} + c_{i,j+1}^{t+\Delta t}\right)\right] + (\Delta t/2)R(c_{i,j}^{t+\Delta t/2}) \end{aligned} \tag{4.55}$$

In the first substep, the x direction is treated implicitly and the y direction is treated explicitly, and vice versa in the second substep. Because only one direction is treated implicitly at each step, the resulting linear system is tridiagonal and may be easily solved (for details, see Ref. 15).

4.4.2 Case 2: $\tau_{\text{Rxn}} \lll \tau_{\text{Diff}}$

When reaction processes are significantly faster than diffusion, the schemes described above often become inefficient. This is because the reaction terms are treated only to second-order accuracy (or even first-order accuracy in the case of nonlinear reaction terms) and to achieve an acceptable approximation it would be necessary to take many small time steps of size $\Delta t \ll \tau_{\text{Rxn}} \ll \tau_{\text{Diff}}$. Furthermore, as spatial locations are coupled by diffusion over the time scale of $\tau_{\text{Diff}} = (\Delta x)^2/D$, it is not always necessary to consider this coupling over very short time scales associated with a 'fast' reaction $\tau_{\text{Rxn}} \ll \tau_{\text{Diff}}$. In other words, if slow diffusion can be integrated accurately with a large time step but fast reaction requires time steps that are much shorter, it is not necessary to update diffusion at every reaction step. This suggests a multiscale approach, in which the reaction terms are integrated using the methods discussed below.

4.4.2.1 Operator splitting method

The operator splitting method may be used to solve any time-dependent equation of the form $\partial c/\partial t = O_1(c) + O_2(c)$, where $O_1(\,)$ and $O_2(\,)$ are some operators. If only one of these operators were present on the right-hand side of the equation, one could update c from time step t to $t + \Delta t$ using an appropriate differencing scheme.

Let these updates be written as $c^{t+\Delta t} = U_1(c^t, \Delta t)$ for $O_1(\,)$ and $c^{t+\Delta t} = U_2(c^t, \Delta t)$ for $O_2(\,)$. In the operator splitting method, one applies these updating schemes sequentially as (i) $c^{t+\Delta t/2} = U_1(c^t, \Delta t)$ and (ii) $c^{t+\Delta t} = U_2(c^{t+\Delta t/2}, \Delta t)$. Importantly, as the updates U_1 and U_2 are performed independently and sequentially, it is possible to treat each process (i.e., 1 and 2) using different numerical schemes, each of which is optimized for the specific process.

In the context of RD systems, $O_1(\,)$ and $O_2(\,)$ represent the diffusion and reaction terms, respectively. Thus, this method assumes the processes of reaction and diffusion are independent or uncoupled over a single time step, Δt. Because the processes are uncoupled, one may treat diffusion and reaction by the numerical scheme most appropriate/efficient for each process. For the diffusive component, the recommended scheme is the Crank–Nicholson method (or ADI method in two or three dimensions), in which the time step is typically $\Delta t \sim D/2(\Delta x)^2$. For the reaction component, in the absence of the spatial derivatives, $\mathrm{d}c_{s,i}/\mathrm{d}t = R_s(c_1 \ldots c_n)_i$, and one is left with a set of n coupled, ordinary differential equations (ODEs) for each grid point – thus, we no longer have to deal with partial differential equations. To integrate these ODEs from time $t + \Delta t/2$ to $t + \Delta t$, there exist a variety of methods[15] from which one may choose depending on the complexity of the reaction kinetics as well as the desired accuracy.

It is important to note that the integration of reaction terms from $t + \Delta t/2 \rightarrow t + \Delta t$ may proceed by an arbitrary number of 'substeps' to achieve the desired accuracy (discussed further below); thus, Δt refers only to the time scale on which the processes of reaction and diffusion are coupled. Specifically, since $\tau_{\mathrm{Rxn}} \ll \tau_{\mathrm{Diff}}$, the diffusive time step $\Delta t \sim D/2(\Delta x)^2$ is long relative to the characteristic reaction time scale, and it would be very inaccurate to integrate the reaction component using the diffusive time step. Instead, one needs to treat the fast reactions at a finer temporal resolution – i.e., use smaller time steps for the reaction terms. The simplest approach is to apply the forward difference scheme using an 'auxiliary' time step $\Delta t'$ much smaller than Δt – i.e., $\Delta t' \ll \Delta t$. A more accurate but more complex alternative is to use a higher order integration scheme, such as the fourth-order Runga–Kutta method, which can even be implemented with adaptive time stepping.[15] If the reaction kinetics contain more than one first-order differential equation, the possibility of *stiffness* arises (specifically, stiffness occurs when different dependent variables – here, concentrations – change on very different time scales). For stiff equations, the Runga–Kutta method may become very inefficient, and one should instead use a method designed for handling stiff equations, such as Rosenbrock methods.[15] Fortunately, these methods are readily available from Numerical Recipes libraries[15] or in mathematics packages such as Matlab.

4.4.2.2 Method of lines

For one-dimensional systems, there is an alternative to the operator splitting method known as the method of lines,[16,17] in which the relevant partial differential equations are transformed into a larger set of ODEs using the finite difference approximation for the diffusion term:

$$\frac{dc_{s,i}}{dt} = \frac{D_s}{\Delta x^2}(c_{s,i-1} - 2c_{s,i} + c_{s,i+1}) + R_s(c_1 \ldots c_n)_i \tag{4.56}$$

where index $s = 1, \ldots, n$ distinguishes the n chemical species involved, and index $i = 2, \ldots, N-1$ corresponds to different, discrete spatial locations (grid points) in the 'bulk' of the domain (the boundary points $i = 1$ and $i = N$ must be treated separately; see below). Since each location i has its own equation, we are now dealing with a set of nN equations, which, however, do not contain partial derivatives. At the same time, these equations capture the diffusive transport between the grid points, since $dc_{s,i}/dt$ depends not only on the concentrations of species at location i, but also on the concentrations at neighboring locations ($i \pm 1$). The boundary conditions are treated as described previously for the finite difference solution to the diffusion equation. For example, Dirichlet conditions at $x = 0$, $c_s(x = 0) = \alpha_s$ correspond simply to $c_{s,1} = \alpha_s$ (such that $dc_{s,1}/dt = 0$) and von Neumann conditions at $x = L$, $(\partial c_s/\partial x)_{x=L} = \beta_s$ can be written as

$$\frac{dc_{s,N}}{dt} = \frac{2D_s}{\Delta x^2}(c_{s,j-1} - c_{s,j} + \beta_s \Delta x) + R_s(c_1 \ldots c_n)_i \tag{4.57}$$

With these preliminaries, the set of ODEs may be solved directly by any stiff equation integrator.[15] As an example of such an integrator combining the accuracy of explicit methods with the stability of implicit schemes, we consider the so-called *semi-implicit Euler method*. Rewriting the original set of ODEs in a simplified, vector form, $\mathbf{dc}/\mathbf{d}t = \mathbf{f}(\mathbf{c})$ (where the bold type indicates a vector, e.g., $\mathbf{c} = [c_{11}c_{21} \ldots c_{n1}c_{12} \ldots c_{nN}]$ is the vector of concentrations of all species at all grid points), we can then write

$$\mathbf{c}^{t+\Delta t} = \mathbf{c}^t + \Delta t \left[\mathbf{f}(\mathbf{c}^t) + \left.\frac{\partial \mathbf{f}}{\partial \mathbf{c}}\right|_{\mathbf{c}^t} \cdot (\mathbf{c}^{t+\Delta t} - \mathbf{c}^t) \right] \tag{4.58}$$

which requires not only an evaluation of $\mathbf{f}(\mathbf{c}^t)$ at every time step, but also of the Jacobian matrix, $\partial \mathbf{f}/\partial \mathbf{c}$ (this is true of all stiff solvers). In the method of lines, the Jacobian is an $nN \times nN$ matrix, which must be evaluated and multiplied at each time step making it the most time-consuming part of the algorithm. Fortunately, because only neighboring cells are coupled by diffusion, the Jacobian matrix is sparse and band-diagonal. Thus, the calculations can be accelerated significantly by using linear algebra routines designed specifically for band-diagonal matrices. An example of an efficient implementation of the method of lines can be found in the 'brussode' program in Matlab, which models reaction and diffusion in the Brusselator chemical system (Section 3.5).

4.4.3 Dealing with Precipitation Reactions

As a special case, we consider a situation when the time scale of reaction is very short compared to that of diffusion ($\tau_{\text{Diff}} \gg \tau_{\text{Rxn}}$ *and* $\tau_{\text{Rxn}} \rightarrow 0$), in which case the details of the reaction kinetics are often of little importance. Experimentally, such a scenario corresponds to many ionic precipitation reactions, which we will see very often in later chapters. As an example of how to model such reactions, consider precipitation of ions A and B into an insoluble precipitate AB. Even though the reaction is very rapid and assumed to be shifted toward the product, some degree of reversibility is always present and the equilibrium constant can be written as $K_{\text{eq}} = k/k_{-1} = 1/[\text{A}][\text{B}]$ (the concentration of the precipitate does not enter because it is insoluble; Section 3.2). Since $K_{\text{sp}} = [\text{A}][\text{B}] = 1/K_{\text{eq}}$, the forward rate is $k[\text{A}][\text{B}]$, the reverse rate is $k_{-1} = kK_{sp}$ and the one-dimensional ($0 \leq x \leq L$) equation describing these reactions coupled to diffusion becomes

$$\frac{\partial c_s}{\partial t} = D_s \frac{\partial^2 c_s}{\partial x^2} - k(c_{\text{A}} c_{\text{B}} - K_{\text{sp}}) \tag{4.59}$$

where $s = \text{A}$ or B. Notice that the above equation is valid for any value of k – i.e., not just for fast reactions for which $\tau_{\text{Rxn}} \ll \tau_{\text{Diff}}$. Thus, when the reaction rate is comparable or slightly greater than the diffusive rate ($kc_o \geq D_{\text{A}}/L^2$, where c_{o} is a characteristic concentration, or equivalently $\tau_{\text{Rxn}} \leq \tau_{\text{Diff}}$), it is possible to apply the numerical methods described previously in Sections 4.1 and Section 4.4.2.

If, however, the reaction rate is much faster than diffusion ($kc_{\text{o}} \ggg D_{\text{A}}/L^2$, or equivalently $\tau_{\text{Rxn}} \lll \tau_{\text{Diff}}$), it is often helpful to consider the reaction process as instantaneous on the time scale of diffusion. In this case, the exact form of the reaction kinetics (e.g., forward rate $= kc_{\text{A}}c_{\text{B}}$) becomes irrelevant, and one can rewrite the governing equation as follows:

$$\frac{\partial c_s}{\partial t} = D_s \frac{\partial^2 c_s}{\partial x^2} - k'\Theta(c_{\text{A}} c_{\text{B}} - K_{\text{sp}}) \tag{4.60}$$

Here, $\Theta(\,)$ is the Heaviside step function (i.e., $\Theta(x) = 0$ for $x \leq 0$, and $\Theta(x) = 1$ for $x > 0$), and $k' \sim kC_o^2$ is a characteristic reaction rate, which must be much greater than the diffusive rate for the approximation to hold.

To solve the above equations numerically, one may apply the operator splitting method (Section 4.1), in which the diffusion term is integrated using the Crank–Nicholson method (or a simpler FD scheme), and the reaction term proceeds instantaneously to equilibrium (i.e., until $c_{\text{A}} c_{\text{B}} = K_{\text{sp}}$) at each time step. In this way, one would first apply the diffusion operation for time $t \rightarrow t + \Delta t/2$, after which the concentration at each location i would be updated by solving the following algebraic equations for $c_{\text{A},i}^{t+\Delta t}$ and $c_{\text{B},i}^{t+\Delta t}$: $c_{\text{A},i}^{t+\Delta t} c_{\text{B},i}^{t+\Delta t} = K_{\text{sp}}$ and $c_{\text{A},i}^{t+\Delta t} - c_{\text{A},i}^{t+\Delta t/2} = c_{\text{B},i}^{t+\Delta t} - c_{\text{B},i}^{t+\Delta t/2}$. The first equation accounts for the solubility relation and the second for the 1:1 stoichiometry of the reaction. In the limit as $K_{\text{sp}} \rightarrow 0$, which is not a bad approximation for many precipitation reactions (e.g., $K_{\text{sp}} \sim 10^{-40}\,\text{M}^5$ for silver(I) hexacyanoferrate,[18] $Ag_4[Fe(CN)_6]$), the solution to the above equations is especially simple, namely $c_{A,i}^{t+\Delta t} = 0$ and $c_{B,i}^{t+\Delta t} = c_{B,i}^{t+\Delta t/2} - c_{A,i}^{t+\Delta t/2}$ for $c_{B,i}^{t+\Delta t/2} > c_{A,i}^{t+\Delta t/2}$; likewise $c_{A,i}^{t+\Delta t} = c_{A,i}^{t+\Delta t/2} - c_{B,i}^{t+\Delta t/2}$ and $c_{B,i}^{t+\Delta t} = 0$ for $c_{A,i}^{t+\Delta t/2} > c_{B,i}^{t+\Delta t/2}$. In other words, the less-abundant reactant is completely consumed (or partially consumed in the case of finite K_{sp}) by the more-abundant reactant to give the insoluble precipitate, AB.

4.5 HEURISTIC RULES FOR SELECTING A NUMERICAL METHOD

When choosing a numerical method, it is first necessary to examine the characteristic time scales of the RD processes one is trying to describe. As we have already seen, the diffusion time scale is given by $\tau_{\text{Diff}} \sim L^2/D$, where L is a characteristic length of the system. The definition of the reaction time scale depends on the order of the reactions involved, but may be approximated as $\tau_{\text{Rxn}}^m \sim 1/k_m c^{\beta_m - 1}$, where k_m and β_m are, respectively, the rate constant and order of reaction m, and c is a characteristic concentration scale. Typically, the concentration scale (there may be more than one) is taken as the initial concentration of a given species or concentration prescribed at the domain boundary (e.g., for Dirichlet boundary conditions). Although there are as many reaction time scales as there are reactions in the system, the most important one corresponds to the fastest reaction – that is, to the reaction with the smallest characteristic time, $\tau_{\text{Rxn}} = \min(\tau_{\text{Rxn}}^m)$. Having identified the relevant time scales, one may then choose from one of the appropriate methods described in Section 4.4. As a rule of thumb, we recommend the Crank–Nicolson method (or ADI method for two and three dimensions) for systems in which $\tau_{\text{Rxn}} \geq \tau_{\text{Diff}}$; the operator splitting method combining Crank–Nicolson diffusion integration and Runga–Kutta reaction integration is most suitable for systems with $\tau_{\text{Rxn}} \ll \tau_{\text{Diff}}$. Of course, for quick-and-dirty solutions, it is hard to beat the simplicity of the explicit FTCS scheme (Section 4.1), but one should be wary of numerical instabilities and/or long run times. Also, for systems with complicated geometries, FD methods become cumbersome to implement, and prepackaged finite element methods (e.g., FEMLAB) provide a convenient and efficient alternative.

4.6 MESOSCOPIC MODELS

The description of RD processes by partial differential equations is a macroscopic one, and is based on average quantities, in which the microscopic nature of a system's elementary constituents does not appear explicitly. Operating on averages does not require analysis of all the microscopic degrees of freedom and significantly reduces the complexity of the problem we wish to model. At the same time, however, the averages cannot be used to analyze how large-scale phenomena are triggered by local and/or transient deviations from these averages, or to model very small RD systems, in which such fluctuations play important roles (see Chapters 7 and 10).

Since RD simulations accounting for the exact dynamics of individual molecules are in most cases computationally prohibitive, a family of intermediate ('mesoscopic') methods – usually referred to as lattice gas automata (LGA)[19] – have been developed to approximate molecular motions and reactions as moves on a regular grid (Figure 4.8). Because the rules governing the dynamics of particles on the lattice are probabilistic, LGA can account for the fluctuations while reproducing a system's macroscopic behavior. LGA methods thus provide a convenient, easy-to-implement tool with which to bridge the micro- and macroscopic descriptions of RD.

In its simplest form, a LGA algorithm approximates the domain of a RD process as a regular lattice with coordination number m (e.g., $m = 3$ for a triangular lattice, 4 for square, and 6 for hexagonal). The molecules of all types reside at the lattice nodes and undergo displacements along the links connecting the nodes. The time evolution of the system occurs at discrete time steps and follows the iterated application of an evolution operator, Θ, such that [state at time $t + 1$] = Θ[state at time t]. The evolution operator typically consists of three basic operations often referred to as 'automaton rules'.

1. ***Propagation*** (Figure 4.8(b)), in which all particles move to the neighboring sites of the lattice (along the directions of the individual velocity vectors). The motions conserve both particle numbers as well as momenta.
2. ***Randomization*** (Figure 4.8(c)), in which the directions of the velocity vectors of particles at each node are changed randomly. This step conserves the number of particles but changes their individual momenta (total momentum is conserved). Randomization of the velocities effectively 'mimics' the effects of elastic collisions between the modeled particles and those of the solvent (which is not explicit in the simulation).

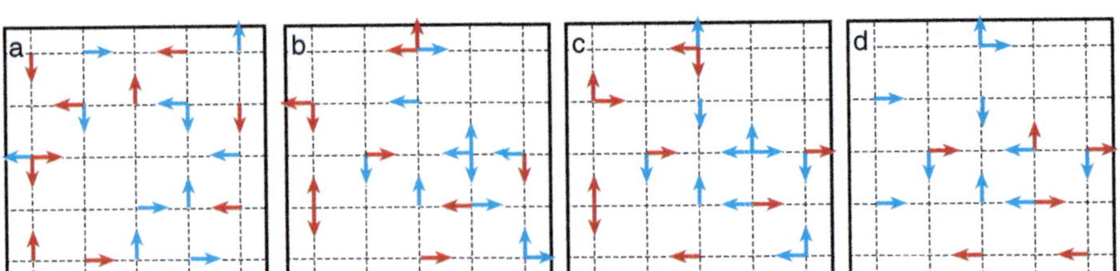

Figure 4.8 In this illustration of LGA, locations of two different types of particles and their corresponding velocity vectors (red and blue arrows) are indicated. The lattice is delineated by dashed lines; solid black lines represent periodic boundaries. (a) Initial configuration for a LGA iteration. (b) The particles propagate by moving one node in the direction of their velocity vectors. (c) Directions of the velocity vectors at each node are randomized. (d) The particles are then reacted according to specific chemical kinetics (here, 2 red → 1 blue and 2 blue → 1 red). New particle counts, locations and velocity vectors are used as the initial configuration for the next iteration

3. ***Reaction*** (Figure 4.8(d)), where the particles at each node are either created or annihilated with a certain probability reflecting the 'macroscopic' stoichiometry and rate of the particular reaction. Reactions at each node are performed independently of one another, and conserve neither particle numbers nor particle momenta.

It can be shown rigorously that repeated application of these simple rules reproduces diffusive phenomena exactly and, with judicious choice of reaction probabilities, can approximate well a system's chemical kinetics. Although proper calibration of LGA against experimental diffusion constants and reactions stoichiometries is sometimes cumbersome, the automaton algorithms are very easy to code, 'intuitive' in their chemical content, and computationally efficient in at least simple RD systems. In addition to their ability of treating fluctuations and stochastic events, LGA methods are useful is solving RD problems with complex boundary/initial conditions, as illustrated in Example 4.3 and later in Chapter 6.

Example 4.3 How Reaction–Diffusion Gives Each Zebra Different Stripes

According to Rudyard Kipling, animals like zebras, giraffes and leopards might have developed their skin patterns as a result of uneven shadows cast by the 'aboriginal flora'. However imaginative and picturesque, this hypothesis has not proven correct, and formation of skin patterns is nowadays attributed to more mundane chemical phenomena. In the middle of the twentieth century, a British mathematical genius named Alan Turing (also known for his contributions to the theory of computation and for breaking the Naval Enigma code during World War II) proposed that the patterns emerge as a result of a reaction and diffusion of chemical 'morphogens'. Since the stripes on each animal are different, Turing postulated that the chemical kinetics underlying their formation should have a stochastic/random element responsible for these idiosyncrasies. Here, we illustrate how such randomness emerges from a LGA model describing reaction and diffusion of morphogens X and Y on a square grid. The equations we use are somewhat similar to those originally proposed by Turing:[20]

$$\begin{aligned}
0 &\xrightarrow{11X} \mathrm{X} \\
0 &\xrightarrow{1} \mathrm{Y} \\
X &\xrightarrow{6X} \mathrm{Y} \\
Y &\xrightarrow{7Y} 0 \\
X &\xrightarrow{6Y} 0
\end{aligned}$$

and the process starts from an array of parallel lines (see leftmost picture below). Some of the reaction rates (indicated above the arrows) depend on the concentrations of X and Y, and the probability that a reaction will occur at a given node is proportional to the rate divided by some maximal rate (here, taken as 150). At a single time step, each reaction can create (0 → X or 0 → Y), annihilate (X → 0 or Y → 0) or interconvert (X → Y) only one molecule of X or Y. To account for the fact that reactions are more rapid than diffusion, propagation of X and Y at every time step occurs with a certain probability, p (here, Y diffuses more rapidly than X and $p_Y = 0.3 > p_X = 0.1$). Randomization at each node is performed after propagation by rotating particle velocities by 0°, 90°, 180°, or 270° with equal probabilities.

The figures below illustrate how this algorithm evolves the initial configuration (five parallel lines, each five nodes wide and with two X morphogens per node; total lattice dimensions 250 × 250 nodes) through 5000 propagation–randomization–reaction cycles. The middle and right pictures are two realizations of 'skin patterns' (visualized by the normalized concentration of the X morphogen; white regions correspond to X > 0.5, dark regions to X < 0.5) obtained in two separate runs. Note that although these two runs were started from the same initial configuration, they produced markedly different end results reflecting the probabilistic nature of the process.

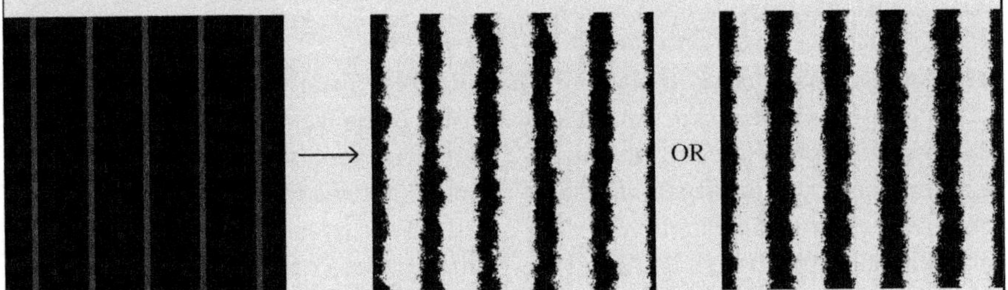

Note. A commented C++ code for the algorithm is included in Appendix C at the end of the book.

REFERENCES

1. Turek, T. (1996) Kinetics of nitrous oxide decomposition over Cu-ZSM-5. *Appl. Catal. B: Environ.*, **9**, 201.
2. Hayes, R.E., Kolaczkowski, S.T., Li, P.K.C. and Awdry, S. (2001) The palladium catalysed oxidation of methane: reaction kinetics and the effect of diffusion barriers. *Chem. Eng. Sci.*, **56**, 4815.
3. Habuka, H., Suzuki, T., Yamamoto, S. *et al.* (2005) Dominant rate process of silicon surface etching by hydrogen chloride gas. *Thin Solid Films*, **489**, 104.
4. Tejedor, P. and Dominguez, P.S. (1995) Fabrication of patterned (311)A GaAs substrates by ArF laser-assisted Cl-2 etching. *Microelectron. J.*, **26**, 853.
5. Baronas, R., Ivanauskas, F. and Kulys, J. (2007) Computational modelling of the behaviour of potentiometric membrane biosensors. *J. Math. Chem.*, **42**, 321.

6. Garvie, M.R. (2007) Finite-difference schemes for reaction–diffusion equations modeling predator–prey interactions in MATLAB. *Bull. Math. Biol.*, **69**, 931.
7. Mikkelsen, A. and Elgsaeter, A. (1995) Density distribution of calcium-induced alginate gels: a numerical study. *Biopolymers*, **36**, 17.
8. Davis, S. (1991) Unsteady facilitated transport of oxygen in hemoglobin-containing membranes and red-cells. *J. Membr. Sci.*, **56**, 341.
9. Chandrupatla, T.R. and Belegundu, A.D. (2002) *'Introduction to Finite Elements in Engineering'*, 3rd edn, Prentice Hall, Upper Saddle River, NJ.
10. Desai, C.S. and Kundu, T. (2001) *Introductory Finite Element Method*, CRC Press, Boca Raton, FL.
11. Logan, D.L. (2007) *A First Course in the Finite Element Method*, 4th edn, Thomson, Toronto.
12. Zienkiewicz, O.C. and Morgan, K. (2006) *Finite Elements and Approximation*, Dover, Mineola, NY.
13. Gockenbach, M.S. (2006) *Understanding and Implementing the Finite Element Method*, SIAM, Philadelphia, PA.
14. Ames, W.F. (1970) *Numerical Methods for Partial Differential Equations*, Barnes and Noble, New York.
15. Press, W.H. (1992) *Numerical Recipes in C: The Art of Scientific Computing*, Cambridge University Press, New York.
16. Reusser, E.J. and Field, R.J. (1979) Transition from phase waves to trigger waves in a model of the Zhabotinskii reaction. *J. Am. Chem. Soc.*, **101**, 1063.
17. Schiesser, W.E. (1991) *The Numerical Method of Lines: Integration of Partial Differential Equations*, Academic Press, San Diego, CA.
18. Mizerski, W. (2003) *Tablice Chemiczne*, Adamantan, Warsaw.
19. Boon, J.P., Dab, D., Kapral, R. and Lawniczak, A. (1996) Lattice gas automata for reactive systems. *Phys. Rep. Rev. Sect. Phys. Lett.*, **273**, 55.
20. Turing, A. (1952) The chemical basis of morphogenesis. *Phil. Trans. R. Soc. Lond. Ser. B*, **237**, 37.

5

Spatial Control of Reaction–Diffusion at Small Scales: Wet Stamping (WETS)

After the thorough review of reaction–diffusion (RD) fundamentals in previous chapters, we are now in a position to discuss and apply these phenomena at micro- and nanoscales. First, we have to make a choice of media in which these processes are to occur. Given that solids do not generally permit efficient diffusion (unless molten[1] or highly pressurized[2]) while liquids are prone to hydrodynamic disturbances disrupting RD patterns and/or structures we wish to create, we focus our attention on an 'intermediate' class of materials – gels (see Section 3.6). Depending on specific composition, gels can support various types of solvents and provide environments for both inorganic and organic reactions. At the same time, their porous network structure makes them permeable to diffusive transport while minimizing hydrodynamic flows. Also, varying the degree of crosslinking changes a gel's elastic properties and alters the average pore size thus allowing regulation of the diffusion coefficients of the reactants participating in a RD process.

Having chosen a medium, we must develop an experimental method of initiating RD. We have already seen that the details of RD patterns and structures depend crucially on the initial/boundary conditions from which the chemicals involved in the process diffuse and subsequently react with one another. Translating into experimental requirements, this means that if we were to prepare micro- and nanoscopic RD patterns, we should be able to deliver RD reagents to the gel medium from locations defined with micro-/nanoscopic resolution. How can this be done? Placing droplets of reagent solutions onto a gel is not an option at such small scales, since the droplets can spread (or bead-up) on the gel surface,

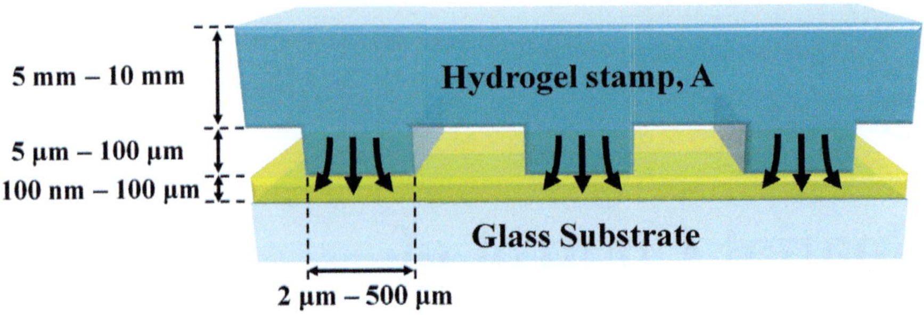

Figure 5.1 Typical configuration of wet stamping. A hydrogel stamp carries chemical A and delivers it into a film of another gel (yellow). The outflow of A from the stamp's features is indicated by red arrows. Numbers give typical ranges of dimensions used in wet stamping experiments

evaporate, slide, or coalesce. Some types of reagents could be first dispersed uniformly throughout the gel in an inactive state and then activated using spatially selective irradiation through a patterned photomask/transparency – this method, however, is limited to only a narrow class of photoactivated chemical transformations. In this chapter, we describe a simple and general alternative to these approaches that is applicable to two-dimensional RD systems and is based on 'injection' of the reagent(s) from one gel into another. This method – called wet stamping (WETS; Figure 5.1) – is nothing more than a small-scale variation of Gutenberg's printing press, save that the 'printing press' is made of a gel and delivers its chemical 'ink' *into* a gel 'paper'.

In its simplest variation, WETS uses (i) a micropatterned hydrogel stamp soaked in a solution of one or more chemicals and (ii) an unpatterned hydrogel film containing the 'partner' chemicals for a RD process. To ensure directional transport from the stamp into the film without any backflow of chemicals, the latter is usually drier than the former. In extreme cases, thin films of completely dehydrated gels can be used. When the stamp is placed onto the film, the regions of contact between the two phases are restricted to the stamp's microfeatures which direct the flow of the chemicals into the film. In other words, the geometry of the microfeatures defines the initial/boundary conditions of a RD process occurring in the patterned film.

5.1 CHOICE OF GELS

Although many types of gels can be used for WETS, practical considerations (ease of preparation, price, structural robustness) suggest the use of the most popular ones: gelatin, agarose, polyacrylamide and silica gels (Figure 5.2).

Gelatin is a protein rich in glycine (almost one in three residues), proline and hydroxyproline amino acids (Figure 5.2(a)) and is produced by partial hydrolysis

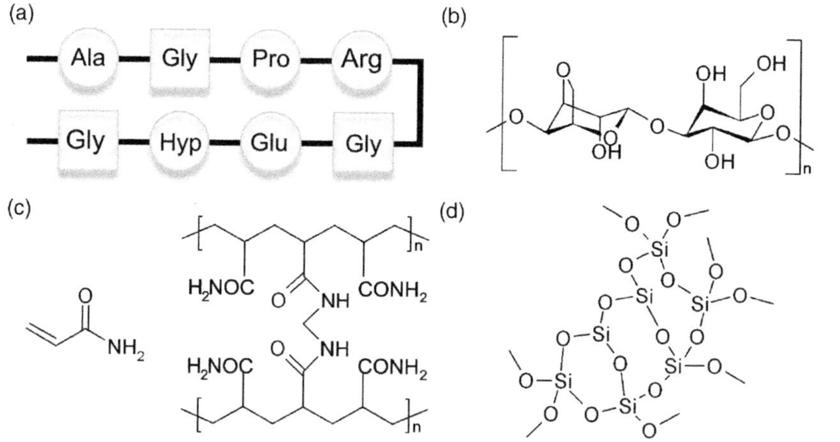

Figure 5.2 Gels for WETS. (a) A typical peptide sequence of gelatin (Ala = alanine; Gly = glycine; Pro = proline; Arg = arginine; Glu = glutamic acid; Hyp = hydroxyproline). Approximately every third residue is glycine. (b) Agarose monomer. (c) Acrylamide monomer (left) and crosslinked polyacrylamide (right). (d) Silica gel

of collagen fibers extracted from the bones and connective tissues of animals such as cattle, pigs, and horses. Upon hydrolysis, the triple helices of collagen are partly destroyed and so the processed gelatin is a mixture of single- and multi-stranded polypeptides. In aqueous solutions and at elevated temperatures, a significant portion of these polypeptides exist in disordered 'coil' conformation; when cooled, however, they wind up into helices, which then aggregate to form collagen-like, right-handed, triple-helical proline/hydroxyproline-rich junction zones. Higher levels of these junctions result in stronger gels.

In the context of WETS applications, gelatin is very suitable for the formation of thin films that support various types of RD processes. Importantly, when a layer of gelatin is cast against a hydrophilic (glass, oxidized polystyrene) flat surface and then slowly dried, it does not buckle/deform and can give very thin (down to ~100 nm), uniform films. Although gelatin is not very stable against harsh chemical agents (i.e., it cannot be gelated in the presence of acids, bases and heavy metal cations), it can be doped with many types of salts useful in RD experiments – halides, chromates, and hexacyanoferrates. Finally, gelatin doped with chromates (e.g., $K_2Cr_2O_7$) can be crosslinked using light (Section 3.6), which provides a facile way of controlling pore size and diffusion coefficients of the species migrating through the gel. There are multiple commercial suppliers of various types of gelatin; most experiments described in this book were performed using inexpensive material available from Sigma-Aldrich (about \$8 for 100 g).

Agarose. Owing to its mechanical and chemical robustness and relatively high degree of elasticity, agarose is the gel of choice for the stamps used in WETS. Chemically, agarose is an unbranched, uncharged polysaccharide (i.e., sugar) consisting of (1 → 3)-β-D-galactose and (1 → 4)-3,6-anhydro-α-L-galactose alternating repeat units (Figure 5.2(b)) and obtained from cell membranes of some species of red algae or seaweed. Depending on its specific origin, the degree of methoxylation and bulk concentration, agarose gelates between 25 and 42 °C, though it is necessary to heat to 70–95 °C to disperse the gel. The gelation process is reversible and involves transition from a fluctuating, disordered coil conformation in solution to a rigid, ordered structure with junction zones held by hydrogen bonds (and, to a lesser extent, hydrophobic interactions). The most generally accepted model of the ordered structure is a coaxial double helix.[3,4]

Once gelated, agarose is relatively stable toward many organic and inorganic substances. Agarose tolerates very well inorganic pH buffers and solutions of proteins and/or DNA. It is often used in gel electrophoresis or gel filtration chromatography. When blocks of agarose are soaked in aqueous solutions of inorganic salts ($AgNO_3$, $CuSO_4$, K_2CO_3, etc.), they remain stable for periods of days to weeks. With acidic substances, degradation is more rapid, and is due to proton-mediated disruption of hydrogen bonds between polysaccharide chains. Yet, even in a relatively concentrated (2 M) HCl, agarose is stable for several hours and can withstand less concentrated solutions for over a week. Agarose tolerates weaker acids relatively well and can be soaked with such harsh chemical agents as HF (commonly used for etching glass). Finally, gelated agarose can accept organic molecules that are not necessarily soluble in water. This can be achieved by partial exchange of water for another miscible solvent (e.g., methanol, ethanol, dimethylsulfoxide, dioxane) that contains an organic compound(s) of interest. Examples of such exchange and infusion of agarose blocks ('stamps') with alkyl thiols, disulfides, siloxanes or amines are described in Campbell *et al.*[5] Experiments described in this reference as well as the majority of WETS systems discussed later in the book used high-strength, Omni-pur® agarose from EMD Biosciences (Darmstadt, Germany; about $80 per 100 g).

Polyacrylamide. Unlike gelatin or agarose, polyacrylamide (PAAm) gels (Figure 5.2(c)) are held together by covalent bonds and are solidified not upon cooling but through a chemical process. PAAm is prepared in water or buffer by copolymerization of a monomer, acrylamide (AAm), and a crosslinker, *N,N′*-methylene-bisacrylamide (Bis), in the presence of a radical initiator, ammonium persulfate (APS), and a catalyst, *N,N,N′,N′*-tetramethylethylenediamine (TEMED).

Once crosslinked, PAAm gels are chemically inert (since their amide groups are relatively unreactive) and stable for prolonged periods of time. They are biocompatible, tolerate inorganic buffer solutions over a wide range of pH values and can accommodate DNA, RNA and proteins. Owing to these characteristics, PAAm gels are used widely for electrophoretic separation of biomolecules. In these applications, the ability to separate macromolecules of different molecular weights can be optimized by adjusting the gel pore size which decreases with increasing ratio of crosslinker to monomer and/or with the total content of solid precursors. For example, 5% PAAm gels are typically used to separate proteins of mass ranging from 60 000 to 200 000 daltons, 10% gels are optimal for the range 16 000–70 000 daltons and 15% gels for 12 000–45 000 daltons.

The major disadvantage of PAAm in WETS is the brittleness of the solidified gels. This makes fabrication and handling of WETS stamps more cumbersome, especially in cases when the micropatterns embossed on the stamp surface are very small (micrometers). On the other hand, PAAm allows for different types of covalent chemical modifications, either by copolymerization of the native AAm with functionalized AAm monomers, or by coupling to gel amide groups. These modifications are chemically better defined than in agarose gels, and in most cases do not change appreciably the mechanical properties of the gel.

Silica gels are chemically inert, easy to functionalize and processable materials typically obtained by polymerization of alkoxysilanes ($Si(OR)_4$, where R stands for an alkyl group) or neutralization of sodium metasilicate. Polymerization of alkoxysilanes is usually performed in water/alcohol mixtures with an acid or base catalyst, and involves hydrolysis of Si–OR groups to Si–OH followed by the formation of siloxane (Si–O–Si) bonds (Figure 5.2(d)). After the initial, rapid gelation, silica gels continue to evolve due to additional crosslinking between the remaining Si–OH and Si–OR groups. This slow process increases shear and elastic moduli, and eventually terminates to give gels with stable physical properties. Silica gels can be functionalized by incorporating organic groups (e.g., $R'Si(OR)_3$, in which the R' does not hydrolyze), other metals (e.g., using $Ti(OR)_4$ or $Al(OR)_3$) or by occluding nanoparticles, clays, or liquid crystals.

Once prepared, silica gels are chemically stable, and can withstand most weak acids/bases and organic solvents. They are also stable toward most inorganic salts and support growth of large, high-quality salt crystals (Section 3.6). On the other hand, silica gels are degraded in the presence of strong acids/bases, strong oxidants and fluorine-containing compounds (HF, F_2, OF_2, ClF_3) that compete with oxygen to form Si–F bonds. Silica gels can be structurally evolved by hydrothermal treatment, and autoclaving them in the presence of water increases the average pore size.

The uniqueness of silica gels as media supporting RD is that they can be processed into dry, porous materials with the retention of shape and size – in this way, it is possible to first execute a RD process in a wet silica gel and then make the developed structure permanent by the removal of solvent. This removal can be effected either by simple evaporation to yield xerogels or by so-called supercritical extraction to give aerogels. Xerogels are usually characterized by low porosity but can be baked to form dense coatings or ceramics. Aerogels[6,7] are more porous and have several fascinating and sometimes ‘extreme’ physical properties – ultralow density ($\sim$1 mg cm^{-3}), the lowest known thermal conductivity (Figure 5.3), refractive index near one (tunable using various chemical additives) and an ability to slow down sound to very small speeds. They are used in many important applications including scratch-resistant coatings, optical elements and thermal insulators for space shuttles.[6,7]

Figure 5.3 Owing to its extremely low thermal conductivity, a piece of inflammable aerogel can be held in a butane-torch flame for prolonged periods of time without burning one's fingers. In this photograph, a student has been holding the slab of aerogel for over a minute

5.2 FABRICATION

While the gel films supporting reaction-diffusion process are trivial to prepare by simple casting or spin-coating, micropatterned stamps require a more elaborate fabrication procedure (Figure 5.4 and Table 5.1).

Photomask. The first step is to define a pattern of interest using photolithography. This process begins by designing the so-called photomask with transparent and opaque regions (Figure 5.4(a)). The design can be carried out in any higher-end graphical program such as Macromedia Freehand, Corel-Draw, or CleWin. The pattern is then printed on a transparency using a high-resolution printer. With professional 20 000 dpi printers, it is possible to print sharp features with resolution down to about 10 μm at a cost of about $50 per transparency (e.g., CAD/Art Services, Poway, CA). If still smaller features are desired, the computer design has to be engraved into a thin, quartz-supported metal film using a focused electron or ion beam. This method can produce submicrometer features, but it is significantly more expensive, and one photomask can cost several thousand dollars.

Table 5.1 Fabrication of wet stamps

Step	Time involved	Typical cost	Reusable?
Photomask preparation	2 h	\$50 on transparency, >\$1000 for chrome masks	*Yes*
Photolithography	5–8 h	If not in-house, \$500–1000 per one 3.5-inch wafer	*Yes*, but patterned features can delaminate after several months. Store in dry place! Do not wash with acetone, hot ethanol orf hot water
PDMS molding	30 min + 4–12 h for curing	~\$10 (cost of PDMS and $CF_3(CF_2)_6(CH_2)_2SiCl_3$ silanizing reagent)	*Yes*, but with repeated casting of gels hard-to-remove residues accumulate in the features. Wash in warm water after every gel casting, do not expose to acids, do not sonicate. Oxidize surface prior to each use
Gel casting and stamp preparation	1–2 h	\$1–2 per stamp (cost of gels)	Usually *no*, although with substances that do not degrade the gel matrix, multiple (tens) stampings can be performed. Pure gels can be stored for months in deionized water

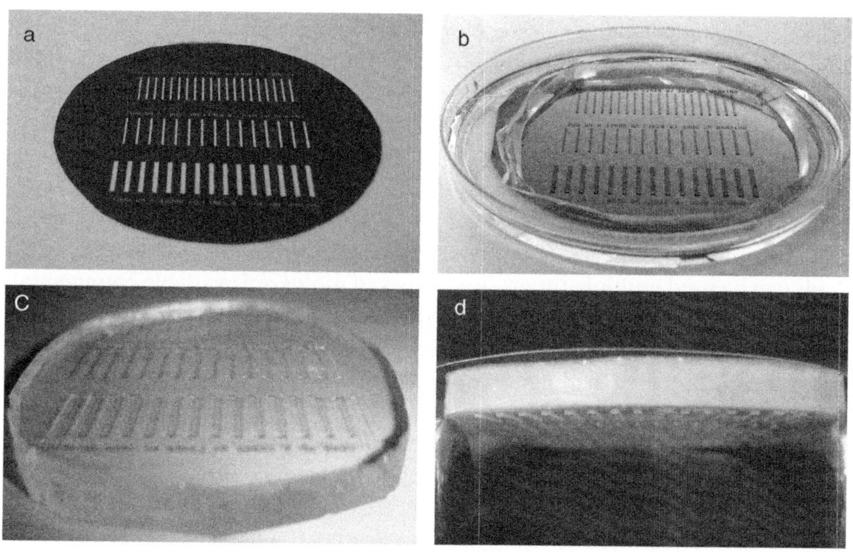

Figure 5.4 Fabrication of wet stamps. (a) A transparency defines microfeatures to be patterned (here, arrays of lines of thickness between 250 μm and 1 mm). (b) Using photolithography, the pattern is transferred into a photoresist supported on a silicon wafer. Here, the patterned lines protrude above the plane of the silicon. (c) PDMS molded against the silicon master. The replicated lines are now *depressions* in the PDMS surface. (d) A layer of agarose cast and cured against PDMS has protruding lines embossed on its surface. This layer can be cut into smaller pieces to give wet stamps

Photolithographic master. Using the photomask, one prepares a photolithographic 'master' (Figure 5.4(b)). This step starts with spin-coating of a thin layer of photocurable polymer ('photoresist') onto a flat silicon wafer. The photomask is then placed onto the photoresist (and gently pressed against it to ensure conformal contact), and the assembly is exposed to a uniform source of UV light. The transparent regions of the photomask allow the UV to pass through; the opaque ones block passage of light. The regions of photoresist exposed to UV undergo a chemical reaction and polymerize/crosslink (in the so-called 'negative' photoresists, the exposed regions degrade). The mask is then taken off, and the photoresist from noncrosslinked regions (or degraded regions in negative photoresists) is washed away to leave behind a pattern of small protrusions (depressions) corresponding to the design of the mask. The surface of the developed master is then exposed to vapors of $CF_3(CF_2)_6(CH_2)_2SiCl_3$ overnight to reduce its tendency to adhere to the polymers that are cast against it (see next step). Note that in order to avoid contamination by dust and to ensure uniform UV exposure of the master, the entire procedure should be performed in a clean-room using specialized optical equipment. If a clean-room facility is not available, photolithographic procedures can be contracted out to companies like NanoTerra (www.nanoterra.com; large-volume orders) or ProChimia (www.prochimia.com). Although such custom fabrication might be quite expensive, the masters can be used multiple times and many copies of stamps can be made from them.

PDMS mold. Next, the photolightographic master is replicated[8] into a silicone elastomer called polydimethylsiloxane (PDMS; commercial name Sylgaard 184, Dow Corning; Figure 5.4(c)). Although one might be tempted to replicate the master immediately into hydrogel, most photoresists are not water- and/or temperature-resistant and casting a hot, water-based gel solution against them would ruin the master in short order. In contrast, PDMS does not contain water, can be cast and solidified at room temperature and is exceptionally stable to common chemicals. After curing, PDMS is gently peeled off the master, and oxidized in a plasma cleaner for about a minute to make its surface hydrophilic. Like photolithographic masters, PDMS molds can be used multiple times; some types of molds are available commercially from Platypus Technologies (www.platypustech.com) and ProChimia.

Hydrogel stamps. Hydrogel precursor solution can now be cast against the PDMS mold. In the case of agarose, this solution (typically, 8% w/w) is hot and solidifies upon cooling at room temperature. To avoid trapping of air bubbles in the gel, it is advisable to perform the procedure in a desiccator (about 3 L volume) under mild vacuum and to warm the PDMS master immediately prior to use. Once gelation is complete ($\sim$30 min), agarose (typically 0.5–1 cm thick) is removed from the PDMS to give a freestanding layer with microfeatures embossed on its lower surface (Figure 5.4(d)). This layer is cut into smaller pieces (1–2 cm $\times$ 1–2 cm $\times$ 0.5–1 cm), which constitute the 'wet stamps'. These stamps are subsequently soaked in a solution of a specific reagent to be used in a RD process. The soaking usually takes 2–24 hours to ensure uniform distribution of reagents within the stamp. Immediately prior to use, the patterned surface of a stamp is blotted dry on a tissue paper ($\sim$5 min) and under a stream of air or nitrogen (15 s). The stamp is then placed onto a gel substrate to initiate RD from its microscopic features. This is where the fun of RD begins!

APPENDIX 5A: PRACTICAL GUIDE TO MAKING AGAROSE STAMPS

While PDMS molding and agarose casting are not sky-rocket technology, these procedures can sometimes be finicky. For best results, we recommend following these step-by-step instructions.

5A.1 PDMS Molding

1. In order to mold PDMS against a photolithographically patterned silicon wafer, it is necessary to first eliminate adhesion between the two materials. To this end, the wafer is cleaned with oxygen plasma (2–5 min) followed by silanization of its surface. Silanization is achieved by placing an oxidized wafer and a small, open vial containing up to five droplets of trichloro (1H,1H,2H,2H-perfluorooctyl) silane (Sigma Aldrich 448931-10G) in a small ($\sim$3 L) plastic or glass chamber connected to a house vacuum (i.e., in a desiccator such as Bel-Art, model F420200000, www.belart.com). The chamber should be evacuated and then kept closed for $\sim$12 h.
2. PDMS is prepared by vigorously stirring a 10:1 w/w mixture of Sylgard 184 prepolymer and crosslinking agent (available as a kit from Dow Corning, www.dowcorning.com) until very well mixed, then degassing under vacuum at room temperature to remove any remaining air bubbles ($\sim$30–60 min).
3. Next, PDMS is poured onto the silanized wafer, to a depth of 5–10 mm (roughly 40 g of solution for a 3-inch wafer in a 100 mm Petri dish), and then degassed until no bubbles remain in contact with the silicon surface and/or with the micropatterned features.
4. The PDMS is then cured by heating in a 65 °C oven until it is no longer tacky, typically for 4 h.
5. Finally, the PDMS mold is removed from the silicon master by cutting the PDMS layer around the wafer's perimeter, and peeling it off gently. For easiest removal, the cut piece should not extend over the edge of the wafer. Some ethanol should be poured between the PDMS and the master to facilitate their separation without mechanical damage; larger masters might require several additions of ethanol and very gentle peeling.

5A.2 Agarose Molding

1. The PDMS mold is first oxidized (again, in a plasma cleaner for 2–5 min) to render its surface hydrophilic.

2. Approximately 100 g of a 4–10% by mass mixture of Omnipur High Gel Strength Agarose (EM-2090, available through VWR, vwr.com) in water is made in a 500–800 mL beaker. It is important to use this specific brand and type of agarose, as other products give stamps that are much less sturdy.
3. The mixture is stirred to break up large agarose clumps, sealed with Saran-wrap (or even a nitrile or latex glove) and heated in a microwave oven until it is vigorously boiling, and the cover has either popped off or significantly expanded. Heating typically takes 60–90 s and results in a very hot solution.
4. The cover is rapidly removed from the beaker, and the oxidized PDMS mold is placed feature-side-up on the agarose surface (this allows the PDMS to warm up and minimizes formation of air bubbles on its surface). After 10 s, the mold is slowly pushed into the still hot agarose, which flows over and covers the features of the PDMS mold.
5. The beaker containing agarose/PDMS is placed in a vacuum desiccator, and is held under vacuum for 90 s. This removes most of the bubbles from the agarose solution. If the PDMS mold rises during degassing, it needs to be pushed back before the agarose gelates.
6. After degassing, a thermally insulating cover (e.g., a terrycloth glove or cloth towel) is placed on top of the beaker, and agarose is allowed to gel at room temperature. This cooling process typically takes 30–45 min.
7. Once the agarose has cooled, a piece of agarose containing the entire PDMS mold is removed from the beaker. This piece is trimmed to the size of the mold, and the mold is then gently peeled off the solidified gel.
8. The agarose stamp is ready for use, and can be stored in water if not needed immediately. The PDMS mold can be cleaned by soaking in hot water, and can be reused multiple times.

REFERENCES

1. Mehrer, H. (1996) Diffusion in intermetallics. *Mater. Trans. JIM*, **37**, 1259.
2. Goeb, O., Herrmann, M. and Siegel, S. (1999) Thermodynamic and microstructural studies of coloration and decoloration processes during gas-pressure sintering of silicon nitride ceramics. *Sci. Eng. Cer. Key Eng. Mater.*, **2**, 217.
3. Arnott, S., Fulmer, A. and Scott, W.E. *et al.* (1974) Agarose double helix and its function in agarose gel structure. *J. Mol. Biol.*, **90**, 269.
4. Arnott, S., Scott, W.E., Rees, D.A. and McNumb, C.G.A. (1974) I-Carrageenan: molecular structure and packing of polysaccharide double helices in oriented fibres of divalent cation salts. *J. Mol. Biol.*, **90**, 253.
5. Campbell, C.J., Smoukov, S.K., Bishop, K.J.M. and Grzybowski, B.A. (2005) Reactive surface micropatterning by wet stamping. *Langmuir*, **21**, 2637.
6. Sanchez, C., Julian, B., Belleville, P. and Popall, M. (2005) Applications of hybrid organic–inorganic nanocomposites. *J. Mater. Chem.*, **15**, 3559.
7. Pierre, A.C. and Pajonk, G.M. (2002) Chemistry of aerogels and their applications. *Chem. Rev.*, **102**, 4243.
8. Whitesides, G.M., Ostuni, E. and Takayama, S. *et al.* (2001) *Annu. Rev. Biomed. Eng.*, **3**, 335.

6

Fabrication by Reaction–Diffusion: Curvilinear Microstructures for Optics and Fluidics

6.1 MICROFABRICATION: THE SIMPLE AND THE DIFFICULT

Miniaturization is the 'arrow of time' of modern technology. Computer chips, flat panel displays, micromirror-based laptop projectors, RFID tags, inkjet valves, microfluidic systems and countless other useful devices are all based on microscopic components that are often invisible to the naked eye. To make such structures, even the proverbial patience and precision of a Swiss watchmaker would not be sufficient and new techniques (collectively known as microfabrication) are needed to manipulate, pattern and structure materials at the scale of micrometers and below. Some of these techniques have achieved accuracy and versatility that is simply breathtaking – for example, in microfabrication of microelectronic components where a fully automated combination of photolithography, vapor deposition and etching[1,2] can produce microprocessors comprising hundreds of millions transistors with the overall process yield up to 93%.[3,4] Yet, for some architectures, even the photolithographic wonders of Silicon Valley may be inadequate. Photolithography is very efficient in delineating microstructures with 'straight' side walls (Figure 6.1(a)), but it is not readily extended to curvilinear topographies (Figure 6.1(b)). In photolithography, it is easy to fully

Chemistry in Motion: Reaction–Diffusion Systems for Micro- and Nanotechnology Bartosz A. Grzybowski

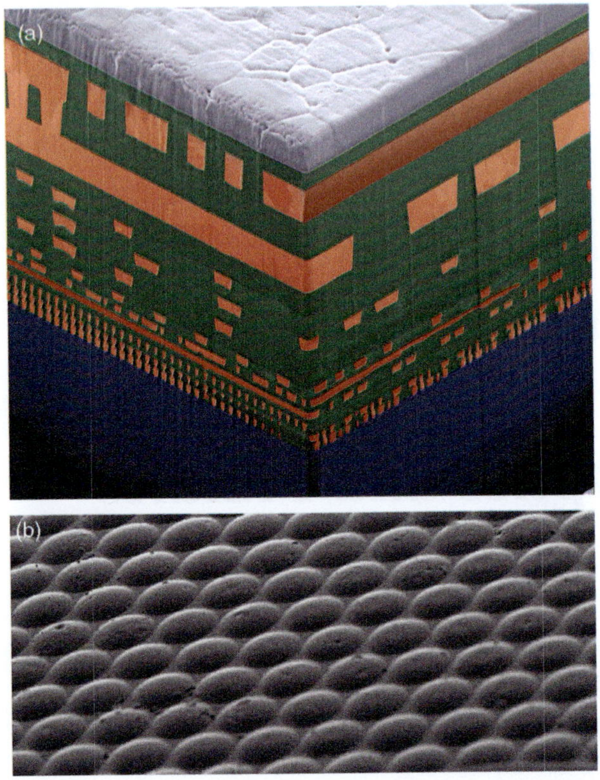

Figure 6.1 (a) Cross-section of 64-bit high-performance microprocessor chip built in IBM's 90 nm Server-Class CMOS technology. Though remarkably complex, PC microprocessors are nowadays manufactured in large quantities, and for as little as tens of dollars. (b) Surprisingly, fabrication of an apparently simple array of microlenses is more involved. A 2×2 cm array can cost several hundred dollars. (a) Reprinted courtesy of International Business Machines Corporation, © International Business Machines Corporation.

UV-degrade/crosslink a polymer substrate beneath a 'transparent/opaque' photolithographic mask and obtain a 'binary' surface relief, but it is hard to set up precise gradients of light intensity to develop a curvilinear or multilevel surface profile. Although one could try to make such structures via serial techniques (localized electrodeposition,[5] proton and ion beam machining,[6,7] laser ablation,[8] powder blasting[9] or electrochemical micromachining[10]), these procedures are usually laborious, expensive and often unsuitable for very small structures. Luckily, there is reaction–diffusion (RD). As we remember from earlier chapters, RD processes are inherently linked with spatial concentration gradients. If we could somehow control these gradients in space and time and translate them into local deformations of a material supporting RD, we could fabricate such coveted nonbinary microstructures. In this chapter, we show how such translation can be achieved by surprisingly simple experimental means.

6.2 FABRICATING ARRAYS OF MICROLENSES BY RD AND WETS

Our first example of wet stamping (WETS) in action deals with arrays of microscopic lenses for uses in optical data storage,[11] imaging sensors,[12] optical limiters[13] and confocal microscopy[14] to name just a few. To make these curvilinear structures we will use an inorganic reaction between silver nitrate ($AgNO_3$) and potassium hexacyanoferrate ($K_4[Fe(CN)_6]_3$), in which silver cations and hexacyanoferrate anions precipitate according to $4Ag^+ + [Fe(CN)_6]^{4-} \rightarrow Ag_4[Fe(CN)_6]$ ($\downarrow$). Importantly, it has been shown that if this precipitation occurs in a gelatin matrix, it causes pronounced and permanent (i.e., persisting after the gel is dried) gel swelling of magnitude linearly proportional to the amount of precipitate generated.[15] This coupling between reaction products and gel dimensions/topography is central to translating RD controllably into the shapes of the objects we wish to fabricate.

Our microfabrication process begins by placing a WETS stamp soaked in a solution of $AgNO_3$ and micropatterned in bas relief with an array of cylindrical depressions ('wells') onto a thin film of dry gelatin uniformly loaded with $K_4Fe(CN)_6$ (Figure 6.2(a)). When the surface of the stamp comes into conformal contact with gelatin, water and ions diffuse from the stamp into the dry gel (Figure 6.2(b) and Example 6.1). In doing so, the silver cations – constantly resupplied from the stamp – precipitate all hexacyanoferrate anions they encounter and cause the gel to swell. The interesting and relevant part of this process is the diffusion from the edges of the circular features radially inwards (in the $-\boldsymbol{r}$ direction; Figure 6.2(b,c)). Here, as the reaction (precipitation) front propagates, the unreacted $Fe(CN)_6{}^{4-}$ experiences a sharp concentration gradient at this front and diffuses in its direction – that is,

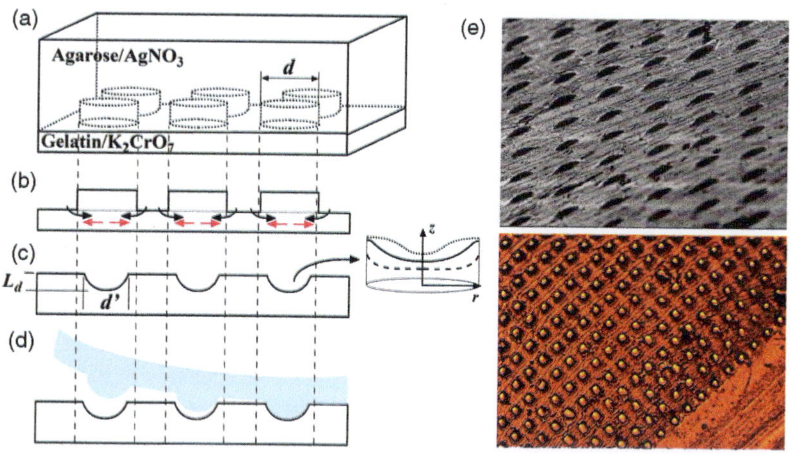

Figure 6.2 (a–c) Scheme of microlens fabrication using WETS. (d) Replication into a polymeric material cast against the lenses. (e) SEM (top) and optical (bottom) images of an array of 50 μm lenses replicated into PDMS. Reprinted from reference 17, with permission © 2005 American Institute of Physics.

radially outward (+ *r* direction, red arrows in Figure 6.2(b)). This 'outflow' decreases the concentration of $[Fe(CN)_6]^{4-}$ anions near the circles' centers so that by the time Ag^+ cations diffuse therein, they find fewer precipitation 'partners' than they had near the patterned edges. In other words, the amount of precipitate and the degree of gel swelling, $z(r)$, both decrease monotonically with decreasing r. The RD phenomena continue until one of the reagents runs out. When RD comes to a halt, the patterned regions are convex depressions in an otherwise flat surface (Figure 6.2(c)). These depressions do not change shape when the gel is dried (drying only compresses wet gelatin uniformly by ~20%) and the developed topography can be replicated[16] into other materials (e.g., optically transparent polydimethylsiloxane, PDMS) to give regular arrays of convex microlenses that focus light efficiently (Figure 6.2(d) and also insets to Figure 6.4(a)).[17]

Example 6.1 A Closer Look at Gel Wetting

Consider a 'wet' agarose stamp containing a solution of a chemical species A (e.g., an ionic salt such as $AgNO_3$). When this stamp is applied onto a thin layer of dry gelatin, doped with another chemical B (e.g., $K_4Fe(CN)_6$), water and A are transported into the gel, where A and B react to form an insoluble product. The transport of water and chemical species proceeds from the stamp and into the gelatin by a two-step kinetic mechanism.

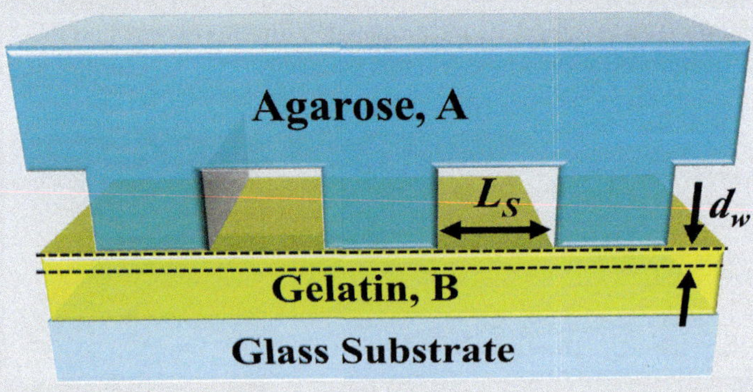

In the first step, water from the stamp rapidly spreads onto the surface of the dry gelatin by capillary wetting. For the case of water wetting a hydrophilic surface such as gelatin, the maximum velocity of the spreading front may be approximated as[18] $V = \gamma\theta_E^3/9\sqrt{3}\eta l$, were γ is the surface tension of the liquid, θ_E is the equilibrium contact angle, η is the dynamic viscosity of the liquid and l is a dimensionless parameter (typically, $l = 15$–20). Thus, for water ($\gamma \sim 70$ mJ m^{-2} and $\eta \sim 10^{-3}$ Pa s) spreading towards an equilibrium contact angle of $\theta_E \sim 5°$, this characteristic velocity is $V \sim 100$ μm s^{-1}. Denoting the characteristic spacing between the edges of the stamp's microfeatures by L_s (typically ~100–500 μm for the examples discussed in this chapter), the characteristic time of the wetting process may be estimated as $\tau_{wet} \sim L_s/V \sim 1$–5 s.

Whereas wetting hydrates the surface layer of gelatin rapidly, water transport into the bulk of the dry gel proceeds slowly by a diffusive process characterized by a diffusion coefficient, $D_w \sim 10^{-7}\,cm^2\,s^{-1}$.[19] Meanwhile, species A and B are free to diffuse (with typical diffusion coefficient $D_s \sim 10^{-5}\,cm^2\,s^{-1}$) and react within the hydrated portion of the gel. The characteristic time required to form any RD patterns/structures over distance L_s is $\tau_{diff} \sim L_s^2/D_s \sim 10$–$250\,s$. During this time, water diffuses a distance $d_w \sim (\tau_{diff} D_w)^{1/2} \sim 10$–50 μm into the bulk of the gel. Importantly, the initial water transport into the gel may be considered independent of that of the chemical species, A, which is effectively 'filtered' from the water due to its reactive consumption by species B. With these considerations, one may treat RD phenomena as occurring in a thin (~10–50 μm) layer of a hydrated gel. This approximation allows us to neglect migration of water within the substrate and variation of diffusion coefficients of salts with gel depth. These simplifications make theoretical treatment of our RD systems considerably easier, and we will use them frequently throughout the book.

For the RD fabrication to be truly useful, it should, as mentioned in Chapter 1, allow for the 'programming' of the shapes and curvatures of the lenses one wishes to prepare. Since local deformations of the gel surface depend on the underlying concentrations of the precipitate, the shapes of the developed lenses can be controlled by the concentrations of the salts used and/or the dimensions of the stamped features. Trends based on experimental surface profilograms (Figure 6.3) illustrate how the depth, L_d, of the lenses changes with changing [$AgNO_3$], [$K_4Fe(CN)_6$] and the diameter of the stamped features, d.

For a given value of d and with [$AgNO_3$] kept constant, L_d increases with increasing [$K_4Fe(CN)_6$] (up to 1% w/w; above this concentration, the gelatin does not gelate properly and tends to 'melt' upon stamping). This trend (Figure 6.3(a)) reflects the fact that when more potassium hexacyanoferrate is available in the patterned film, the more precipitate can form, and the higher the degree of swelling. At the same time, increasing the concentration of $K_4Fe(CN)_6$ decreases the effective diameter of the lens, d', and increases the surface curvature. These effects can be explained by the extent of the propagation of the precipitation front inwards: when [$K_4Fe(CN)_6$] is low, all ions are precipitated below or in the vicinity of the stamped ring and only these regions develop curvature; when more hexacyanoferrate is available, the front travels further, curving the center of the circle and leaving behind itself a flat region of uniform precipitation.

When d and [$K_4Fe(CN)_6$] are kept constant (Figure 6.3(b)), L_d increases with increasing [$AgNO_3$] up to ~10% w/w concentration, but the lenses are relatively flat-bottomed. Curved lenses approximated by sections of a sphere are obtained

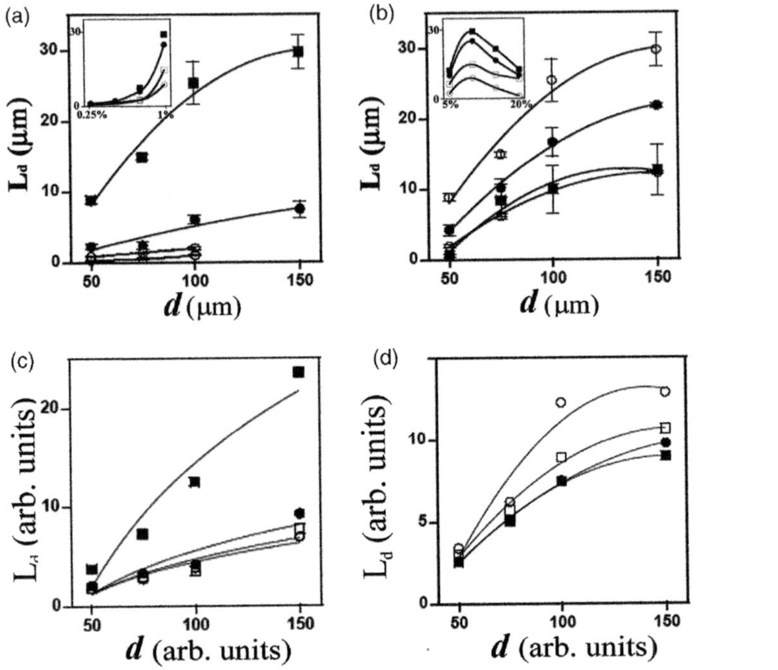

Figure 6.3 (a) Experimental dependence of the depth, L_d, of circularly symmetric depressions on feature size, d, for varying concentrations of $K_4Fe(CN)_6$ in gelatin (○, 0.25%; □, 0.5%; ●, 0.75%; ■, 1.0%) and for a constant concentration of $AgNO_3$ in the stamp (10%). Inset shows the same data plotted against $[K_4Fe(CN)_6]$ for different values of d (○, 50 μm; □, 75 μm; ●, 100 μm; ■, 150 μm). (b) L_d as a function of d for $[K_4Fe(CN)_6] = 1\%$ w/w and for variying $[AgNO_3]$ (□, 5%; ○, 10%; ●, 15%; ■, 20%). Inset shows the same data plotted against $[AgNO_3]$ for different values of d (○, 50 μm; □, 75 μm; ●, 100 μm; ■, 150 μm). Graphs in (a) and (b) were created based on profilometric measurements of the gelatin masters. Standard deviations of L_d were collected from at least three independent stampings and two profilometric scans (averaged over ten times each) spanning two to five features for each stamping. For lenses with $d < 100$ μm, standard deviations were less than 1%. (c, d) Dependencies of L_d on d modeled using the lattice gas method and corresponding to the experimental trends in, respectively, (a) and (b). The units on both axes are arbitrary but linearly proportional to the experimental ones – that is, in (c), relative concentrations for curves ○:□:●:■ = 1:2:3:4; in (d) ■:○:□:● = 1:2:3:4. For further details, see Ref. 17 (Reprinted with permission from *Appl. Phys. Lett.* (2004), **85**, 1871. © 2005 American Institute of Physics.)

for $[AgNO_3] > 10\%$. In this regime, however, L_d decreases with increasing concentration of silver nitrate. This trend is due to 'flooding' of the gelatin with a large amount of $AgNO_3$, so that the reaction front reaches the center of a lens rapidly, precipitates the $Fe(CN)_6^{4-}$ still present therein and lifts this region up with respect to the perimeter of the lens.

Finally, for given concentrations of the participating chemicals, L_d increases with increasing diameter of the stamped circles d (Figure 6.3(a,b)) until it plateaus at $d \sim 150$–$200\,\mu\text{m}$. In this limit, the precipitation front from the perimeter of the circle does not propagate all the way to the circle's center (remember that diffusive transport `slows down' with distance traveled), and L_d depends only on the degree of swelling directly below the features. In addition, increasing d decreases the curvature of the lenses: in large circles, the central area where no precipitation occurred remains flat.

The technique can be straightforwardly extended to arrays of microlenses of arbitrary base shapes. Figure 6.4 shows optical micrographs of 'pyramidal' lenses obtained by wet stamping arrays of triangles (Figure 6.4(a)), squares (Figure 6.4(b)), hexagons (Figure 6.4(c)) and stars (Figure 6.4(d)). As in the case of circularly symmetric patterns, the topographic details of these reliefs can be controlled by the concentrations of the participating chemicals, and can be faithfully reproduced into optically useful polymers.

6.3 INTERMEZZO: SOME THOUGHTS ON *RATIONAL* DESIGN

Our method's sensitivity to the RD process parameters (salt concentrations and dimensions of the stamped features) is in some sense a double-edged sword. On the one hand, by changing these parameters we can fabricate a continuum of microlens shapes (hemispheres, 'copulas' of different curvatures, pyramids, etc.); on the other hand, we cannot readily guess *a priori* the values of parameters that would produce a *particular* lens (say, a section of sphere with $L_d = 20\,\mu\text{m}$ and diameter $d = 600\,\mu\text{m}$). Of course, we could standardize the method by recording the experimental trends (such as those described in Section 6.2) for different parameter values, and then extrapolate/approximate the values needed to build a desired structure. This brute-force approach, however, would be time-consuming and laborious and would certainly detract from the appeal of supposedly `effortless' RD fabrication. The more rational approach is through modeling.

For conventional patterning/microfabrication techniques, modeling is often of only ornamental value, as these methods can be successfully practiced without any theoretical background. In sharp contrast, RD fabrication without some theoretical guidance is rather hopeless since the participating chemicals evolve into final structures via nontrivial and sometimes counterintuitive ways. In other words, simple 'what you pattern is what you get' heuristics simply do not apply. Fortunately, in Chapter 4 we learned that there are many theoretical tools with which one can model RD. For practical applications of RD microfabrication, where the initial/imposed geometries might be quite complex, numerical rather than analytical approaches are better suited. In this chapter, we will use arguably the simplest of these methods – the discrete lattice gas (LG) approach – which is easy to set up and thus accessible to

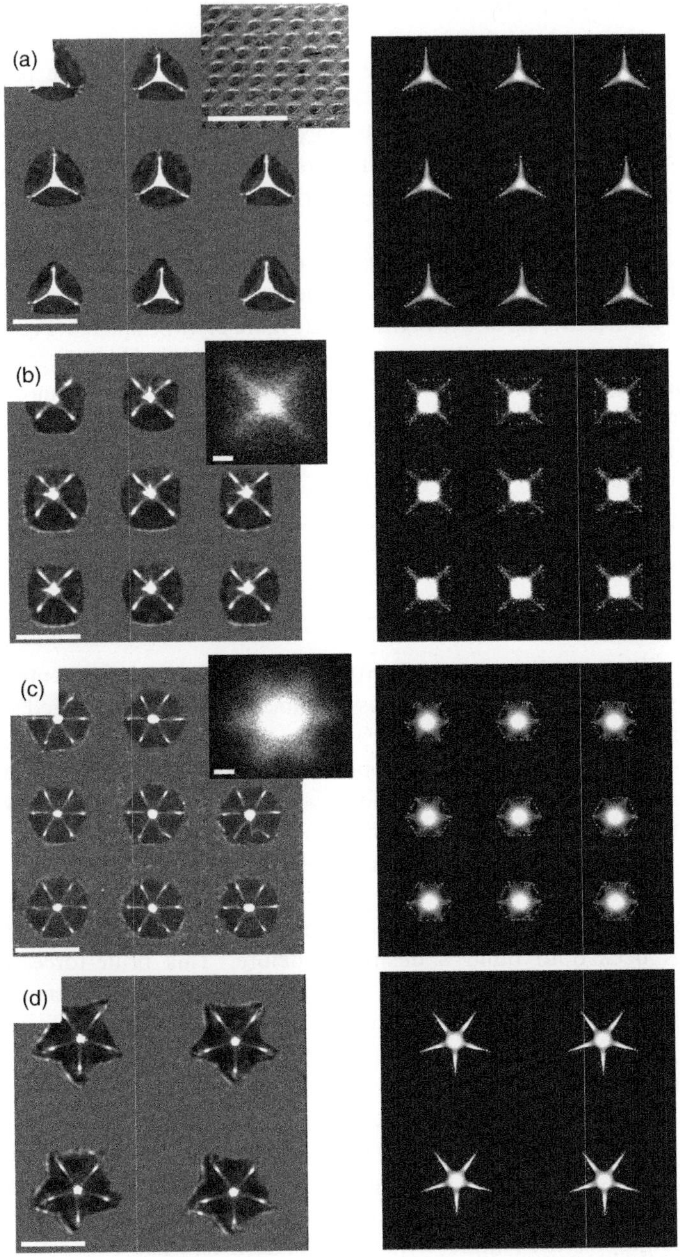

Figure 6.4 Experimental (left) and modeled (right) images of RD-fabricated microlenses having polygonal base shapes. Inset in (a) illustrates a long-range order in the stamped array of triangles. Insets in (b) and (c) are the images of the focal plane of the corresponding lens arrays replicated into PDMS. Scale bars in the primary images in (a–d) correspond to 150 μm; those in the insets are 1 mm in (a) and 2 μm in (b) and (c). (Reprinted with permission from *Appl. Phys. Lett.* (2004), **85**, 1871. © 2005 American Institute of Physics.)

experimentalists, rapid and, given the simplifications it entails, surprisingly accurate.

6.4 GUIDING MICROLENS FABRICATION BY LATTICE GAS MODELING

Recall from Section 4.6 that, in a LG-type model, the domain of a RD process (here, gelatin substrate) is represented as a discretized grid, and the molecules/ions placed onto it are subject to several basic rules describing reaction and diffusion events. For our experimental system, each node on the lattice has the same initial concentration of hexacyanoferrate ions (B) while silver cations (A) are delivered from the features of the stamp. There are three simulation steps.

1. A is added to the nodes of the gel directly beneath the stamped features. The number of A molecules at the nodes corresponding to the gelatin/agarose interface is kept constant (that is, the stamp is approximated as an infinite, constant-concentration reservoir of A).
2. A and B are allowed to perform a diffusion move on the square lattice, with the experimentally determined[15] relative diffusion coefficients of the two salts as $D_B/D_A = 0.3$. To give physical meaning to these coefficients on a lattice (where all the steps are of the same length), the particles of the two types are assigned certain probabilities of moving in each of the diffusive steps. These probabilities are determined such that A always moves to an adjacent node, and B moves with probability $p_B = D_B/D_A = 0.3$. Once the particle is chosen to move, it has equal chance of migrating to each of the nearby nodes (e.g., $1/4$ for the square lattice).
3. In cells where A and B are present in sufficient quantities, they react according to the specific reaction stoichiometry – in our system: $4A + B \rightarrow A_4B\ (\downarrow)$.

These three steps are repeated until one of the reagents (usually the limiting reagent, B, in the gelatin) runs out. At this point, RD terminates, and the topography of the surface is reconstructed by multiplying the local concentrations of the precipitate by a scaling factor (determined experimentally from one 'test' pattern).

Despite its simplicity, the LG model reproduces well the experimental profiles of the microlenses fabricated using features of different dimensions/shapes and different concentrations of participating chemicals (Figures 6.4 and 6.5). The same set of parameters is used to simulate all these structures and the execution of the program (available for download at http://dysa.northwestern.edu/Research/ Progreactions.dwt) takes only tens of seconds on a standard desktop PC. This speed combined with the ease of coding initial conditions (a bitmap picture is sufficient as input) makes this program quite helpful in the

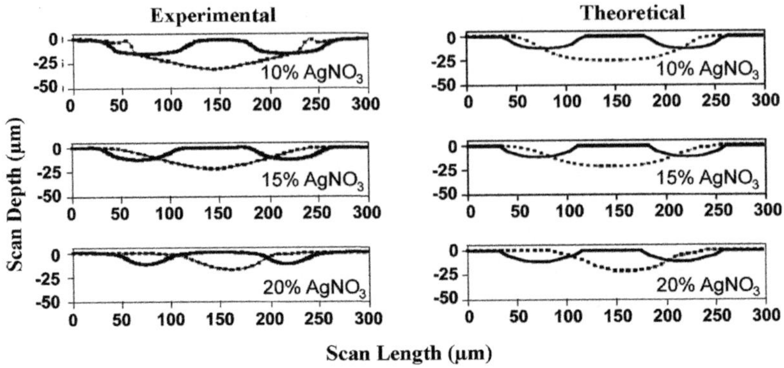

Figure 6.5 Experimental (left) and modeled (right) surface profiles of microlenses developed from circular features ($d = 75$ μm, solid line; $d = 150$ μm, dashed line) for varying concentrations of silver nitrate and for 1% w/w $K_4Fe(CN)_6$ in dry gelatin. (Reprinted with permission from *Appl. Phys. Lett.* (2004), **85**, 1871. © 2005 American Institute of Physics.)

design of not only microlenses but also other types of structures that we will discuss shortly.

Before we proceed, however, one more important comment is due. A curious reader might have noticed that while the LG model reproduces experimental results for particular sets of parameters, it does not back-track parameters that would lead to a desired structure. Although it would be ideal to propagate the RD process from a desired structure 'backwards', all the way to the initial conditions, RD is not time reversible (Example 6.2) and such reverse-engineering is not possible. In its absence, we will optimize the parameters with the help of the so-called Monte Carlo method. Say we wish to make a circular lens having a profile z_{ij} (i, j subscripts specify lattice locations). For the lack of better ideas, we begin with a 'guess' choice of concentrations and dimensions of the stamped pattern (our parameter set, P^{curr}), with which we run the LG model as described before to obtain a surface profile z_{ij}^{curr} (the superscripts on z and P remind us this is our 'currently-best' guess). Unless we are incredibly lucky or clairvoyant, our trial profile is probably quite different from the target z_{ij}. This difference can be quantified as $\delta^{\text{curr}} = \sum_{i,j}(z_{ij} - z_{ij}^{\text{curr}})^2$, and our task is to minimize it. In an effort to do so, we change one or more parameters (new set, P^{new}) and calculate a corresponding profile, z_{ij}^{new}, and difference $\delta^{\text{new}} = \sum_{i,j}(z_{ij} - z_{ij}^{\text{new}})^2$. Now we have two options. First, if $\delta_{\text{new}} < \delta_{\text{curr}}$, we are getting closer to the optimal solution and so we keep the adjusted parameter set as our best guess so far – in other words, P^{new} becomes P^{curr}. Second, if $\delta_{\text{new}} \geq \delta_{\text{curr}}$, we are not making progress toward the target structure, and it is tempting to reject the change in parameters unconditionally and keep the previous set, P^{curr}, as 'currently-best'. However tempting, such unconditional rejection is not a great idea, for by accepting only the changes that decrease δ

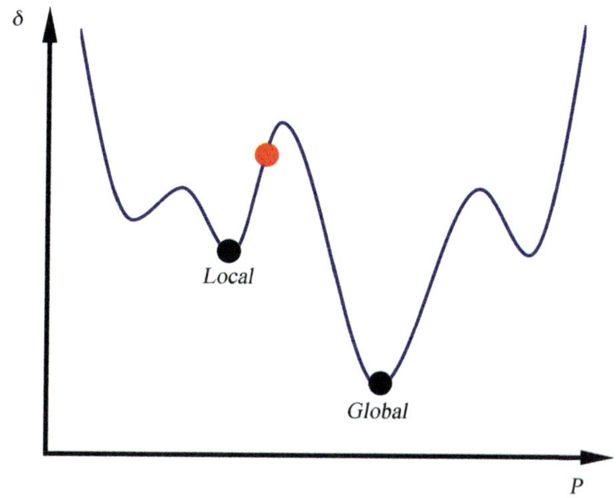

Figure 6.6 Optimization procedure starting from a trial set of parameters (red circle). Acceptance of parameter sets always decreasing error, δ, leads to a locally optimal solution. To reach a global optimum, it is first necessary to move 'uphill' by accepting some parameter changes that increase δ

monotonically, we move to a locally optimal solution and might never find the global one (Figure 6.6). Sometimes, one needs to move uphill before descending into a deeper valley – which in technical jargon means that it is sometimes advisable to accept a parameter set that increases δ to escape from a local minimum. Of course, such uphill excursions should not be taken too frequently, and there are rigorous methods that describe probabilities with which one should reject or accept a parameter set that increases δ (Example 6.3).

Overall, after deciding whether to accept or reject the new set of parameters, a new trial solution is generated and compared with the 'currently-best' one, and this cycle is repeated until ultimately converging to a solution that is close (within desired accuracy) to the target RD structure. Because this process might take quite a few steps, especially if many parameters are being optimized, computational speed is essential. With rapid LG algorithms where simulation of each RD structure takes only a few seconds on a low-end laptop, one can perform of the order of a thousand optimization cycles per day, which is not mind-boggling but at least reasonable. If more accurate numerical methods are needed to capture the details of a RD process, each optimization step might take hours to days, and optimization procedures become practically unfeasible since waiting several months for an answer can hardly be considered helpful in the design process. For now, let us stay with some more fabrication tasks where the models are quick and helpful.

Example 6.2 Is Reaction Diffusion Time-Reversible?

A process is said to be time-reversible if the governing equation of the system remains unchanged under the time transformation from $t \rightarrow -t$.[20] Time-reversible processes are commonly found in classical Newtonian mechanics of the specific form $m(d^2x/dt^2) = F(x)$, where m is the mass of an object, x is the displacement and the force $F(x)$ is any function of x. Since $F(x)$ is independent of t, applying the transformation $t' = -t$ yields $m(d^2x/dt'^2) = F(x)$. Because the transformed equation is identical to the initial equation, this system is considered to be time-reversible. As an example, consider a ball falling freely in a uniform gravitational field, for which $m(d^2x/dt^2) = g$, where x is the elevation and g is the acceleration due to gravity. If we took a movie of the ball's motion and then played it backwards, we would see that although it is now moving upwards, its trajectory is perfectly realistic and obeying all the laws of mechanics.

With this definition of time-reversibility, is it possible for RD processes to be time-reversible? The short answer is no, since applying the same transformation, $t' = -t$, to a general reaction-diffusion equation, $\partial c/\partial t = D\nabla^2 c - R$, yields a significantly different equation, $\partial c/\partial t' = -D\nabla^2 c + R$.[21] Note that this equation now has a negative diffusion coefficient, which is certainly unphysical (from Chapter 2, Example 2.4 we know that diffusion coefficient is related to the mean squared displacement of a random walker, which is strictly a positive quantity). Despite the lack of physical meaning of the time-reversed equation, however, one might hope this equation to be sound as a solely mathematical entity – indeed it is! Therefore, one might expect that a RD system can be run 'backwards' in time (in a purely mathematical sense, of course) in order to recover a suitable set of initial conditions and 'reverse-engineer' the structure one wishes to make. Unfortunately, this hope is justified only as long as there is no *noise* in the system.

To see why this is so, we first consider a 'noise-free' analytical solution to a simple, pure diffusion problem, in which the initial concentration profile on a one-dimensional domain, $0 \leq x \leq L$, is given by $c(x, 0) = c_0[1 + \cos(2\pi x/L)]$, where c_0 is the average concentration over the domain. When this profile is evolved via the diffusion equation, $\partial c/\partial t = D\nabla^2 c$, with no flux conditions at the boundaries, $(\partial c/\partial x)_{0,L} = 0$, it yields a time-dependent solution, $c(x, t) = c_0[1 + \cos(2\pi x/L)\exp(-4\pi^2 Dt/L^2)]$, which asymptotically approaches the steady-state solution, $c(x,\infty) = c_0$. The reader is encouraged to verify that for any finite time t this equation can be time-reversed to converge back onto $c(x,0)$.

Now, let us take one of the evolved profiles – for instance, $c(x, t) = c_0[1 + 0.1\cos(2\pi x/L)]$ corresponding to $t = \ln(10)L^2/4\pi^2 D$ – and add to it a small-amplitude 'noise' of length scale smaller than the problem's domain, L (see figure below). For the sake of argument, let this 'noise' be quite regular and expressed as $\varepsilon_0 \cos(n\pi x/L)$, where $n \gg 2$ means that its period is small compared to L. If we now run the diffusion equation 'backwards' ($D \rightarrow -D$) in an attempt to recover the initial conditions, we obtain the reverse solution

$$c(x,t) = c_0[1 + 0.1\cos(2\pi x/L)\exp(4\pi^2 Dt/L^2)] + \varepsilon_0 \cos(n\pi x/L)\exp(n^2\pi^2\ Dt/L^2).$$

Notice that while both the 'real' and the 'noise' parts of the concentration profile grow exponentially in time, the growth rate of the latter is much more rapid (since $n \gg 0$). Thus after a short period of time, the noise will grow much larger than the real profile, and our hopes of recovering the initial conditions are ruined.

Importantly, the same behavior would be observed with any other form of noise provided that its characteristic length scale ('wavelength') is small. Recall now that in all numerical simulations – and, in particular, simulations of RD systems – the modeled trends (e.g., time-dependent concentration profiles) are always to some degree noisy, which is an unavoidable consequence of the limited precision of calculations and round-off errors. If we were to propagate such numerical solutions backwards in time, the noise would grow back exponentially with the shortest 'wavelength' noise increasing most rapidly. In other words, the backward equations would be unstable against small perturbations.

In summary, the mathematical 'trick' of time-reversibility can be used only for these few RD systems that have analytical solutions.

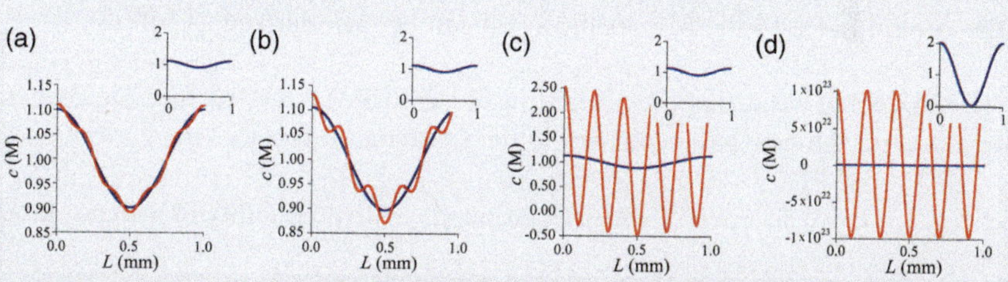

The figure above shows concentration profiles for a time-reversed, one-dimensional diffusion system described in the text. The graphs correspond to different times run backward: (a) $t = 0$ s, (b) $t = -1$ s, (c) $t = -5$ s and (d) $t = -58$ s. Blue curves represent exact analytical solutions; red curves have the analytical solutions with noise. The insets show analytical solutions plotted on the same vertical scale, and are included to illustrate the recovery of the original $c(x, 0)$ profile, which is not discernible in the main graphs (where these curves seem to flatten out due to the rapidly increasing noise). Parameters used to plot these figures were $\varepsilon_0 = 0.01$ M, $n = 10$, $L = 1$ mm, $D = 1 \times 10^{-5}$ cm^2 s^{-1}, $c_0 = 1$ M.

Example 6.3 Optimization of Lens Shape Using a Monte Carlo Method

The Monte Carlo (MC) method is a popular numerical algorithm for finding global minima in complicated parameter spaces. In this example, we will combine it with a lattice gas (LG) simulation to find optimal concentrations of reagents that produce 150 μm wide lenses with 80 μm radius of curvature from 150 μm diameter wet stamped circles.

The LG part of the algorithm is as described in Section 6.3 in the main text, and is based on a square grid with 2.5 μm lattice spacing, and ratio of diffusion coefficients $D_A/D_B = 0.3$. The swelling of the gelatin layer is calibrated against experimental profiles from Figure 6.5 such that one molecule of C at a given node causes $\alpha = 0.54$ μm gel swelling therein.

The MC optimization starts with a trial set of parameters, P^{curr} – for example, [A] = 60 at the nodes beneath the stamp (corresponding to 20% w/w solution of $AgNO_3$ in the stamp) and [B] = 25 at all nodes of the substrate (corresponding to 1% w/w $K_4Fe(CN)_6$ in dry gelatin). Using this parameter set, the LG simulation is averaged over three runs to yield a 'current' lens profile z_{ij}^{curr} (e.g., top, blue curve in the figure below) and the error – which we ultimately wish to minimize – between this curve and the target lens (top, dashed curve), δ^{curr}. A new set of parameters is then generated by modifying [A] and [B] by adding a random integer to each (here, from −3 to + 3) with the constraint that neither concentration can become negative. The new profile of the lens, z_{ij}^{new}, and difference from the target, δ^{new}, are then calculated and the so-called Boltzmann criterion is applied to the quantity $\Delta = \delta_{new} - \delta_{curr}$. This criterion is the central piece of the MC method and can be summarized as follows:

(i) if $\Delta < 0$ and we are apparently getting closer to the target structure, we always 'accept' the new parameter set, in the sense that it replaces the current one, $P^{new} \Rightarrow P^{curr}$;
(ii) if $\Delta \geq 0$, we accept P^{new} only conditionally, with probability $\Pr = \exp(-k\Delta)$.

Technically, the conditional acceptance involves generating a random number between 0 and 1 and comparing it to the value of Pr. If the number is smaller than Pr, the new parameter set is accepted; if it is larger, the new set is rejected. Note that if the value of Δ is large, Pr is very small, and it is very unlikely to accept configurations that increase the difference with the target structure substantially. At the same time, the Boltzmann criterion permits *some* moves 'uphill', which help us get out of local minima (cf. Figure 6.6). In this respect, the positive parameter k plays an important role as it specifies the degree to which we 'penalize' uphill excursions. In a very efficient variant of the MC method called simulated annealing, the value of k is gradually increased so that at the beginning of the simulation the system is not trapped in any local minimum, but, as the solutions become better and better, it becomes harder to escape deep – and hopefully, global – minima.[22]

After the Boltzmann criterion is performed, a new parameter set is generated and tested, and the cycle is continued until a certain halt condition. In our simple example this condition is that 25 consecutive changes in the parameter set do not change P^{curr} – if this happens, we regard P^{curr} as our optimal solution.

The figure below illustrates how simulated annealing MC (with $k = 5 \times 10^{-5} \times n$, where n is the number of MC iterations) optimizes the concentrations of [A] and [B] to converge onto the target lens shape.

The simulation starts from the initial ([A] = 60, [B] = 25) parameter set (corresponding to the top, surface profile colored blue) and converges to the optimal solution (bottom, green profile) in 166 MC iterations. The optimal parameters are found to be [A] = 38 and [B] = 39, corresponding to 1.56% w/w concentration of $K_4Fe(CN)_6$ in the gelatin and 12.7% w/w concentration of $AgNO_3$ in the stamp, respectively. The C++ source code of the simulation is available for download at http://dysa.northwestern.edu/RDbook/index.html.

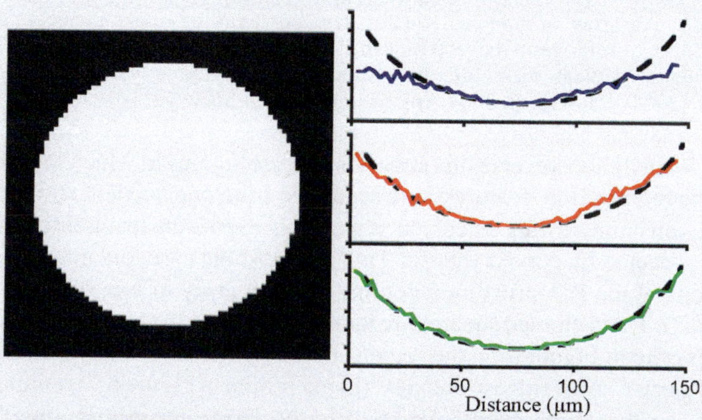

The left-hand panel of the figure above shows initial geometry. Stamped regions where B is delivered to gelatin loaded with A are colored black. Right: simulated profiles corresponding to the initial ([A] = 60, [B] = 25) guess (blue), an intermediate ([A] = 54, [B] = 33) solution after 20 MC iterations (red) and optimal solution ([A] = 38, [B] = 39) after 166 MC steps (green). Dotted lines correspond to the target structure. Ticks on the vertical axes are spaced by 10 μm.

6.5 DISJOINT FEATURES AND MICROFABRICATION OF MULTILEVEL STRUCTURES

In the fabrication of lenses, the region of the stamp in contact with the gelatin surface was continuous, and the areas where the lenses ultimately developed were

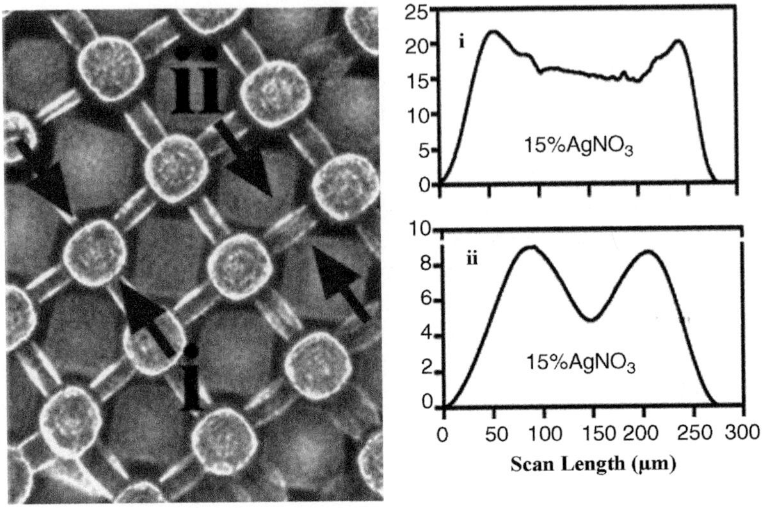

Figure 6.7 An array of stamped squares develops pairs of ridges connecting nearby features. Surface profilograms have: (i) a scan over one 'swollen' square and (ii) a scan over a pair of ridges. Vertical scales are in micrometers. (Reprinted with permission from *Langmuir* (2004), **21**, 418. © 2004 American Chemical Society.)

disjoint. We will now reverse this situation and use stamps in which the substrate-contacting regions (the 'features') are separated from one another. At first glance, this seems an uninteresting extension as we might expect the features to swell most and be connected by curved valleys. This is almost the case, but not quite. Let us have a look at the RD structure emerging from an array of wet stamped squares (Figure 6.7). The patterned squares are indeed swollen to the highest degree (upper surface profile in Figure 6.7), the regions between them are valleys, but there are also unexpected small ridges/'buckles' (lower profile in Figure 6.7) running across the valleys – that is, perpendicular (sic!) to the RD fronts propagating from the squares. Figure 6.8 shows that this phenomenon is no idiosyncrasy of square patterns, and intricate ridge arrangements are also seen in arrays of stamped circles, triangles, crosses, or undulating lines.

An interesting and potentially useful microfabrication opportunity immediately presents itself: if we could somehow control these ridge formations, we would be able to fabricate in one step surfaces that have several levels – tall, swollen features, intermediate-height ridges and deep valleys. With this in mind, we will first try to understand the underlying RD mechanism.

Consider an arrangement of four stamped squares (Figure 6.9) delivering $AgNO_3$ into dry gelatin loaded uniformly with $K_4Fe(CN)_6$. As before, Ag^+ cations diffuse outwards from the features (yellow arrows in the left panel of Figure 6.9), and precipitate $Fe(CN)_6^{4-}$ anions to swell the gel. In sharp contrast to the microfabrication of circular lenses, however, the unreacted potassium hexacyanoferrate moves in response to more complex, noncentrosymmetric gradients.

Figure 6.8 Multilevel 'buckled' surfaces forming from wet stamped arrays of features of different geometries: (a) circles; (b) triangles; (c) crosses; (d) array of 'teeth'; (e) undulating lines. Left column shows large-area photographs; center column shows corresponding close-ups. Topographies modeled by the LG method are shown in the right column. In all cases, the stamped regions are the most elevated ones (i.e., tallest), and the intermediate-height ridges bisect the 'valleys' between them. In all experimental images, the scale bars correspond to 200 μm. In the modeled patterns, the stamped regions are colored orange, precipitate is blue and the regions where ridges form are indicated by red lines. (Reprinted with permission from Langmuir (2005), **21**, 418. © 2005 American Chemical Society.)

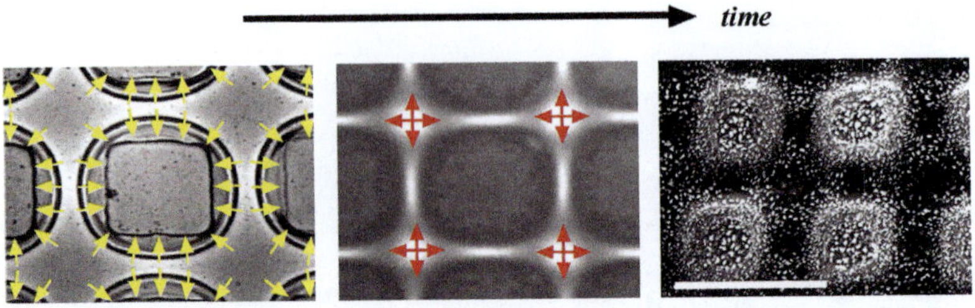

Figure 6.9 Mechanism of ridge formation in an array of squares. Initially, reaction fronts (visible as a concentric ring around the squares) propagate outwards from the stamped features while $[Fe(CN)_6]^{4-}$ ions migrate in the opposite directions (yellow arrows). This migration rapidly reduces the concentration of $[Fe(CN)_6]^{4-}$ between the nearby squares and sets up secondary flows (red arrows) of $[Fe(CN)_6]^{4-}$ along the concentration gradients. These secondary flows give rise to the ridges connecting the squares. In the rightmost picture, the ridges were visualized by reacting the $Ag_4[Fe(CN)_6]$ precipitate with water to obtain Ag_2O grains clearly visible under an electron microscope. Scale bar = 200 μm. (Reprinted from Grzybowski, B.A. and Campbell, C.J. (2007) *Materials Today*, **10**, 38–40. © 2007, with permission from Elsevier.)

In the pattern of squares, these gradients are initially most pronounced between the neighboring squares, and $[Fe(CN)_6]^{4-}$ is rapidly cleared from these regions (clear areas in the left panel of Figure 6.9) – at this point, the lines become valleys connecting raised regions of the swollen features. This structure is not final though, as there is still unused $Fe(CN)_6{}^{4-}$ between the four squares (darker, diamond-shaped area in the left panel of Figure 6.9), which diffuses not only in the directions of the incoming reaction fronts, but also 'horizontally' and 'vertically' towards the valleys devoid of $Fe(CN)_6{}^{4-}$ (red arrows in the centre panel of Figure 6.9). The Fe $(CN)_6{}^{4-}$ ions diffusing along these directions are 'intersected' by Ag^+ diffusing from the edges of the squares. As a result, 'secondary' precipitation – and concomitant swelling – occurs approximately along the lines joining the nearest vertices of neighboring squares, and gives rise to small ridges running across the deep valleys (right panel of Figure 6.9).

For other types of wet stamped patterns, the mechanism is qualitatively similar and the formation of ridges is due to the migration of hexacyanoferrate ions along the gradients transverse to the direction of RD propagation. While the locations and relative heights of the ridges can be determined precisely by modeling (cf. http://dysa.northwestern.edu/programs/BuckleFinder.exe for downloadable 'Buckle Finder' software), the rule of thumb is that they connect the closest and most curved regions of the nearby features (cf. Figure 6.8). Also, it has been shown experimentally that for typically used,[15] approximately 50 μm thick gelatin films, the difference between the lowest and the highest points of the developed RD patterns is between 20 and 40 μm, and the heights of the ridges (with respect to the nadir of the valley) are 2–10 μm. Precise dimensions depend predominantly on the concentration of $AgNO_3$ delivered from the stamp and on the periodicity of the stamped arrays. Specifically, for a given geometry of the array, the heights of the ridges increase roughly linearly with increasing concentration of silver nitrate solution up to 15% w/w; more concentrated solutions are impractical to use as they destroy the gelatin matrix. For a given concentration of the stamping solution and for the spacing between the features smaller than roughly twice the feature's diameter, heights of the ridges increase with increasing spacing up to ~150 μm. When the spacing is larger, no ridges appear, reflecting the fact that not all of the wide space between the features is wetted and swollen, and the precipitation zones responsible for the ridge formation do not form. Spacing between the nearest, parallel ridges increases linearly with feature spacing. Lastly, ridges form from features as small as 50 μm and separated by as little as 25 μm.

6.6 MICROFABRICATION OF MICROFLUIDIC DEVICES

One of the most appealing applications of the multilevel surfaces developed by RD and replicated into polymeric materials is in the fabrication of multilevel microfluidic systems. Microfluidics[23,24] is a booming field of research focusing on the development of systems of miniaturized channels, reactors, valves, pumps, and detection elements with which to manipulate and analyze minuscule (down to femtoliter) amounts of fluid samples. Over the last decade, integration of individual components has led to the development of microfluidic circuits that begin to match the complexity of computer chips (Figure 6.10),[25] and can perform complex tasks such as high-throughput protein crystallization, drug screening, synthesis of multicomponent colloidal particles, and more.

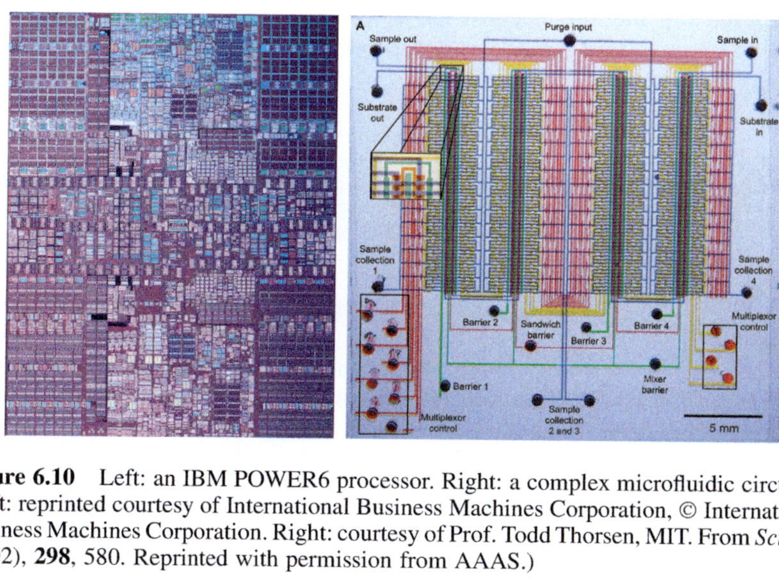

Figure 6.10 Left: an IBM POWER6 processor. Right: a complex microfluidic circuit.[26] (Left: reprinted courtesy of International Business Machines Corporation, © International Business Machines Corporation. Right: courtesy of Prof. Todd Thorsen, MIT. From *Science* (2002), **298**, 580. Reprinted with permission from AAAS.)

At the same time, microfluidics faces some challenges that are specific to this regime of fluid flow[26] and, sometimes, hard to overcome. In particular, since fluids flowing in microscopic channels do not develop any turbulences and flow neatly side-by-side ('laminarly'), they mix only very slowly via molecular diffusion across the interface that separates them. Since mixing of the reaction substrates is a necessary precondition to subsequent chemical reactions, considerable experimental and theoretical effort has been devoted to the design of devices/architectures that enhance mixing in microchannels. Of particular interest to us is the class of the so-called passive microfluidic mixers, which contain no moving parts and achieve mixing by virtue of channel geometry alone. In their classic 2002 *Science* paper, Stroock *et al.*[27] showed that if the bottom of a microfluidic channel is decorated with small ridges, the fluids flowing in such a channel advect and 'twist' with respect to one another, effectively increasing the area of interface and the degree of mixing (Figure 6.11). The downside to this elegant approach, however, is that the device microfabrication is relatively complicated and requires either several rounds of photolithography and precise registration[28] or the use of serial techniques. It would thus be desirable to develop multilevel patterning techniques that are both parallel and one-step – fabrication with RD can meet both of these criteria.

To fabricate a channel-with-ridges micromixer, consider the wet stamped pattern shown in the top row of Figure 6.12. Here, the stamped regions (colored orange) swell uniformly, so that the spaces between them are valleys in an otherwise flat surface. At the same time, the spiked, triangular protrusions on the edges of the stamped pattern cause migration of $[Fe(CN)_6]^{4-}$ ions along the long axis of the valley; these migrations, in turn, translate into the accumulation of precipitate (blue color in Figure 6.12) and formation of ridges along the lines joining the spikes' tips. Overall, the structure produced by RD comprises an approximately 50 μm deep channel with transverse ridges that – depending on the specific concentration of reactants used – are 5–10 μm high.

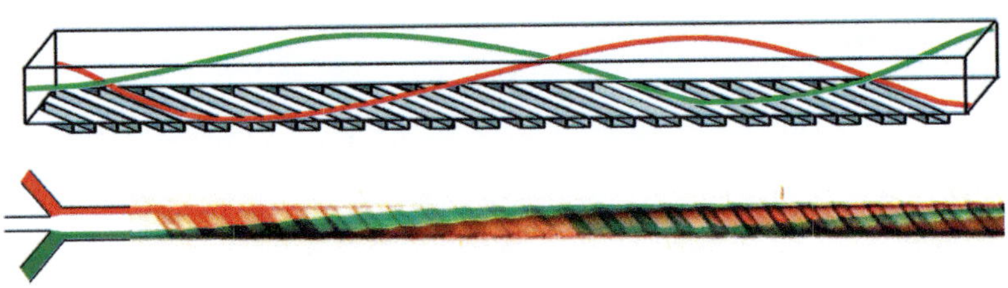

Figure 6.11 Top: scheme of a passive microfluidic mixer, in which ridges at the bottom of the channel cause chaotic advection of otherwise laminarly flowing fluids (here, red and green curves). Bottom: the experimental realization of advective mixing in a device fabricated in PDMS by molding from a photolithographically patterned master. (Courtesy of Prof. Abraham D. Stroock, Cornell University. From *Science*, Chaotic Mixer for Microchannels, Abraham D. Strook *et al.*, copyright (2002), **295**, 647. Reprinted with permission from AAAS.)

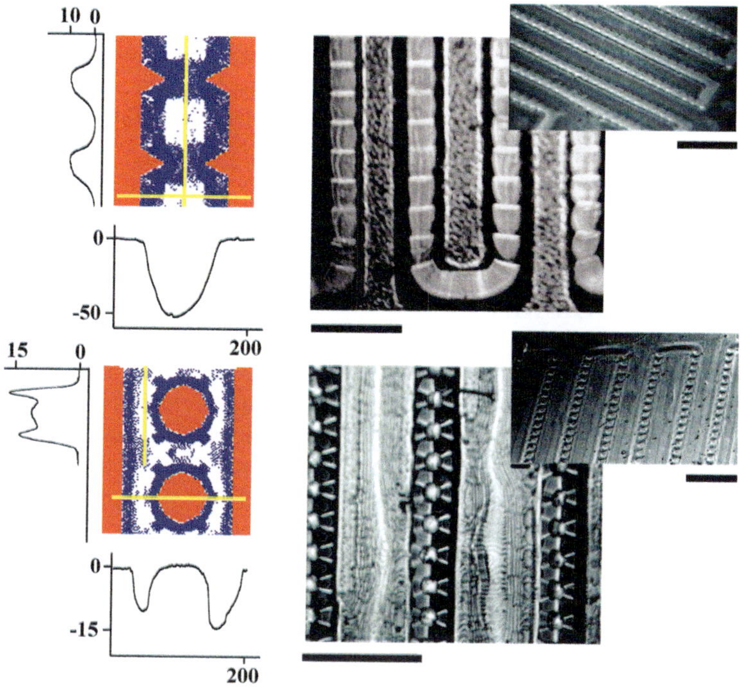

Figure 6.12 Reaction–diffusion fabricates parallel-buckle (top row) and 'caterpillar' (bottom row) passive microfluidic mixers. The left column shows results of simulations. Orange color delineates the stamp geometry; blue corresponds to precipitate; yellow lines give the directions of profilometric scans shown next to the images. All dimensions are in micrometers. The right column shows optical micrographs of channel systems replicated into PDMS. Scale bars are 500 μm in large-magnification images and 1 mm in the insets. (Reprinted from Grzybowski, B.A. and Campbell, C.J. (2007) *Materials Today*, **10**, 38–40. © 2007, with permission from Elsevier.)

In the second design we pursue (Figures 6.12, bottom row), the arrangement of ridges is more complex and is inspired by the caterpillar flow mixer developed by Ehrfeld and co-workers.[29] In this case, the target structure has crossed ridges at the bottom of the channel. One way to fabricate such ridges is to stamp channels with small islands (orange circles in the lower left picture in Figure 6.10). LG modeling indicates that these islands cause diffusion of the chemicals between narrow and wide regions of the channels and, ultimately, formation of four precipitation bands connecting each circle to the walls of the channel. Experimental images in the lower row of Figure 6.12 confirm that the precipitation bands translate into surface ridges to produce a caterpillar-mixer architecture. Interestingly, because the stamped circular islands swell to almost the same level as the edges of the channel, they 'pull up' the bottom of the valley and the channels that emerge are shallower that those we prepared in the first example.

Both the parallel-ridge and caterpillar topographies developed in gelatin by RD can be easily replicated into PDMS (Figure 6.12, right column) and, upon sealing their top surfaces, can be used as efficient micromixers. From a practical point of view it is important to note that the method can fabricate not only individual channels but entire circuits over areas as large as 3×3 cm. And it can do so on the bench-top, without specialized equipment and in a matter of tens of minutes.

6.7 SHORT SUMMARY

Throughout this chapter we have dealt with only one chemical reaction and yet managed to fabricate a variety of microstructures that would otherwise be hard to make. The ease with which we prepared them is encouraging and illustrates the inherent flexibility of RD as a microfabrication tool. It must be remembered, however, that this flexibility comes hand-in-hand with the complexity of chemical and diffusive processes underlying RD, and that even apparently simple fabrication tasks should – at least in principle – be guided by modeling. While for the precipitation–swelling system we considered a simple LG approach proved a sufficient modeling aid, the more complex phenomena we will consider in later chapters will require a higher-end theoretical description.

REFERENCES

1. Bratton, D., Yang, D., Dai, J.Y. and Ober, C.K. (2006) Recent progress in high resolution lithography. *Polym. Adv. Technol.*, **17**, 94.
2. Clemens, J.T. (1997) Silicon microelectronics technology. *Bell Labs Tech. J.*, **2**, 76.
3. Chung, J.W. and Jang, J.J. (2007) The integrated room layout for a semiconductor facility plan. *IEEE Trans. Semicond. Manufact.*, **20**, 517.
4. Leachman, R.C., Ding, S.W. and Chien, C.F. (2007) Economic efficiency analysis of wafer fabrication. *IEEE Trans. Automat. Sci. Eng.*, **4**, 501.
5. Madden, J.D. and Hunter, I.W. (1996) Three-dimensional microfabrication by localized electrochemical deposition. *J. Microelectromech. Syst.*, **5**, 24.
6. Fu, Y.Q., Bryan, N.K.A. and Shing, O.N. (2000) Calibration of etching uniformity for large aperture multilevel diffractive optical element by ion beam etching. *Rev. Sci. Instrum.*, **71**, 1009.
7. Watt, F., van Kan, J.A. and Osipowicz, T. (2000) Three dimensional microfabrication using maskless irradiation with MeV ion beams: proton-beam micromachining. *MRS Bull.*, **25**, 33.
8. Schaffer, C.B., Brodeur, A., Garcia, J.F. and Mazur, E. (2001) Micromachining bulk glass by use of femtosecond laser pulses with nanojoule energy. *Opt. Lett.*, **26**, 93.
9. Belloy, E., Sayah, A. and Gijs, M.A.M. (2002) Micromachining of glass inertial sensors. *J. Microelectromech. Syst.*, **11**, 85.
10. Schuster, R., Kirchner, V., Allongue, P. and Ertl, G. (2000) Electrochemical micromachining. *Science*, **289**, 98.

11. Nikolov, I.D., Goto, K. Mitsugi, S. *et al.* (2002) Nanofocusing recording probe for an optical disk memory. *Nanotechnology*, **13**, 471.
12. Daemen, N. and Peek, H.L. (1994) Microlenses for image sensors. *Philips J. Res.*, **48**, 281.
13. Gryaznova, M.V., Danilov, V.V., Belyaeva, M.A. *et al.* (2002) Optical limiters based on liquid-crystal microlenses. *Opt. Spectrosc.*, **92**, 614.
14. McCabe, E.M., Taylor, C.M. and Yang, L. (2003) Novel imaging in confocal microscopy. *Proc. Soc. Photogr. Instrum. Eng.*, **4876**, 1.
15. Campbell, C.J., Klajn, R., Fialkowski, M. and Grzybowski, B.A. (2005) One-step multilevel microfabrication by reaction–diffusion. *Langmuir*, **21**, 418.
16. McDonald, J.C., Duffy, D.C., Anderson, J.R. *et al.* (2000) Fabrication of microfluidic systems in poly(dimethylsiloxane). *Electrophoresis*, **21**, 27.
17. Campbell, C.J., Baker, E., Fialkowski, M. and Grzybowski, B.A. (2004) Arrays of microlenses of complex shapes prepared by reaction–diffusion in thin films of ionically doped gels. *Appl. Phys. Lett.*, **85**, 1871.
18. de Gennes, P.G., Brochard-Wyart, F. and Quere, D. (2003) *Capillarity and Wetting Phenomena: Drops, Bubbles, Pearls, Waves*, Springer Science, New York.
19. Campbell, C.J., Fialkowski, M., Klajn, R. *et al.* (2004) Color micro- and nanopatterning with counter-propagating reaction–diffusion fronts. *Adv. Mater.*, **16**, 1912.
20. Strogatz, S.H. (1994) *Nonlinear Dynamics and Chaos*, Westview Press, Cambridge.
21. Ghez, R. (2001) *Diffusion Phenomena, Cases and Studies*, Kluwer Academic/Plenum, New York.
22. Landau, D.P. and Binder, K. (2005) *A Guide to Monte Carlo Simulations in Statistical Physics*, 2nd edn Cambridge University Press, Cambridge.
23. Song, H., Chen, D.L. and Ismagilov, R.F. (2006) Reactions in droplets in microflulidic channels. *Angew. Chem. Int. Ed.*, **45**, 7336.
24. Stone, H.A., Stroock, A.D. and Ajdari, A. (2004) Engineering flows in small devices: microfluidics toward a lab-on-a-chip. *Annu. Rev. Fluid Mech.*, **36**, 381.
25. Thorsen, T., Maerkl, S.J. and Quake, S.R. (2002) Microfluidic large-scale integration. *Science*, **298**, 580.
26. Squires, T.M. and Quake, S.R. (2005) Microfluidics: fluid physics at the nanoliter scale. *Rev. Mod. Phys.*, **77**, 977.
27. Stroock, A.D., Dertinger, S.K.W., Ajdari, A. *et al.* Chaotic mixer for microchannels. *Science*, **295**, 647.
28. Walsby, E.D., Wang, S., Xu, J. *et al.* (2002) Multilevel silicon diffractive optics for terahertz waves. *J. Vac. Sci. Technol. B*, **20**, 2780.
29. Ehrfeld, W., Hartmann, J. Hessel, V. *et al.* (2000) Microreaction technology for process intensification and high throughput screening. *Proc. Micro Total Anal. Syst. Symp.*, 33.

7

Multitasking: Micro- and Nanofabrication with Periodic Precipitation

7.1 PERIODIC PRECIPITATION

In the previous chapter, we discussed and applied reaction–diffusion (RD) processes that generate only one (e.g., a lens) or few (e.g., ridges) distinct microstructures for each wet-stamped feature. We will now make a useful addition to our microfabrication toolbox and consider a class of chemical reactions that generate multiple structures from one feature.[1] These reactions are collectively termed 'periodic precipitation' and involve select pairs of inorganic salts which – while diffusing through a gel matrix – create regular arrays of precipitation zones. Since their discovery more than a hundred years ago,[2] periodic precipitation (PP) phenomena have attracted considerable scientific interest for their relevance to the issues of nonlinear chemical kinetics,[3,4] wide occurrence in nature[5–7] and aesthetic appeal as in the case of alternating defect-rich chalcedony and defect-poor quartz zones in iris agates.[8] Most of the research on PP has so far focused on macroscopic systems and simple geometries[9–11] (quasi-one-dimensional gel columns, single droplets or infinite fronts), for which several empirical laws characterizing the emerging patterns have been formulated.[12,13] At the same time, relatively little is known about PP at the microscale and in complex geometries, where the localized and controllable delivery of participating chemicals becomes technically challenging. Wet stamping is well suited to explore these interesting regimes and harness PP for micro- and even nanofabrication. Before we discuss the specific applications, however, let us first focus on the key characteristics and the origin of the phenomenon.

Chemistry in Motion: Reaction–Diffusion Systems for Micro- and Nanotechnology Bartosz A. Grzybowski

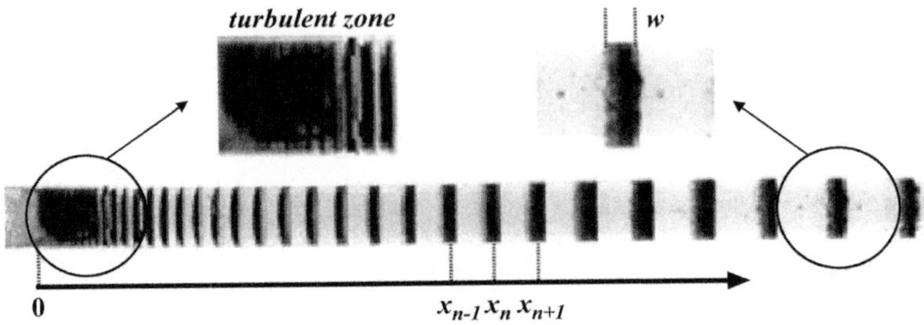

Figure 7.1 Typical, macroscopic setup for periodic precipitation. The outer electrolyte, A, is applied at $x < 0$ to a long ($\sim$1 m) tube filled with gel and inner electrolyte B. Beyond the initial turbulent zone, discrete bands appear at locations x_n

7.2 PHENOMENOLOGY OF PERIODIC PRECIPITATION

Consider a long column (Figure 7.1) of wet hydrogel (e.g., agar, gelatin) loaded uniformly with one of the inorganic chemicals listed in Table 7.1[14–27] as an 'inner electrolyte', B. When a concentrated solution of the corresponding 'outer electrolyte', A, is placed onto the gel, A starts diffusing into the substrate where it precipitates B. Close to the gel/outer electrolyte interface (position $x = 0$), the

Table 7.1 Common periodic precipitation systems: gels, inner (B) and outer (A) electrolytes, and literature sources

Gel	B	A	Reference
Agar	Zn^{2+}	OH^-	14
Agar	Fe^{2+}	NH_3	15
Agar	I^-	Pb^{2+}	16
Agar	F^-	Pb^{2+}	17
Agar	Mn^{2+}	S^{2-}	17
Agar	Cu^{2+}	S^{2-}	17
Agar	Cd^{2+}	S^{2-}	17
Agarose	Al^{3+}	OH^-	18
Gelatin	Ba^{2+}	SO_4^{2-}	19
Gelatin	Co^{2+}	NH_3	20
Gelatin	Ni^{2+}	NH_3	21
Gelatin	$Cr_2O_7^{2-}$	Ag^+	22
Gelatin	$Cr_2O_7^{2-}$	Pb^{2+}	22
Gelatin	OH^-	Mg^{2+}	16
Gelatin	Co^{2+}	OH^-	23
Gelatin	Ni^{2+}	NH_3	17
Gelatin	Cd^{2+}	NH_3	17
Gelatin	Mg^{2+}	NH_3	17
Silica	HPO_4^{2-}	Ca^{2+}	24,25
Poly(vinyl alcohol)	Cu^{2+}	OH^-	26
Poly(vinyl alcohol)	Co^{3+}	OH^-	27

precipitation is spatially continuous, but at larger distances distinct precipitation bands separated by clear regions appear. Interestingly, irrespective of the salt pair chosen from Table 7.1 and the exact geometry/dimensions of the system,[28] the bands obey a common set of scaling laws as follows.

1. **The spacing law** first described by Jablczynski in 1923 relates the positions x_n and x_{n+1} of the consecutive precipitation bands n and $n+1$ (Figure 7.1).[12] Jablczynski noticed that for a given PP system and concentrations of salts, the ratio x_{n+1}/x_n is constant for all bands. In other words, the positions of the bands form a power series, $x_n = p^n x_0$, where $p = x_{n+1}/x_n$ is the so-called spacing coefficient, which for most systems is larger than 1 but usually does not exceed 1.5.
2. **The Matalon–Packter law**[13] relates the spacing coefficients to the initial concentrations of the inner ($[B]_0$) and outer ($[A]_0$) electrolytes via $p([A]_0,[B]_0) = F([B]_0) + G([B]_0)/[A]_0$, where $F([B]_0)$ and $G([B]_0)$ are decreasing functions of $[B]_0$.
3. **The width law** in its simplest form stipulates that the ratio of widths of two consecutive precipitation bands is constant, $w_{n+1}/w_n = q$. This result has been confirmed both in experiments and in simulations. An interesting and much more contentious extension is that width and spacing coefficients are related through a power law, $q = p^k$. Several workers have found that the exponent k is 'universal' to various systems, and its value is between 0.9 and 0.95. At the same time, there is no rigorous explanation of the origin of this universality.
4. **The time law** discovered by Morse and Pierce[29] states that if the time elapsed until the formation on the n-th band is t_n, then the value of x_n^2/t_n approaches a constant value as n increases. The reader will no doubt notice a quantitative similarity between this expression and the definition of diffusion coefficient, which is a direct consequence of the diffusive transport of salts in the gel (still, it is not entirely trivial why this law holds when diffusion is coupled to very fast precipitation).

Because these trends hold for virtually all systems exhibiting PP, they likely reflect some common mechanism underlying the phenomenon. The first model attempting to explain PP mechanism was published in 1897 by a famous German chemist, Wilhelm Ostwald. Ostwald based his explanation on the well-known fact that although mixtures of inorganic salts become thermodynamically unstable if the product of their concentrations exceeds the solubility product, K_{sp} (Section 3.3), their precipitation can be kinetically hindered. As a result, the salts start precipitating and nucleating into microcrystals only at some level of 'supersaturation', $K > K_{sp}$. In the case of PP, supersaturation means that the A and B salts migrating through the gel can remain stable until, at some location, x, their concentrations reach the threshold value of K. At that point, the salts nucleate into microcrystals, which trigger an 'explosion' precipitating the supersaturated solution in the crystals' vicinity. Precipitation continues until the concentrations of salts drop to thermodynamically stable (equilibrium) threshold, K_{sp}. Importantly, since the reaction is very fast, slow diffusion cannot feed A and B immediately into the depleted zone to restore supersaturation. By the time supersaturation can be re-established, the front of A moves forward (in the $+x$ direction in Figure 7.1), and the next precipitation event occurs at a new location giving rise to another band therein.

Since Ostwald's times, several more detailed scenarios have been proposed that can be classified as either prenucleation or postnucleation depending on the sequence of the elementary events. Prenucleation models are, in one way or another, descendants of Ostwald's supersaturation theory, and assume that the formation of precipitation bands is a result of a supersaturation wave, which leads to precipitation. While specific implementations of prenucleation differ in the details of the nucleation/precipitation events[28] they generally reproduce the scaling laws mentioned above and several other experimental trends. What they do not capture, however, is the presence of precipitate *between* precipitation bands that is observed in some experimental systems. Postnucleation models can account for these observations by assuming that before the bands emerge, the salts form a homogeneous sol of solid particles, which subsequently loses its stability and 'focuses' into periodic striations. Although the debate as to the most accurate model of periodic precipitation is far from being resolved (cf. Example 7.1), the prenucleation models appear to be more popular and in some sense more chemically intuitive. Here, we will use one of the more advanced of such models (called the 'diffusive intermediate model') that has been shown to reproduce the essential features of PP.

7.3 GOVERNING EQUATIONS

Consider an ionic reaction of the form $nA^{m+} + mB^{n-} \rightarrow C$. Diffusion of the ions through the gel is described by diffusion coefficients D_A and D_B, and rapid formation of C occurs when the product of concentrations $[A]^n[B]^m$ exceeds the solubility product, K_{sp}. The created C molecules, however, do not immediately precipitate but are instead free to diffuse (with diffusion coefficient D_C) until their local concentration reaches some saturation threshold $[C]^*$. At that point, nucleation occurs followed by aggregation of C into an immobile precipitate, D. Although the introduction of the C intermediate might seem redundant, notice that without this free-to-diffuse species, the bands would form only at points where supersaturation is reached. Consequently, the precipitation would be 'point-like' and the resultant bands would be infinitely sharp, in clear contrast to experiments. The ability of C to diffuse before it converts to immobile precipitate, D, effectively 'smears' the bands and gives them finite thickness.

Let us write the first-round RD equations describing the process:

$$\begin{aligned}
\partial[\mathrm{A}]/\partial t &= D_A\nabla^2[\mathrm{A}] - nk\Theta([\mathrm{A}]^n[\mathrm{B}]^m - K_{\mathrm{sp}})\\
\partial[\mathrm{B}]/\partial t &= D_B\nabla^2[\mathrm{B}] - mk\Theta([\mathrm{A}]^n[\mathrm{B}]^m - K_{\mathrm{sp}})\\
\partial[\mathrm{C}]/\partial t &= D_C\nabla^2[\mathrm{C}] + k\Theta([\mathrm{A}]^n[\mathrm{B}]^m - K_{\mathrm{sp}}) - k'\Theta([\mathrm{C}] - [\mathrm{C}]^*)\\
\partial[\mathrm{D}]/\partial t &= k'\Theta([\mathrm{C}] - [\mathrm{C}]^*)
\end{aligned} \tag{7.1}$$

Here, k are the characteristic reaction rates which are much greater than the diffusive rates, and $\Theta(x)$ is the Heaviside step function (i.e., $\Theta(x) = 0$ for $x \leq 0$, and $\Theta(x) = 1$ for $x > 0$) that reflects very rapid formation of C (if $[\mathrm{A}]^n[\mathrm{B}]^m > K_{\mathrm{sp}}$) and D (if $[\mathrm{C}] > [\mathrm{C}]^*$; see Section 4.4.3 for more details).

Example 7.1 Periodic Precipitation via Spinodal Decomposition

While virtually all models of periodic precipitation agree on the description of reaction–diffusion processes governing the outer and inner electrolytes (A and B, respectively), they differ in describing the dynamics with which the forming species C 'partitions' between C-rich (i.e., the precipitate bands) and C-deficient regions. One way of looking at this process is that C undergoes a phase transition, in which a single phase of C initially dispersed in solution separates into two phases (C-rich and C-deficient). This scenario can be described by a class of models known collectively as spinodal decomposition. Although for PP these models may not be as intuitive as the kinetic equations and nucleation-and-growth formalism, they are becoming quite popular and offer several unique insights into the nature of the PP process.

To get a better feel for the physics of phase transitions, consider first a mixture of two liquids (L_1 and L_2), which are miscible at high temperatures, but separate into two phases when the temperature is lowered. This system may be described by the so-called 'regular solution model,'[30] the phase diagram of which is as follows:

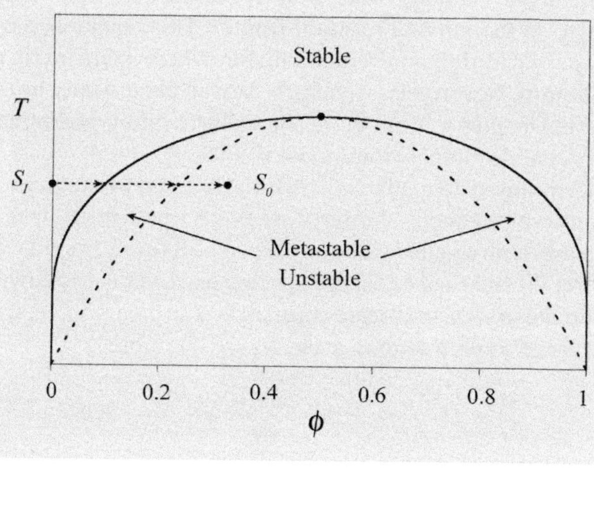

(In the figure, T stands for the temperature, and ϕ is the volume fraction of liquid L_1. The solid and dashed curves are the coexistence and spinodal curves, respectively.)

Below the critical temperature, a binary mixture may be in one of three distinct states depending on the volume fraction, ϕ, of L_1 (cf. path from S_I to S_O in the figure above). For small ϕ, the mixture is *stable*; however, upon increasing ϕ (analogous to increasing [C] in the context of PP) the system enters a *metastable* state, which has a higher free energy than the phase-separated state. Although the mixed state is no longer thermodynamically favored, there is an energy barrier that must be overcome to achieve phase separation. Increasing ϕ further, this energy barrier disappears, and the system becomes *unstable* against phase separation. The dynamics of the phase transition differs for systems in the metastable and unstable states, respectively.

For systems in the metastable state, the transition is an activated process, in which thermal fluctuations are necessary to overcome the energy barrier separating the local free energy minimum of the mixed state from the global minimum associated with the phase-separated state. In PP, this process is described by a nucleation and growth mechanism (Section 7.3), in which nuclei larger than a critical size form spontaneously via thermal fluctuations and subsequently grow by the addition of free C species. Importantly, there is a characteristic rate, J, associated with nucleation, which may be small compared to the rate at which C is produced. If this is the case, the system may move through the metastable regime and into the unstable regime before forming any nuclei. Phase separation then occurs immediately by a process known as spinodal decomposition.

The dynamics of spinodal decomposition may be approximated by the so-called Cahn–Hilliard equation: $\partial m/\partial t = -\lambda\nabla^2[\varepsilon m - \gamma m^3 + \sigma\nabla^2 m] + S$, where $m = 2c - 1$ and c is the mole fraction of C, λ is a kinetic constant, $\varepsilon > 0$ measures the deviation from the critical temperature, γ and σ are positive constants (to ensure stability) and S is a source term that describes the production of C at the moving reaction front.[10] This equation is a nonlinear, fourth-order partial differential equation, for which there exist no general analytical solutions. Nevertheless, it may be solved numerically by the methods we discussed in Chapter 4. With these somewhat lengthy preliminaries, let us apply the spinodal decomposition model to PP.

Consider an infinitely long tube initially divided into two regions: (i) $x < 0$ is filled uniformly with a gelled solution of species A (mole fraction, a_o); (ii) $x > 0$ is filled uniformly with a gelled solution of species B (mole fraction, b_o) (part (a) of the figure on the next page). These species react via the following second-order reaction: $A + B \rightarrow C$, with rate constant k. The governing equations for A and B (in terms of mole fractions a and b) are therefore

$$\frac{\partial a}{\partial t} = D\frac{\partial^2 a}{\partial x^2} - kab, \quad \frac{\partial b}{\partial t} = D\frac{\partial^2 b}{\partial x^2} - kab$$

This type of reaction–diffusion problem has been studied in detail,[31] and it is possible to describe the reaction term $R(x, t) = kab$ by the following analytical expression:

$$R(x,t) \approx \frac{0.3ka_o^2K^{4/3}}{t^{2/3}}\exp\left(\frac{-[x-x_f(t)]^2}{2w(t)^2}\right)$$

Here, $x_f(t) = \sqrt{2D_f t}$ is the position of the reaction front with D_f defined by the relation $\mathrm{erf}(\sqrt{D_f/2D}) = (a_o - b_o)/(a_o + b_o)$, and $w(t) = 2\sqrt{D}t^{1/6}/(ka_oK)^{1/3}$ is the width of the front where $K = (1 + b_o/a_o)(2\sqrt{\pi})^{-1}\exp(-D_f/D)$.

The knowledge of the reaction term allows us to treat the process of PP by a single differential equation. Specifically, $R(x, t)$, enters the Cahn–Hilliard equation as a source term. In dimensionless form, this equation becomes $\partial\bar{m}/\partial\bar{t} = \bar{\nabla}_2[\bar{m}^3 - \bar{m} + \bar{\nabla}_2\bar{m}] + \frac{1}{2}\bar{R}(\bar{x}, \bar{t})$ where length is scaled by $\sqrt{\sigma/\varepsilon}$, time by $\sigma/(\lambda\varepsilon^2)$, and mole fraction by $\sqrt{\varepsilon/\gamma}$. The equation may be solved numerically (here,[10] with $D_f = 21.72$, $w_0 = 2\sqrt{D}/(ka_oK)^{1/3} = 4.54$ and $\Lambda = 0.3ka_o^2K^{4/3} = 0.181$) via the finite difference scheme to give a PP profile shown in part (b) of the figure below.

Proponents of the spinodal decomposition approach argue that this model reproduces many of the experimentally observed spacing laws (e.g., Matalon–Packter law) without the need for 'artificial' thresholds for nucleation and aggregation. While this is certainly attractive, this approach has as many or more unknown parameters (e.g., in the Cahn–Hilliard equation) than alternatives based on the nucleation and growth mechanism. Finally, we note that the distinction between the nucleation and spinodal decomposition regimes (as separated by the spinodal line) is an artificial consequence of the mean-field description; in real PP systems, it is difficult to distinguish between the two mechanisms.[32]

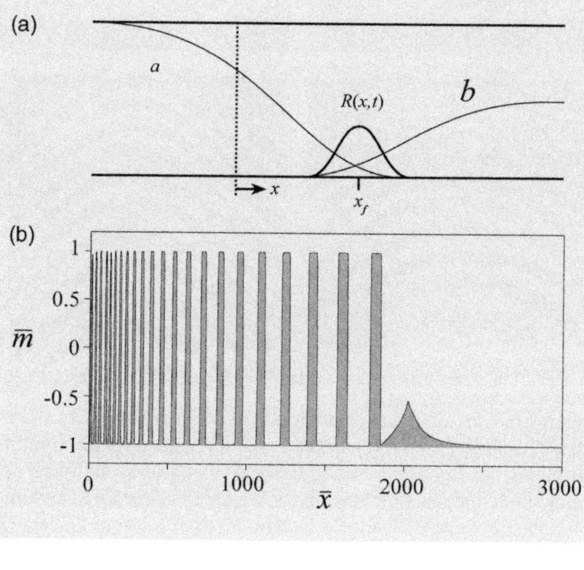

The figure above shows (a) a schematic illustration of the RD system and (b) a periodic precipitation pattern as described by the spinodal decomposition model.

So far, we have treated all reactions as fully deterministic – that is, occurring *always* whenever the local concentrations satisfy certain threshold conditions. Taking a closer look at the PP patterns in Figures 7.2(c) and Figure 7.3(b) we see,

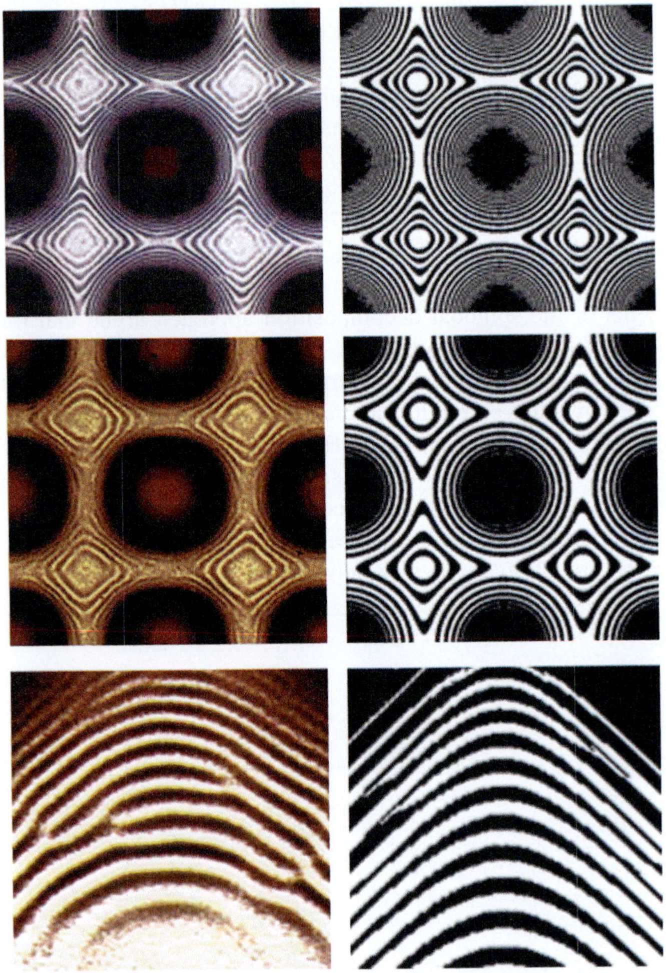

Figure 7.2 PP patterns obtained in experiment (left) and in numerical simulations (right) for (top) an array of $d = 150\,\mu m$ squares separated by $D = 750\,\mu m$, for gel thickness $H = 36\,\mu m$. Middle: same geometry but for $H = 126\,\mu m$. Bottom: wedge-like stamped geometry. Notice defects/dislocations in the bands. Accounting for these effects in the simulations requires the presence of a stochastic component (e.g., term R in Equation (7.3)). (Reprinted with permission from *J. Phys. Chem. B* (2005), **109**, 2774. © 2005 American Chemical Society.)

however, that there is an appreciable degree of randomness in the structures in the form of defects (where the bands 'split') and uneven distribution of precipitates (e.g., larger crystallites in and between the bands in Figure 7.3(b)). At the molecular level, these irregularities reflect the fact that the precipitation/nucleation events are probabilistic/'stochastic' in nature and at times, though in principle permitted, might not take place. You might recall from statistical physics or thermodynamics classes that for large systems, such stochastic effects are relatively unimportant, since many random events effectively 'cancel out'.[33] On the other hand, when the size of the system decreases, chance and noise begin to play a significant role. Because we are most interested in micro- and nanoscale PP patterns in which numbers of ions/molecules are rather low (e.g., down to 1–2 molecules of precipitate per nm^3 for 1.6% w/w $K_4Fe(CN)_6$ and 2.5% w/w $FeCl_3$) we will prudently add stochastic elements to our RD equations. One way of doing so is to multiply some (or even all) reaction terms by a random number between 0 and 1. Here, we will adopt a method introduced by Henisch[34] and multiply the reaction terms during the formation of mobile precipitate by a stochastic term

$$R([\mathrm{A}],[\mathrm{B}]) = \delta[c_r + (1-c_r)r]\exp(-s^2) \tag{7.2}$$

In this expression, r is a random number from the interval $[0, 1]$, $s = |(n[\mathrm{A}] - m[\mathrm{B}])/(n[\mathrm{A}] + m[\mathrm{B}])|$ is the so-called 'near-equality' condition that measures how close the local concentrations of the reactants are to the reaction's stoichiometric coefficients (if they are close, more reaction events can occur), δ is calculated from the equation $([\mathrm{A}] - \delta/m)^n([\mathrm{B}] - \delta/n)^m = K_{sp}$ and measures the extent of reaction (i.e., the number of chemical transformations per unit volume) and the coefficient $0 < c_r < 1$ determines the degree of stochasticity of the reaction ($c_r = 1$ corresponds to fully deterministic nucleation; $c_r = 0$ means that the process is entirely probabilistic).

The last addition we wish to make to the model concerns the presence of the immobile precipitate. A practicing chemist knows well that placing a small crystal of salt (a 'seed') into a salt solution facilitates crystallization. In the case of PP, the immobile precipitate composed of small crystallites[5,34] seeds the aggregation of the surrounding C intermediates (even if $[\mathrm{C}] < [\mathrm{C}]^*$). To account for this effect, we introduce into the third equation a consumption term $[\mathrm{C}]N(\mathrm{C}, \mathrm{D})$ defined after Chopard *et al.*[35] as follows: (i) if some aggregate is already present at a given location $\boldsymbol{x}$, $\mathrm{D}(\boldsymbol{x}) > 0$, then $N(\mathrm{C}, \mathrm{D}) = 1$; (ii) else, if $\mathrm{D}(\boldsymbol{x}) = 0$ but there is some precipitate in the vicinity $d\boldsymbol{x}$ of location $\boldsymbol{x}$, and if the concentration of C is above the aggregation threshold, $\mathrm{C}(\boldsymbol{x}) > \mathrm{D}^*$, then $N(\mathrm{C}, \mathrm{D}) = 1$; (iii) otherwise $N(\mathrm{C}, \mathrm{D}) = 0$. The four RD equations can then be written as:

$$\begin{aligned}
\partial[\mathrm{A}]/\partial t &= D_{\mathrm{A}}\nabla^2[\mathrm{A}] - nkR([\mathrm{A}],[\mathrm{B}])\Theta([\mathrm{A}]^n[\mathrm{B}]^m - K_{sp}) \\
\partial[\mathrm{B}]/\partial t &= D_{\mathrm{B}}\nabla^2[\mathrm{B}] - mkR([\mathrm{A}],[\mathrm{B}])\Theta([\mathrm{A}]^n[\mathrm{B}]^m - K_{sp}) \\
\partial[\mathrm{C}]/\partial t &= D_{\mathrm{C}}\nabla^2[\mathrm{C}] + kR([\mathrm{A}],[\mathrm{B}])\Theta([\mathrm{A}]^n[\mathrm{B}]^m - K_{sp}) \\
&\quad - k'[\mathrm{C}]\Theta([\mathrm{C}] - [\mathrm{C}*]) - k''[\mathrm{C}]N([\mathrm{C}],[\mathrm{D}]) \\
\partial[\mathrm{D}]/\partial t &= k'[\mathrm{C}]\Theta([\mathrm{C}] - [\mathrm{C}*]) + k''[\mathrm{C}]N([\mathrm{C}],[\mathrm{D}])
\end{aligned} \tag{7.3}$$

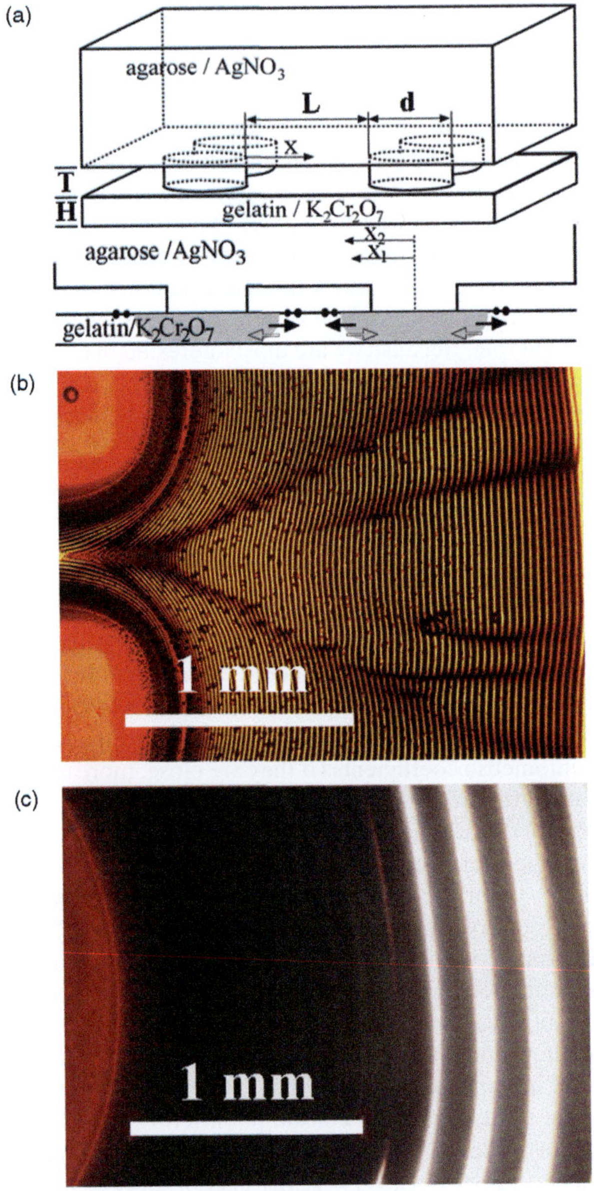

Figure 7.3 (a) Scheme of the experimental arrangement for WETS of PP patterns. Typical dimensions: feature sizes, $d = 50\,\mu m$–1 mm; spacing between the features, $L = 25\,\mu m$–1 mm; feature depth, $T = 40$–$50\,\mu m$; gel thickness, $H = 2$–$120\,\mu m$. Gray areas in the lower diagram correspond to turbulent precipitation zones (not to scale). The arrows give approximate directions of diffusion of $AgNO_3$ (solid arrows) and $K_2Cr_2O_7$ (open arrows). Positions of PP bands are indicated by solid ovals at the gelatin surface. (b) Periodic bands originating from two 500 μm squares. More than 200 bands are resolved within the first millimeter from the feature edges (not all are visible in this image). (c) No bands are resolved within ~1 mm from the edge of a small droplet $AgNO_3$

Of course, these equations are not solvable analytically, and it is necessary to use numerical methods to simulate PP. Throughout this chapter, we use the Crank–Nicholson scheme discussed in detail in Section 4.4.1. Since the patterns we consider are two-dimensional but depend on the thickness of the gel layer (see Section 7.9 below), all concentrations depend on three Cartesian coordinates $\boldsymbol{r} = (x, y, z)$ and on time, t, and the simulation domain is a three-dimensional square grid $N_x \times N_y \times N_z$, with typical dimensions $N_x = N_y$ from 100 to 200 and N_z varying from 20 to 80. For the pair of salts used in most experiments described later, $A = AgNO_3$ and $B = K_2Cr_2O_7$ and for their reaction $2AgNO_3 + K_2Cr_2O_7 \rightarrow Ag_2Cr_2O_7(\downarrow) + 2KNO_3$, the typical concentrations are $[A]_0 \sim 1.0\,M = 16.9\%$ w/w and $[B]_0 \sim 0.1\,M = 2.9\%$ w/w, $K_{sp} = 10^{-11}\,mol^3\,L^{-3}$, and the relative magnitudes of diffusion coefficients $D_A{:}D_B = 1{:}0.06$ were measured experimentally. The remaining model parameters $D_C = 0.04D_A$, $C^* = 0.08$, $D^* = 0$ and $c_r = 0.9$ were optimized against three different experimental patterns. In the simulations, the simulation domain initially contained only species B at concentration $[B](\boldsymbol{r}, t = 0) = [B]_0$; the concentrations of species A, C and D were zero. The effect of stamping was approximated by fixing the concentration of A to a constant value of $[A](\boldsymbol{r}_s, t) = [A]_0$ at those regions of the domain, $\boldsymbol{r}_s$, in contact with the stamp features. The other domain boundaries were considered impermeable to the diffusion of all species (i.e., no flux conditions): $\nabla[\mathbf{A}] \cdot \boldsymbol{n} = 0$, $\nabla[\mathbf{B}] \cdot \boldsymbol{n} = 0$, $\nabla[\mathbf{C}] \cdot \boldsymbol{n} = 0$, $\nabla[\mathbf{D}] \cdot \boldsymbol{n} = 0$, where $\boldsymbol{n}$ is the unit normal to the surface of the domain.

This model reproduces the empirical laws from Section 7.2, and also – to few percent accuracy – the more quantitative features of two-dimensional patterns (Figure 7.2). Its drawback is that it is relatively slow, and one high-quality simulation requires several hours on a desktop PC. The issues of computing times become even more serious when the patterns need to be averaged over several stochastic realizations of the reaction terms R, which is necessary when one is dealing with more subtle effects, especially those depending on the thickness of the gel layer (e.g., change in periodicity, refraction phenomena, etc.; Section 7.4). Therefore, while this model is not rapid enough to allow 'reverse engineering' of the PP patterns via optimization of experimental parameters (see Example 6.3), it is a valuable tool with which various hypotheses can be tested rigorously and experimental trends confirmed. Significantly, all of the 'heuristic' trends we will use in the fabrications tasks described in subsequent sections have been confirmed by simulations based on equations 7.3.3.

7.4 MICROSCOPIC PP PATTERNS IN TWO DIMENSIONS

In principle, the first and thinnest precipitation bands can form at very small distances from the source of the outer electrolyte. In macroscopic PP setups, however, this is not the case, and the first few millimeters or even centimeters are the zone where hydrodynamic flows obscure diffusive transport and prevent formation of regular bands (Fig. 7.1 and 7.3c). By the time bands appear, they are usually of millimeter dimensions and of little interest to us. To overcome this problem and resolve microscopic precipitation zones, we will apply wet stamping (WETS) in which the transport of water and salt through the stamp/substrate interface is diffusive to a very good approximation (see Example 6.1 and Fialkowski *et al.*[36]).

As in previous WETS arrangements, one salt (here, 0.5–25% by weight $K_2Cr_2O_7$) is contained in a thin (2–120 μm) layer of dried gelatin, while the other ($AgNO_3$) is delivered from an agarose stamp (Figure 7.3(a)). The stamp is infused with silver nitrate by soaking in 10–50% w/w salt solution for two hours. Prior to use, the stamp is dried by placing on a filter paper for up to 1 hour – this drying step minimizes the spilling of the outer electrolyte onto gelatin so that $AgNO_3$ is delivered diffusively.

Figure 7.3(b) shows PP bands that develop within about an hour in dichromated gelatin wet-stamped with a pattern of microscopic squares. The total number of bands resolved in this structure is several hundreds, with some bands as thin as 700 nm (Figure 7.4). This is in sharp contrast to bands obtained from a small droplet of $AgNO_3$ solution placed onto the same gelatin film (Figure 7.3(c)), where more than ~1 mm around the drop is covered with structureless precipitate. Clearly, WETS can get us into the microscopic regime of PP.

From a practical perspective, microscopic resolution is only one of the necessary conditions for microfabrication via RD. It is also essential that we can *control* the formation of RD structures to achieve the desired – or at least close to desired – periodicities, band thicknesses, etc. One way to do this is to use different combinations of salts and/or substrate materials in the hope of identifying those that yield the target patterns. This strategy, however, appears very inefficient since the potential parameter space (combinations of salts, nature of the gel) to explore is prohibitively large. A more sensible approach is to identify generic trends that for different salt pairs and gel substrates (e.g., $AgNO_3/K_2Cr_2O_7$/gelatin used here) would allow for controlling pattern characteristics by the dimensions of the system, material properties of the gel substrate or the concentrations/diffusivities

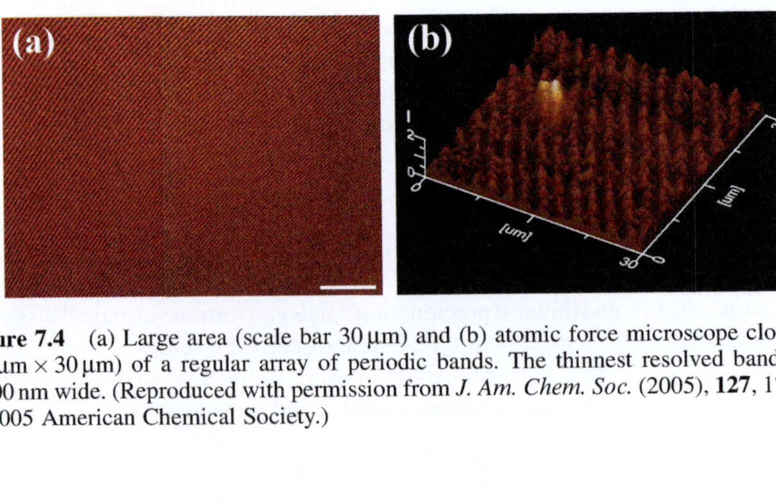

Figure 7.4 (a) Large area (scale bar 30 μm) and (b) atomic force microscope close-up (30 μm × 30 μm) of a regular array of periodic bands. The thinnest resolved bands are ~700 nm wide. (Reproduced with permission from *J. Am. Chem. Soc.* (2005), **127**, 17803. © 2005 American Chemical Society.)

of the precipitating salts. Below, we will list several such dependencies observed in experiments and, whenever possible, rationalize them by scaling arguments and/or numerical simulations.

7.4.1 Feature Dimensions and Spacing (Figure7.5)

For a given gel thickness, H, and feature geometry, the spacing coefficient, p, is linearly proportional to the ratio of feature size to the period of the pattern, $d/(d + L)$; see Fig. 7.3.

Explanation. This dependence is a consequence of the Matalon–Packter law, $p([A]_0, [B]_0) = F([B]_0) + G([B]_0)/[A]_0$ discussed in Section 7.2. To show this, we first invoke an experimental observation that the width of the precipitation zone below and around each feature where no bands are resolved scales with the feature size, d. This result is quite intuitive given that, for constant H and L, larger features 'flood' the surface of the dry gel substrate more rapidly than smaller features (as a spilled bucket of water wets a carpet more rapidly than a spilled glass), and the distance over which hydrodynamic flows subside and transport becomes diffusive is larger for the former. Rapid formation of the precipitate depletes the inner electrolyte within the precipitation zone by a value proportional to $[B]_0 d$. As a result of this depletion, the inner electrolyte left between the features experiences a concentration gradient and diffuses towards the precipitation zone (to equalize concentrations within the gel). Consequently, its effective concentration in the regions where the periodic bands ultimately form decreases by $\Delta[B] \propto [B]_0 d$ or, in terms of relative decrease, by $\Delta[B]/[B]_0 \propto d/(d + L)$. The spacing coefficient corresponding to the reduced concentration $[B]_0 - \Delta[B]$ (Figure 7.5) is then found

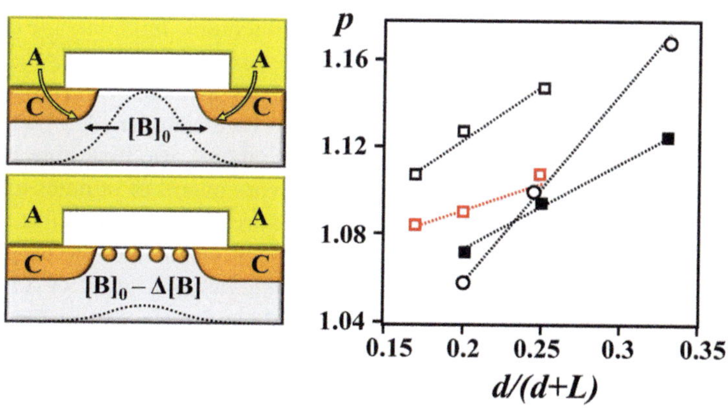

Figure 7.5 Schemes on the left illustrate the outflow of the inner electrolyte, B, in response to the rapid formation of the precipitation zones (C) under and around the features. The graph gives experimental (*black*) and modeled (*red*) dependencies of the spacing coefficient p on $d/d + L$

by series expansion around its reference/idealized value corresponding to no depletion (e.g., for infinite distance between the features):

$$p([A]_0, [B]_0 - \Delta[B]) = p([A]_0, [B]_0) - \Delta[B](\partial p/\partial[B])_{[A]_0,[B]_0} \quad (7.4)$$

Now, by the Matalon–Packter law, the derivative term can be written as $(\partial p/\partial[B])_{[A]_0,[B]_0} = (dF/d[B])_{[B]_0} + (1/[A]_0)(dG/d[B])_{[B]_0}$. Because both F and G are decreasing functions of the concentration of inner electrolyte (Section 7.2), the value of this derivative is negative. Using the scaling relationship for $\Delta[B]/ \propto [B]_0\, d/(d+L)$, we then have

$$p([A]_0, [B]_0 - \Delta[B]) = p_0 + \alpha d/(d+L) \quad (7.5)$$

where p_0 and α are positive constants. The graph in Figure 7.5 shows that this scaling agrees with experimental results for two square arrays of squares ($d = 100\,\mu m$, open square markers; $d = 150\,\mu m$, open circular markers), and for a square array of circles ($d = 150\,\mu m$, filled square markers). The general trend is also reproduced by simulations (e.g., red line in Figure 7.5 corresponds to a square array of $d = 100\,\mu m$ squares) based on Equations (7.3), although the exact values are off by about 2–5%. Note that the different slopes of the lines for geometrically similar patterns (e.g., square array of circles vesus square array of squares with the same d) suggest that this is a rather subtle phenomenon and accurate *a priori* prediction of p is difficult. At present, Equation (7.5) appears to be useful in situations when at least two values of p for a given d or L have been determined experimentally so that other values can be then extrapolated.

7.4.2 Gel Thickness (Figure 7.6)

Whereas the geometry/dimensions of the stamped features modulate the effective concentration of the inner electrolyte, B, the gel thickness, H, affects that of the outer electrolyte, A, delivered from the stamp. As a result, for $H > \sim 15\,\mu m$, the spacing coefficient, p, increases linearly with increasing H.

Explanation. To see how this trend comes about, we first note that the characteristic times in which periodic bands form are in tens of minutes to several hours, and water has enough time to wet the entire depth of the substrate (see Example 6.1 in Chapter 6). Therefore, the cross-sectional area of the propagating front of A is proportional to H. At the same time, the rate of transfer of A is driven by the concentration gradient at the agarose/gelatin interface, and because all delivered ions are constantly being depleted below and around the stamped feature, the flux of these ions entering the gel does not change appreciably as long as the reaction front propagates. It follows that the effective concentration of A at its front propagating in the substrate decreases with increasing thickness of the gelatin layer, $[A]_0^{eff} \approx [A]_0/H$. By the Matalon–Packter law, we then have

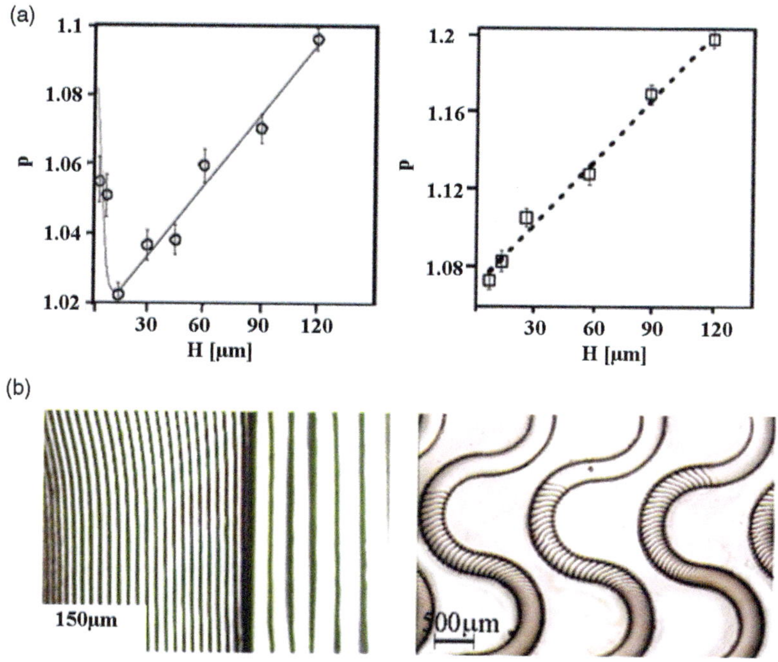

Figure 7.6 (a) Experimental (left) and simulated (right) dependencies of the spacing coefficient, p, on the gel thickness, H, for a square array of circles ($d = 150\,\mu$m, $D = 750\,\mu$m). (b) Left picture: both the absolute spacing and the spacing coefficient change when a PP front travels from a thinner ($H = 10\,\mu$m, $p = 1.023$) into a thicker gel ($H = 150\,\mu$m, $p = 1.125$). Right picture: PP from a planar front (from a rectangular block of agarose) does not propagate on very thin (~2 μm) portions of the gel, and is confined to deeper (~40 μm), wiggly grooves. (Reprinted with permission from *J. Phys. Chem. B* (2005), **109**, 2774. © 2005 American Chemical Society.)

$p(H) \approx \alpha' + \beta' H$, with α' and β' being some constants. This linear dependence is observed in the simulations and also in experiment, albeit in the latter case only down to $H \sim 15\,\mu$m. Recent studies have shown that below $\sim 10\,\mu$m, the salt-doped gelatin layer has an anomalously low absorptivity of ions from solutions of inorganic salts, likely due to the to the existence of a boundary layer near the surface of the support.[22] In other words, when the gelatin layer becomes very thin, the effective concentration of the outer electrolyte $[A]_0^{\mathrm{eff}}$ decreases with decreasing H and, consequently, $p(H)$ increases. In fact, when $H \sim 2\,\mu$m, the concentration is so low that no PP occurs at all. A striking manifestation of this effect is illustrated in the right-hand panel in Figure 7.6(b) which shows a gel cast onto a polymeric support having wiggly grooves embossed on its surface: the thickness of the gelatin layer in the grooves is ~40 μm, while that on the flat portion of the stamp is ~2 μm. The PP patterns form only in the groves but not in a very thin gel (for some more interesting behaviors in gels of varying depth, see Example 7.2).

7.4.3 Degree of Gel Crosslinking

The degree of gel crosslinking is an experimental parameter that is relatively easy to control for most gels and can have a profound effect on the propagation and characteristics of the PP patterns. These effects were studied systematically first by Matsuzawa's[37] and then Zrinyi's[38] groups in gels made of poly(vinyl alcohol) and crosslinked to a desired degree with glutaraldehyde. These studies concluded that while the spacing does not change appreciably with increasing gel crosslinking, the bands become thinner until for very crosslinked gels they do not form at all (instead, a structureless precipitate is observed). While a quantitative treatment of this effect is still lacking, the currently accepted, qualitative explanation is that since the more-crosslinked gels are 'stronger', they offer higher elastic resistance to the forming precipitate, which needs to make room for itself by expanding the gel. Consequently, formation of wide bands in highly crosslinked gels is unfavorable due to the required large increase in the gel's elastic potential energy.

For the purposes of fabrication, the appealing feature of the band thinning phenomenon is that if it were possible to crosslink only the select regions of the gel, the propagated PP patterns would have locally variable band thickness and – in the limit of high degree of crosslinking – would not form over these regions at all. One way to achieve such localized 'thickening' of the gels is by using WETS to deliver an extra crosslinking agent. For the gelatin films doped with potassium dichromate, however, a much more straightforward way is to simply irradiate the gel with UV light through a photomask (Figure 7.7).[39] In the exposed regions Cr(VI) undergoes photoreduction to Cr(III), which then coordinates and crosslinks the electron-donating groups of the gelatin's amino acids.[40]

7.4.4 Concentration of the Outer and Inner Electrolytes

We have already seen that the concentrations of A and B can change the spacing coefficient of the pattern and that this dependence can be expressed by the versatile Matalon–Packter law. In this subsection, we mention briefly another concentration dependence[39] whereby the relative width of the bands, w_n, compared to that of the clear 'slits' between them, s_n, increases with increasing [A] and/or [B]. This effect is illustrated vividly in Figure 7.8. The left image shows a 20 μm thick gelatin film loaded with 10% w/w $B = K_2Cr_2O_7$ and patterned with 10% w/w of $A = AgNO_3$ solution. The bands that form are characterized by the spacing coefficient $p = 1.06$ and are thin relative to their spacing – the average ratio of w_n to s_n is 0.27 (thinnest band is 1.6 μm). The image on the right shows gelatin which has the same thickness and degree of crosslinking, but the concentrations of A and B are both 25%. Although the spacing coefficient is similar ($p = 1.08$), the bands are now thick and the average value of w_n/s_n is 17 with the narrowest slit as thin as 900 nm! Given that the spacing coefficient varies relatively slowly with increasing concentrations,

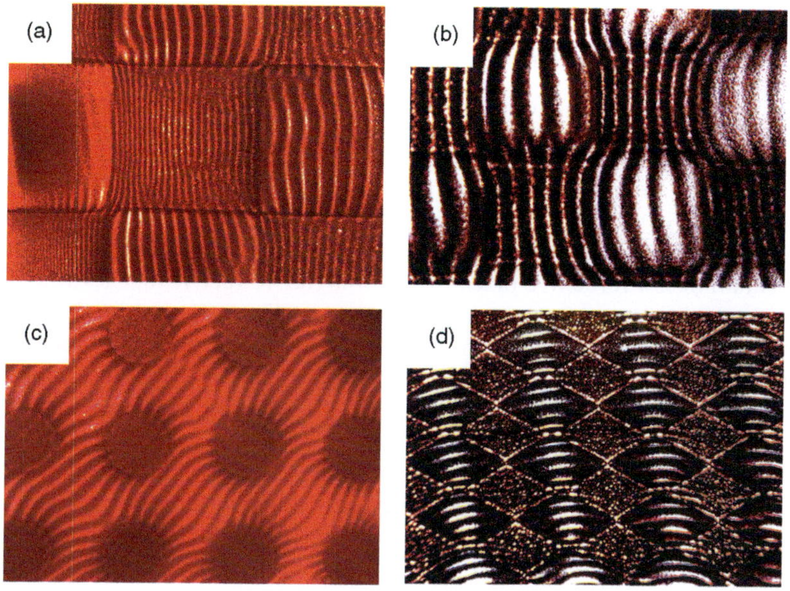

Figure 7.7 PP microstructures in photopatterned gels.[39] (a) Bands in the alternating-square regions irradiated with a large dose of UV ($E \sim 800\,\mathrm{mJ\,cm^{-2}}$) are thinner than those in unirradiated squares; note that the bands are discontinuous at the boundaries between irradiated and unirradiated portions of the surface. (b, c) With smaller doses of UV ($E \sim 300\,\mathrm{mJ\,cm^{-2}}$), the bands have different curvatures and thicknesses, but they preserve continuity at the boundaries. (d) For very large UV dose ($E \sim 1500\,\mathrm{mJ\,cm^{-2}}$), bands do not form at all over the irradiated rhombs. The field of vision in all pictures is ~750 μm. (Reproduced with permission from *J. Am. Chem. Soc.* (2005), **127**, 17803. © 2005 American Chemical Society.)

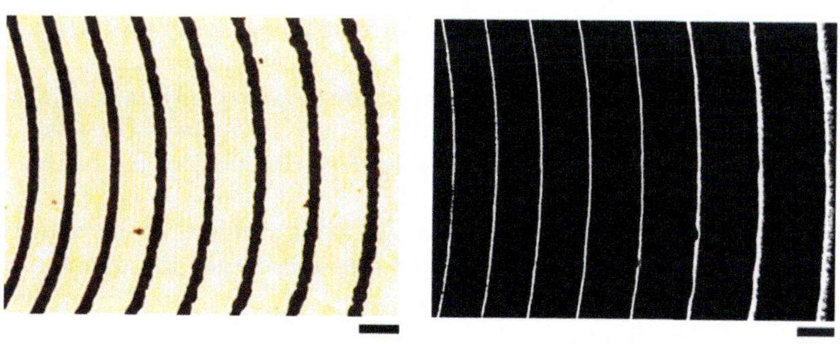

Figure 7.8 PP patterns formed on the same gel, with the same salt pair but for different salt concentrations (low in the left image, high in the right). Scale bars = 10 μm. (Reproduced with permission from *J. Am. Chem. Soc.* (2005), **127**, 17803. © 2005 American Chemical Society.)

these trends can be explained by simple conservation of mass, whereby more concentrated salts produce more precipitate and give thicker bands.

Example 7.2 Wave Optics and Periodic Precipitation[41]

The increase in the 'wavelength' of the bands propagating from a thinner to a thicker gel (Figure 7.6(b)) is reminiscent of the propagation of light waves from a medium of low refractive index to one of higher refractive index. Curiously, this optical analogy can be extended further, and it can be shown that at the locations where gel thickness changes, the PP patterns refract and reflect. In doing so, the incident and the refracted rings obey a law analogous to Snell's law of classical optics, with a reciprocal of the spacing coefficient being a counterpart of the refraction index, $p'_2 \sin\alpha_1 = p'_1 \sin\alpha_2$. In this expression, the subscripts denote the two regions of the gel, the spacing coefficients are defined as $1 + p'_{1,2} = x^{1,2}_{n+1}/x^{1,2}_n$ (note that these coefficients differ by unity from the notation in the main text, $p'_{1,2} = p_{1,2} - 1$) and α_1 and α_2 are the incident and refracted angles (defined as the angles between the normal to the interface and the normal to the 'incident' and 'refracted' rings at the interface, respectively; see Figure at the end of the example).

To derive the law of PP refraction,[41] we begin by noting that the rings are continuous at the interface between the two regions of the gel. Physically, continuity is the consequence of the fact that the precipitate present in one region induces ring formation (nucleation and growth) in the other, and propagates continuously across the interface.

We denote the distance between two subsequent n-th and $(n + 1)$-th rings along the interface as d_n (see figure below) and consider propagation from medium denoted by subscript '1' into that denoted by subscript '2'. By simple trigonometry, $d_n = w^1_n/\sin\alpha_1 = w^2_n/\sin\alpha_2$, where w^1_n and w^2_n denote the distances between the rings in regions 1 and 2, respectively, and α_1 is the angle of incidence, and α_2 the angle of refraction of the reaction front. Using the Jablczynski spacing law in each medium $(x^{1,2}_{n+1}/x^{1,2}_n = 1 + p'_{1,2})$, we approximate $w^{1,2}_n$ as $w^{1,2}_n = x^{1,2}_{n+1} - x^{1,2}_n \approx p'_{1,2}x^{1,2}_n$ to obtain $\sin\alpha_1/\sin\alpha_2 = (p'_1 x^1_n)/\ (p'_2 x^2_n)$. Finally, because $x^1_n = x^2_n$ at the interface, $p'_2 \sin\alpha_1 = p'_1 \sin\alpha_2$ which is a LR counterpart of Snell's law with the 'indices of refraction' of the two media replaced by $1/p'_{1,2}$.

In the figure below the top panel defines the dimensions of the refracting rings. The middle and bottom panels are the experimental illustration of the refraction law. When the precipitation fronts travel from a thinner layer into a thicker gel, the curvature of the rings increases ($\alpha_1 < \alpha_2$) and so $p'_1 < p'_2$. When propagation is from a thick to a thin gel, curvature decreases ($\alpha_1 > \alpha_2$) and $p'_1 > p'_2$. (Reproduced by permission from *Phys. Rev. Lett.* (2005), **94**, 018303. © 2005 American Physical Society.)

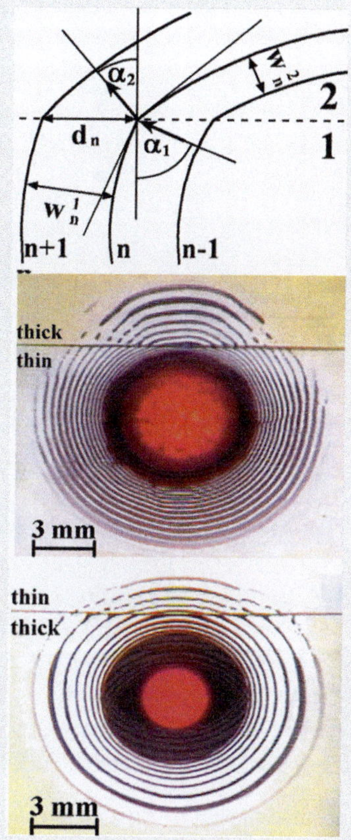

7.5 TWO-DIMENSIONAL PATTERNS FOR DIFFRACTIVE OPTICS

Although the list of trends discussed in the previous section is certainly not exhaustive, it offers enough experimental flexibility to fabricate a variety of microstructures of some interesting optical properties. In particular, PP patterns comprising microscopic and submicroscopic opaque bands separated by transparent slits can diffract light (Figure 7.9).

Although it is quite tempting to call the PP patterns simply 'diffraction gratings' of sorts, this terminology should be used with some caution. In most commonly used gratings, diffracting features are equally spaced and modulation of opacity (or refractive index) is truly periodic. For PP, the spacing and thickness

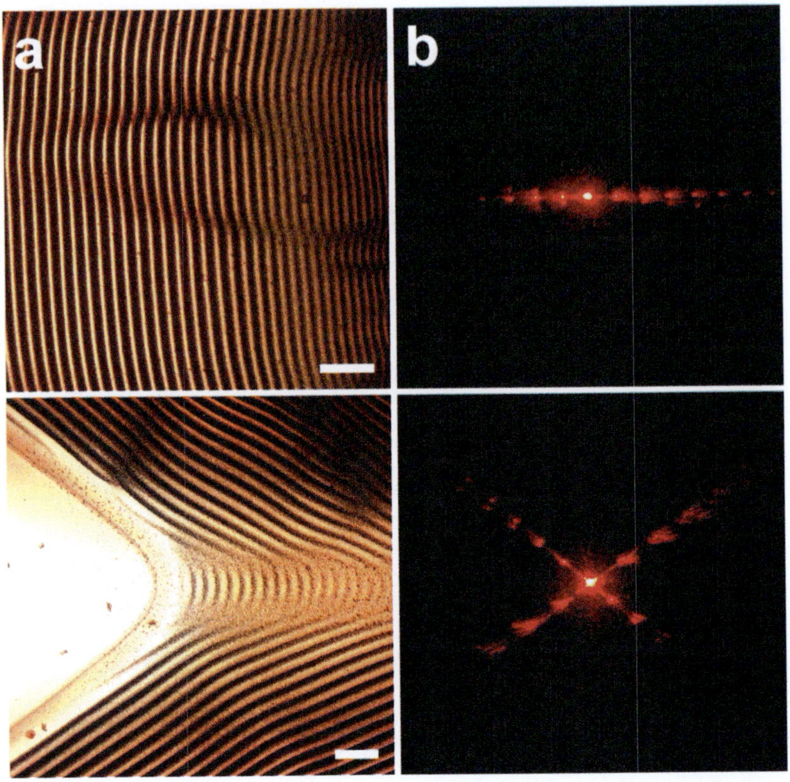

Figure 7.9 (a) Periodic precipitation patterns created in geletin by one (top row) and two (bottom row) overlapping fronts. (b) The corresponding diffraction patterns from a 632 nm HeNe laser. Scale bars are 200 μm

of consecutive precipitation bands change with distance from the source, and the only true periodicity is that imposed by the wet-stamped array. Strictly speaking, the PP patterns become periodic only when the spacing coefficient is adjusted to unity (we have seen in previous sections that values quite close to unity can be achieved by adjusting parameters such as gel thickness or feature spacing). This is not to say that aperiodic structures are not interesting. On the contrary, arrays of parallel lines of different spacings have been shown to increase sharpness of diffraction spots compared to periodic structures,[42] and aperiodic two-dimensional arrangements of microfeatures are useful in optical signal processing (e.g., Dammon lenses[43]), wavefront engineering,[44] and light focusing.

In many of these applications, it is desirable to pattern different regions of the substrate with different types of features or eliminate patterns from some portions of the substrate so that only separate optical elements are formed. The UV gelatin

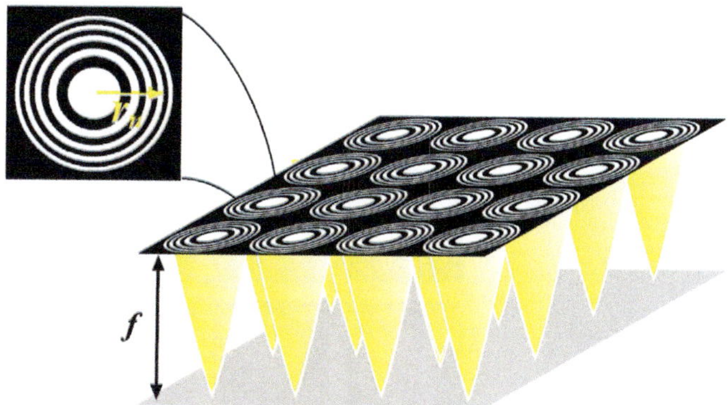

Figure 7.10 Scheme of an individual FZP and an array of FZPs. Such arrays are used for parallel micromanipulation of particles (via optical tweezing,[43] in photolithography,[40,44] and in image processing). Yellow cones represent focused light; f stands for focal length

crosslinking method described in Section 7.4.3. is quite suitable for this task, and Figure 7.7 illustrates its applications to the fabrication of complex two-dimensional surfaces in which the propagation of PP fronts is altered (Figure 7.7(a–c)) or even eliminated (Figure 7.7(d)) over the irradiated, highly crosslinked regions.

With this spatial control of PP and with the ability to adjust band spacings and thicknesses, let us focus on one type of useful[45–47] diffractive element that is composed of concentric, alternating transparent and opaque rings (Figure 7.10). When the radii r_n at which the transparent and opaque zones switch are given by $r_n = \sqrt{n\lambda f + n^2\lambda^2/4}$, this structure is called a Fresnel zone plate (FZP) and focuses light of wavelength λ at a focal point located at a distance f from the center of the structure. The remarkable feature of this lens is that it is entirely planar, and focuses light not by the refraction of the incoming light, but rather by diffraction.[45,48,49]

Since in the FZP both the distances and thicknesses of the opaque zones increase towards the plate's center, this structure is a promising fabrication target for PP propagated from a circular boundary inwards. Indeed, patterns propagated from stamps decorated with circular microwells of radius R (Figure 7.11(a)) have precipitation bands that are qualitatively similar to the opaque zones of the Fresnel plates. These bands are located at radii r_n (counting from the stamped circle's center) which obey a Jablczynski-like spacing law, $\ln(R - r_n) \propto (N - n)$, where N is the total number of resolved bands.

As illustrated in Figure 7.11(b,c), the numbers of the resolved bands and their locations can be regulated by [$AgNO_3$], dimensions of the stamped circles (in particular, the spacing coefficient scales linearly with the circle diameter, so $p \sim d$; Figure 7.5) and the amount of time that the stamp is in contact with the gel (the longer the time, the more bands are resolved). Most importantly, the

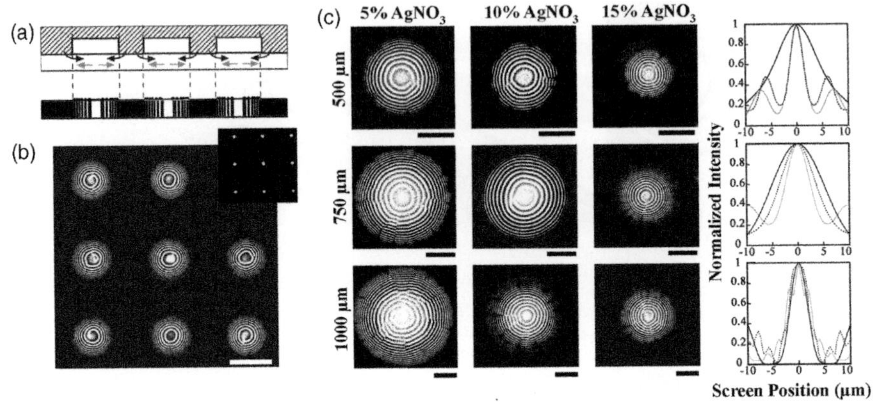

Figure 7.11 Fresnel-like lenses fabricated by PP. (a) Experimental scheme showing a stamp with outlines of the lenses (diameters, $2R = 500$–$1000\,\mu m$) applied onto a $10\,\mu m$ gelatin layer doped with $K_2Cr_2O_7$. The arrows indicate the directions of diffusion of Ag^+ cations (black arrows) towards the centers of the circles and of chromate ions (gray arrows) in the opposite direction. (b) Optical micrograph of a square array of $500\,\mu m$ lenses. The inset shows the image of the intensity distribution of the focused light along the focal plane. Scale bar $= 500\,\mu m$. (c) Optical micrographs of PP lenses obtained from circles of different diameters, and with different concentrations of $AgNO_3$. All scale bars are $250\,\mu m$. The graphs in the rightmost column have calculated distributions of light intensity at the focal planes of the corresponding lenses (solid lines, 5% $AgNO_3$; dashed lines, 10% $AgNO_3$; gray lines, 15% $AgNO_3$). (Reprinted with permission from *J. Appl. Phys.* (2008), **97**, 126102. © 2008 American Institute of Physics.)

developed PP patterns focus visible light efficiently. This is illustrated in Figure 7.11(b), which shows a square array of PP $500\,\mu m$ lenses and the corresponding image (inset) of the focal plane located ~4 mm away from the plane of the patterned film, and with the focal points ~$15\,\mu m$ in diameter. Also, diffraction patterns (see Example 7.3) calculated for various lenses shown in Figure 7.11(c) indicate that these structures focus light to ~$10\,\mu m$ at the focal point, which agrees with experimental results.[50] A curious observation is that these structures have better focusing properties than FZPs of similar number of bands and identical focal distance (for instance, a PP lens with 17 precipitation bands gives the same half-width of light at the focal point as does a FZP plate with ~50 bands). Are not these PP patterns fascinating?

More applications of periodic precipitation are still waiting to be discovered. In optical sensors, an interesting property of the patterns is that the clear spaces between the precipitation bands can be chemically modified with indicator molecules that change absorptivity and/or index of refraction upon external changes, and thus alter the diffraction image of the PP pattern.[51] Also, the ability to reduce silver dichromate to colloidal silver by exposure to vapors of formaldehyde might provide a way (possibly in combination with electroless plating) for the conversion of the PP patterns into electrically conductive wires – while it is somewhat fantastic to dream of PP as a vehicle to self-building electric microcircuitry, would anybody bet his/her savings it cannot be done?

Example 7.3 Calculating Diffraction Patterns

Suppose we have prepared a PP pattern by propagating reaction fronts towards one another from two parallel lines, say 1 mm apart. The PP structure that emerges should look like the one showed in the figure below, where the band spacings and widths increase from both sides toward the midpoint between the lines (ideally, as $x_{n+1}/x_n = p$ and $w_{n+1}/w_n = q$). This structure can be considered a one-dimensional version of PP lenses described in Section 7.5, and one might expect that it is going to focus the incoming light into a narrow 'line' (not a point, as in circular lenses). Our task in this example is to calculate the diffraction pattern from this structure and to find the values of parameters p and q that (for a given number of bands, say, 20) give the highest quality focusing.

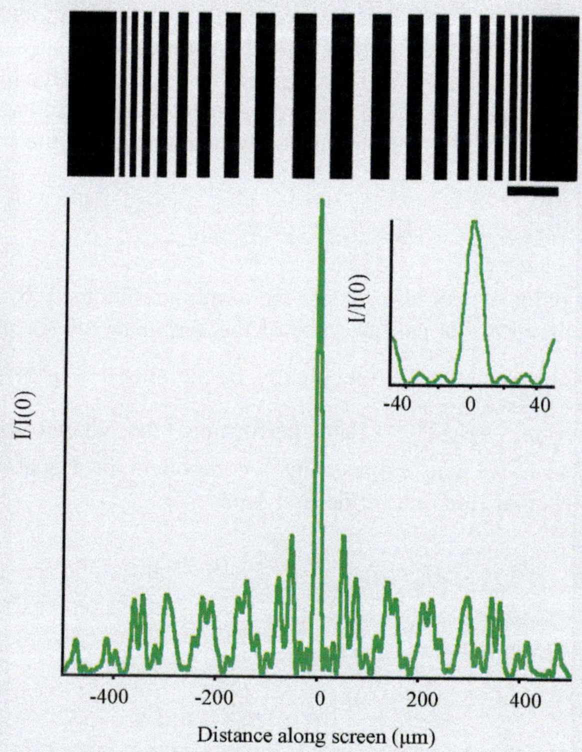

The Figure above shows an optimized PP grating and its calculated diffraction pattern at $z = 2$ cm. Scale bar in the upper picture is 100 μm. The inset in the lower graph magnifies the area around the focused peak (x axis is in μm).

We first define the so-called transmission function of the PP structure such that it is unity for the clear regions and zero for the opaque bands. Since the problem is symmetric with respect to the y coordinate, this function depends only on x:

$$\tau(x) = \begin{cases} 1 & x \text{ is between two bands} \\ 0 & x \text{ is within a band} \end{cases} \tag{1}$$

Using this definition, we now use the Fresnel–Kirchhoff formalism to calculate the distribution of light amplitude, u, over a plane of a 'screen' parallel to the plane of the PP structure and at a distance z from it, as shown in the figure on the next page:

$$u(x_i, y_i) = \frac{1}{j\lambda} \int\int \tau(x_0, y_0) \frac{e^{jkr}}{r} \mathrm{d}x_0 \mathrm{d}y_0 \tag{2}$$

where $j = \sqrt{-1}$, $k = 2\pi/\lambda$, λ is the wavelength of light, x_0 and y_0 are the coordinates in the plane of the grating, x_i and y_i are the coordinates in the focus/screen plane, r is the distance from point (x_0, y_0) to (x_i, y_i), and the integration is over the extent of the PP pattern The importance of the amplitude is that its square modulus gives the distribution of light intensity in the plane of the screen:

$$I(x_i, y_i) = |u(x_i, y_i)|^2 \tag{3}$$

While at this point we could calculate the amplitude/intensity by brute-force numerical integration, let us make use of the symmetry of the problem and simplify the formulas slightly. Specifically, by defining $\rho = \sqrt{z^2 + (x_0 - x_i)^2}$ so that $r = \sqrt{\rho^2 + (y_0 - y_i)^2}$ then performing the change of variables $y_0 - y_i = \rho \cosh t$, we can express the y-component of Equation (2) as a zeroth-order Hankel function of the first kind:

$$H_0^{(1)}(a) = \frac{1}{j\pi} \int_{-\infty}^{\infty} e^{ja\cosh t} \mathrm{d}t$$

so that

$$u(x_i) = \frac{\pi}{\lambda} \int \tau(x_0) H_0^{(1)}(k\rho) \mathrm{d}x_0 \tag{4}$$

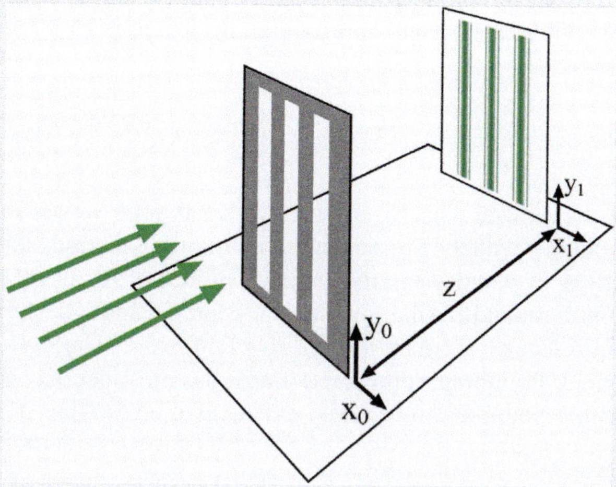

In the Figure above, light passes through the diffraction grating (gray) with the coordinates x_0 and y_0, and creates a diffraction pattern on the screen that is distance z away with coordinates x_1 and y_1.

Assuming that $k\rho \gg 1$ (i.e. distance to the screen is much larger than the wavelength of light), we have

$$H_0^{(1)}(a) \cong \frac{1}{\pi}\sqrt{\frac{\lambda}{\rho}}\,e^{j(a-\pi/4)}$$

and

$$u(x_i) = \frac{1}{\sqrt{\lambda}}e^{-j\pi/4}\int \frac{\tau(x_0)}{\sqrt{\rho}}e^{jk\rho}dx_0$$

In addition, if $z^2 \gg (x_0 - x_i)^2$ (i.e., the distance to the screen is much larger than the width of either the pattern or the image on the screen), we can expand ρ using the binomial expansion as

$$\rho = z + \frac{(x_0 - x_i)^2}{2z} - \frac{(x_0 - x_i)^4}{8z^3}\cdots \approx z + \frac{x_i^2}{2z} - \frac{x_i x_0}{z} + \frac{x_0^2}{2z}$$

From this expansion, we will derive two common approximations. If we keep the first three terms from the expansion, $\rho = z + (x_i^2/2z) - (x_i x_0/z)$, then we are using the Fraunhofer approximation, which gives the amplitude distribution of

$$u(x_i) = \frac{1}{\sqrt{z\lambda}}\exp\left[j\left(kz + k\frac{x_i^2}{2z} - \frac{\pi}{4}\right)\right]\int \tau(x_0)e^{-jkx_ix_0/z}dx_0 \qquad (5)$$

which is valid for the far field, where $kx_0^2/2z < \pi/2$. On the other hand, keeping all four terms yields the Fresnel approximation, which is valid as long as the expansion holds:

$$u(x_i) = \frac{1}{\sqrt{z\lambda}} \exp\left[j\left(kz - \frac{\pi}{4}\right)\right] \int \tau(x_0) e^{jk(x_0 - x_i)^2/2z} dx_0 \tag{6}$$

Since we do not know at what distance, z, the focal point we are looking for is located, we continue with the Fresnel approximation.[47] Searching through different combinations of p and q parameters that produce different PP patterns and screen positions z, we find that the most narrow 'central' band (~13 μm wide at half height) is obtained at $z = 2$ cm for $p = 1.2$, $q = 1.14$, $w_1 = 9$ μm, and $x_1 = 90$ μm. **Note.** Of course, by taking appropriate transmission functions, the formulas developed in this example can be easily extended to other types of PP patterns.

7.6 BUCKLING INTO THE THIRD DIMENSION: PERIODIC 'NANOWRINKLES'

We saw in Chapter 6 that the formation of a precipitate inside of a gel matrix can cause gel swelling. This is also the case for PP where the zones of $Ag_2Cr_2O_7$ precipitate rise above the gel's 'background' level. Interestingly, surface profilograms and atomic force microscope images in Figure 7.12 show that the submicrometer heights, h, of the buckled bands increase roughly linearly with band position, x. Before we see how this scaling can be used in some nontrivial microfabrication, let us first explain its origin.

To do so we make use of four observations. First, PP uses all $K_2Cr_2O_7$ present in the gel, and all $Ag_2Cr_2O_7$ precipitate produced is collected in the precipitation zones (as confirmed experimentally[52]). Second, the n-th ring is formed from silver dichromate precipitated between locations z_{n-1} and z_n, whose relative positions between the rings, $(x_n - z_{n-1})/(z_{n-1} - x_{n-1})$, are constant for all n (Figure 7.12(c)).[53,54] Third, the relative increase in the heights of the buckled bands is larger than in their width, $w_n \approx w$.[22] Fourth, the degree of surface deformation is proportional to the amount of precipitate generated at a given location (Chapter 6). From the first two observations, it follows that $z_n - z_{n-1} \propto x_n$ and the amount of $Ag_2Cr_2O_7$ collected by each ring (per unit length of the ring) is roughly proportional to x_n. Using the third and fourth statements, we can then approximate (admittedly, crudely) the shape of the bands as triangles whose heights scale as $h_n = (z_n - z_{n-1})/w \propto x_n$ – that is, linearly with x_n.

In the WETS experiments, the linear scaling holds remarkably well for buckles of nanoscopic heights prepared under various experimental conditions and lying within ~500 μm from the stamped features (i.e., within the region completely

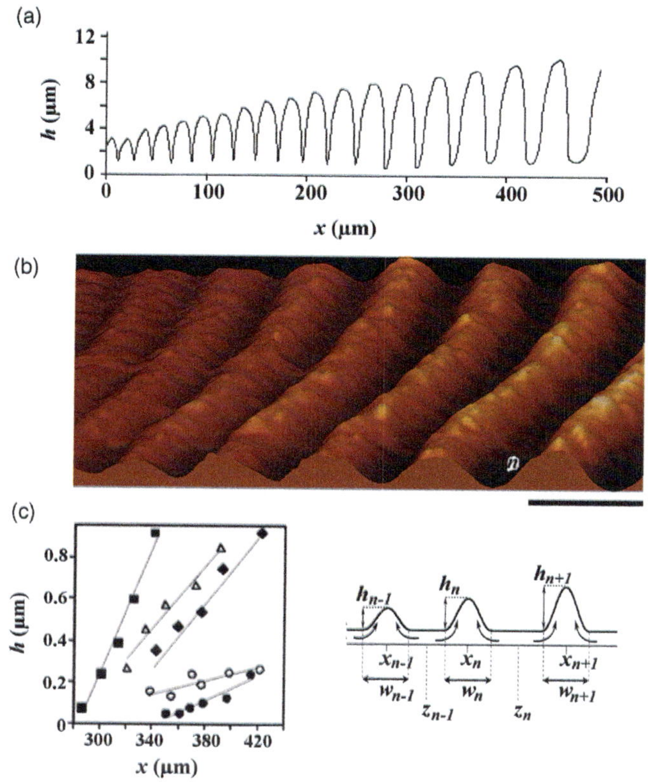

Figure 7.12 (a) Profilogram and (b) an atomic force microscopy scan of an array of periodic precipitation bands of linearly increasing heights (25% $AgNO_3$, 10% $K_2Cr_2O_7$; scale bar in the atomic force microscopy image is 10 μm). (c) Dependence of the height of the *n*-th ridge, h_n, on distance from the center of the feature, x_n, plotted for five gel layer thicknesses: 8 μm (open circles), 25 μm (filled circles), 42 μm (filled squares), 50 μm (filled rhombs) and 65 μm (open triangles). Diagram on the right defines the quantities used to develop pertinent scaling arguments. (Reproduced with permission from *J. Am. Chem. Soc.* (2005), **127**, 17803. © 2005 American Chemical Society.)

wetted during stamping). In addition, the relationship between the amount of precipitate and the surface deformation allows for engineering surface reliefs, in which the heights of the buckles and the slopes, $S = (h_n/x_n)$, of the buckle arrays can be controlled with remarkable precision and reproducibility by several control parameters.[39]

One way to do so is to change the degree to which the gel substrates are hydrated (Figure 7.13(a)). Wetter stamps (Figure 7.13(b), left) hydrate the gel to a higher effective depth, H_{eff}, than drier ones (Figure 7.13b, right), allowing periodic precipitation to occur in a thicker layer. Since the amount of precipitate produced (per unit area of the surface) scales with this layer's thickness, so does the degree of the gelatin deformation along the precipitation bands. Overall, the wetter the gelatin,

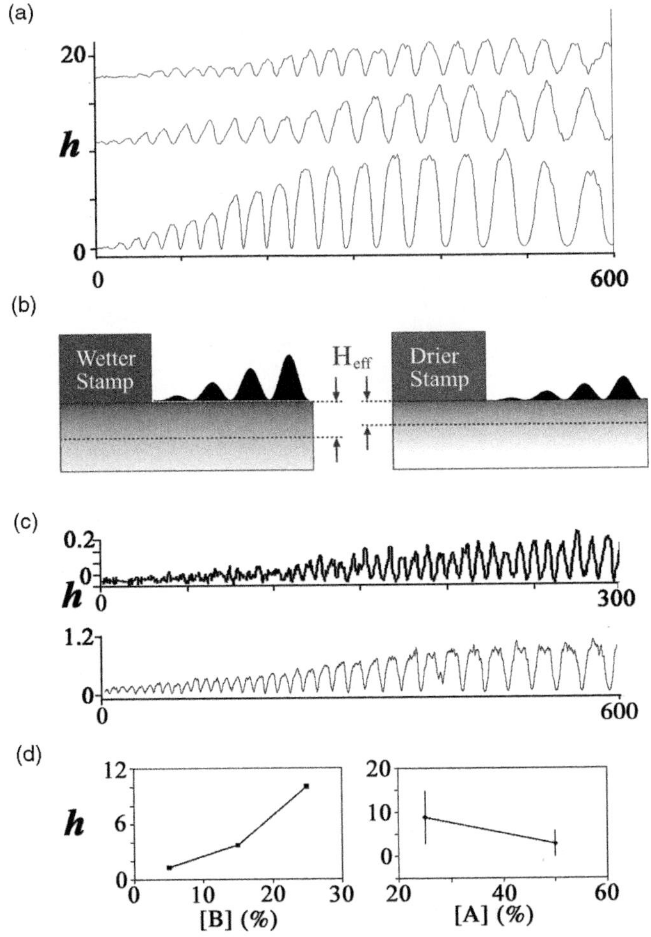

Figure 7.13 Profilograms of wrinkled surfaces. (a) Profilogram of PP bands for various times of stamp drying: 4 h (top), 1 h (middle) and 20 min (bottom). (b) Illustration of the effective hydrated layer. (c) The heights and slopes of the wrinkles decrease with increasing UV-controlled crosslinking of gelatin. Irradiation doses: $E = 180\,\mathrm{mJ\,cm^{-2}}$ for the top profile; $E = 11\,\mathrm{mJ\,cm^{-2}}$ for the bottom one (4.5 μm gels, 12% $K_2Cr_2O_7$, 45% $AgNO_3$, stamps dried for 1 h). (d) Wrinkle heights increase with increasing [$K_2Cr_2O_7$] (left graph) and decrease when [$AgNO_3$] increases (right graph). The trends were obtained by comparing the heights of precipitation bands of the same number $n = 20$ on gels differing only in the concentration of either $K_2Cr_2O_7$ or $AgNO_3$. Standard deviations were taken from three to five independent experiments. Other than the concentrations, all dimensions are in micrometers. (Reproduced with permission from *J. Am. Chem. Soc.* (2005), **127**, 17803. © 2005 American Chemical Society.)

the higher are the PP bands. At the same time, because the positions of the bands depend only weakly on the water content, the slopes of the arrays increase with the degree of substrate hydration. Another possibility is to adjust the degree of gel crosslinking (Section 7.4.3). The more crosslinked the substrate, the stiffer and harder to buckle it becomes, and the smaller the surface undulations (Figure 7.13 (c)). Lastly, the degree of buckling can be adjusted by the concentrations of the electrolytes used. Because the inner electrolyte, $B = K_2Cr_2O_7$, contained in the gel layer is a limiting reagent for the PP process, its concentration in the substrate determines the total amount of precipitate produced and collected into the bands. It follows that the band heights increase with increasing [B] (Figure 7.13(d), left). Increasing the concentration of the outer electrolyte, $A = AgNO_3$, decreases the spacing coefficient (via the Matalon–Packter law) and makes the arrays of buckles more 'shallow' (Figure 7.13(d), right). All in all, there are quite a number of ways in which to control the third dimension of PP – the big question, as in the case of two-dimensional patterns, is whether these structures are of any practical use.

7.7 TOWARD THE APPLICATIONS OF BUCKLED SURFACES

Recall from Chapter 6 that fabrication of nonbinary surface reliefs at the micro- and nanoscales is a tedious task. In this context, the ability to control the slopes of the wrinkle arrays combined with replication of the PP patterns into more durable polymeric materials appears a promising strategy for creating topographic gradients.

One area of research where these microstructured surfaces can be useful is creating surface topographies that control liquid spreading.[55] The upper image in Figure 7.14 shows how the nanowrinkles replicated into polydimethylsiloxane

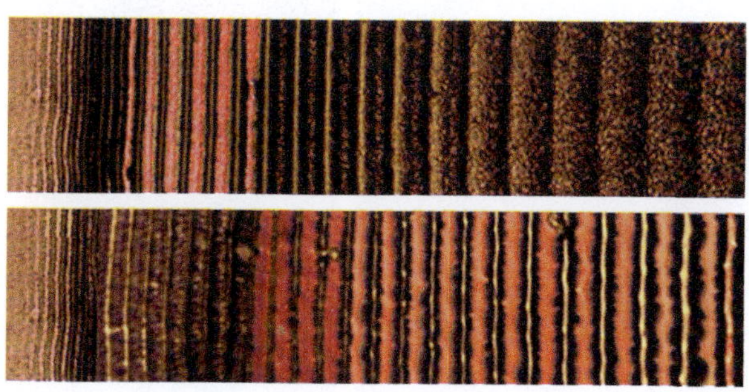

Figure 7.14 Discontinuous dewetting of light mineral oil (top) and pure ethylene glycol (bottom) from an oxidized PDMS replica of an array of parallel undulations, whose heights increase linearly from 4 μm (left) to 8 μm (right). Liquid is colored pink. (Reproduced with permission from *J. Am. Chem. Soc.* (2005), **127**, 17803. © 2005 American Chemical Society.)

(PDMS) 'capture' a dewetting, low-contact-angle (with respect to unstructured PDMS) liquid in high-curvature spaces between small wrinkles. In contrast, high-contact-angle liquids applied onto the same surface (lower image in Figure 7.14), are retained exclusively to the low-curvature 'valleys' between large wrinkles.

The ideas of curvature and tension also underlie the use of wrinkles to control behaviors of cells.[56] It has been known since the late 1970s[57–59] that cells can respond to mechanical stresses mediated by surface roughness by altering their orientation, differentiation and even functionality. The ability to control cell response by topographical cues has been sought in tissue/cellular engineering and in developmental cell biology.[60–62] With binary micropatterns made by photolithography, each fabricated substrate can be used to probe cells' response to features of one specific height – in contrast, the arrays of wrinkles of continuously increasing heights allow for parallel (i.e., simultaneous) monitoring of cellular responses to a continuum of surface topographies. This is illustrated in Figure 7.15, which shows organization of Rat2 fibroblast cells on patterns growing linearly in height from ~20 nm to ~10 μm. Images in the left column show cells on

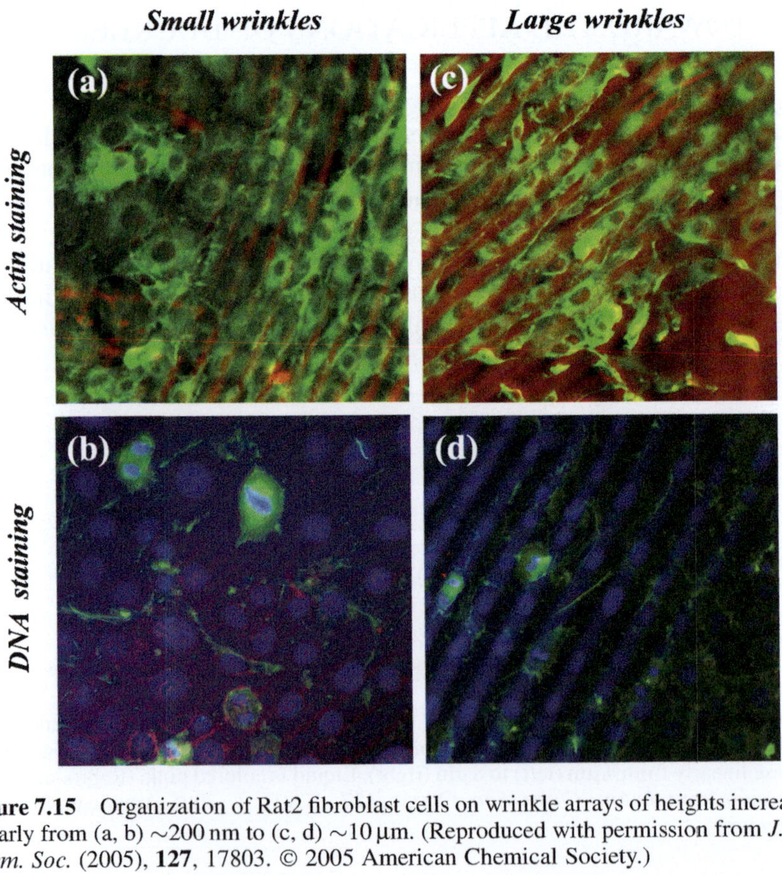

Figure 7.15 Organization of Rat2 fibroblast cells on wrinkle arrays of heights increasing linearly from (a, b) ~200 nm to (c, d) ~10 μm. (Reproduced with permission from *J. Am. Chem. Soc.* (2005), **127**, 17803. © 2005 American Chemical Society.)

the region of the wrinkles where the heights increase from ~200 to 400 nm, and spacing is approximately 10 μm. Figure 7.15(a) is a superposition of a phase-contrast image showing the presence of surface structures (red channel) and fluorescence image visualizing the so-called actin cytoskeleton (loosely speaking, the 'bones' of the cell (see Example 8.2); green channel). As the wrinkles' heights increase (from upper left to lower right in all images), cells progressively orient in the grooves between them. Figure 7.15(b) merges three channels: DNA staining to visualize cell nuclei (blue), actin staining (green) and phase-contrast to visualize the surface (red). As seen, small wrinkles do not influence the shapes of the nuclei, which remain roughly circular. In contrast, cells on larger undulations – in Figure 7.15(c,d) spaced by 30–40 μm and ~10 μm high – are fully oriented along the grooves; interestingly, their nuclei are also elongated in this direction, suggesting that nano-/microtopographies can be used to noninvasively manipulate the organelles within the cells, a topic we will revisit later in Chapter 8.

To close this brief survey of potential uses of buckled surfaces, let us revisit the FZPs. Recall that in the structures discussed in Section 7.5, light focusing was a result of the modulation of transmission function between opaque ($\tau = 0$; Example 7.3) and transparent regions ($\tau = 1$). Although such optically 'binary' PP patterns focused light efficiently, they were supported by gelatin, which is not a sturdy material that one would wish to use in real-life applications. Fortunately, even though the precipitate bands cannot be replicated into more durable supports, the topography of the PP rings can. And, as it turns out, the concentric nanowrinkles replicated into an optically transparent material such as PDMS (Figure 7.16) are also excellent focusing elements! To understand why this happens, note that the

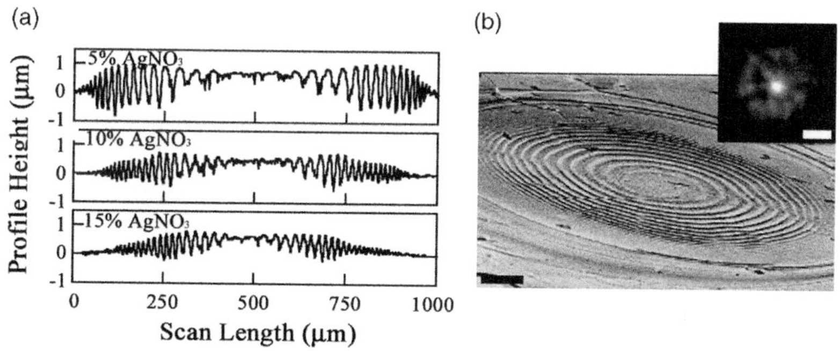

Figure 7.16 (a) Experimental profilograms of PDMS Fresnel-like lenses taken from PP patterns propagated from 1 mm circular boundaries (cf. Figure 7.11). In all cases, the concentration of inner electrolyte ($K_2Cr_2O_7$) is kept constant. As the concentration of $AgNO_3$ increases, the wrinkles appear closer to the center of the lens, and their heights decrease (the reader is encourage to justify this behavior based on the scaling arguments developed earlier in this chapter). (b) A scanning electron microscope image of a lens replicated from gelatin into PDMS (scale bar is 200 μm); the insert shows an optical micrograph of the focal point of the lens (scale bar is 100 μm). (Reprinted with permission from *J. Appl. Phys.* (2005), **97**, 126102. © 2008, American Institute of Physics.)

optical path (and optical phase) of light depends on the index of refraction of the medium through which it travels. When an initially plane light wave travels through the optically transparent PDMS, the local differences in the thickness of PDMS (due to the wrinkles on its surface) translate into differences of the optical phase of the outgoing light. Basic optics tells us that the thickness, h, of PDMS at location (x, y) can be related to the optical phase by $\varphi(x,y) = 2\pi(n_{PDMS} - n_{air})h(x,y)/\lambda$, where n_{air} is the index of refraction in air ($n_{air} = 1.0$), n_{PDMS} is the index of refraction in PDMS ($n_{PDMS} = 1.43$) and λ is the wavelength of light. The phase shift, in turn, can be related to the transmission function as $\tau(x,y) = \exp i\varphi(x,y)$. In other words, although our replicated structures have no opaque regions, they do cause spatial modulation of the transmission function by virtue of varying surface topography. Calculations similar to those outlined in Example 7.3 confirm that this modulation gives rise to efficient light focusing. An interested reader is referred elsewhere[44,63] for further discussion of this and related effects that are sometimes called wavefront engineering – in our case, engineering enabled by PP reactions.

7.8 PARALLEL REACTIONS AND THE NANOSCALE

Though there are many interesting PP systems and applications we have not covered in this chapter, the reader will no doubt be able to conjure new geometries, in which these fascinating reactions can be propagated to microfabricate small structures. What we would like to do in this closing discussion is to touch on two challenging issues we consider important for further development and practical application of PP systems. The first one has to do with integrating periodic PP with other types of chemical reactions and the ability to execute several different fabrication reactions in parallel. As an example, let us consider combination of PP with the fabrication of concave microlenses discussed in Chapter 6. At first sight, this appears a straightforward task, since we could deliver $AgNO_3$ to a gel substrate containing reaction 'partners' for both PP ($K_2Cr_2O_7$) and for surface swelling ($Ag_4[Fe(CN)_6]$). Unfortunately, the two reactions that ensue consume the same $AgNO_3$ 'food' and, depending on the specific concentrations used, one can get either good-quality PP bands or a large degree of substrate swelling – but not both of these effects simultaneously. In a more general context, the difficulty lies in finding chemicals that would react selectively without interference of cross-precipitation ('chemical orthogonality') and would lead to independent surface deformations ('mechanical orthogonality'). While finding appropriate orthogonal chemistries remains a challenge for future research, an alternative strategy can be used that is based on spatial separation of different reactions. One technically straightforward approach is to use gel films composed of multiple layers with each layer supporting a different fabrication process. Figure 7.17 illustrates this procedure in a two-layer system. Here, $AgNO_3$ delivered from the stamp reacts (i) with $Ag_4[Fe(CN)_6]$ contained in

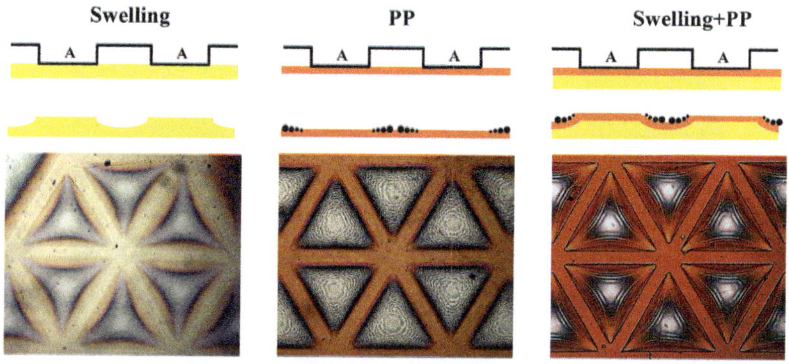

Figure 7.17 Parallel microfabrication. $AgNO_3$ delivered from the stamp reacts with $Ag_4[Fe(CN)_6]$ to cause gel swelling (left); with $K_2Cr_2O_7$ to produce periodic precipitation patterns (middle); or with both of these substances contained in different substrate layers to fabricate a swollen surface decorated with smaller PP bands. The sides of the stamped triangles are 300 μm in all images. (Reprinted with permission from *Materials Today*, (2007), **10**, 38. © 2007 Elsevier.)

the bottom layer to cause gel swelling and (ii) with $K_2Cr_2O_7$ contained in the top layer to produce PP bands. The sum of these processes gives deep curvilinear features (like the lenses discussed in Section 6.2) decorated with smaller PP undulations. Architectures of this type are potentially interesting in the context of microfluidic optical detection, where substances flowing through deep channels could be analyzed by light diffracting from the smaller scale reliefs embossed on the channel wall(s).[64]

Another example is illustrated in Figure 7.18 where PP bands are propagated 'on top of one another' in a wedge-like geometry in two stacked layers of different

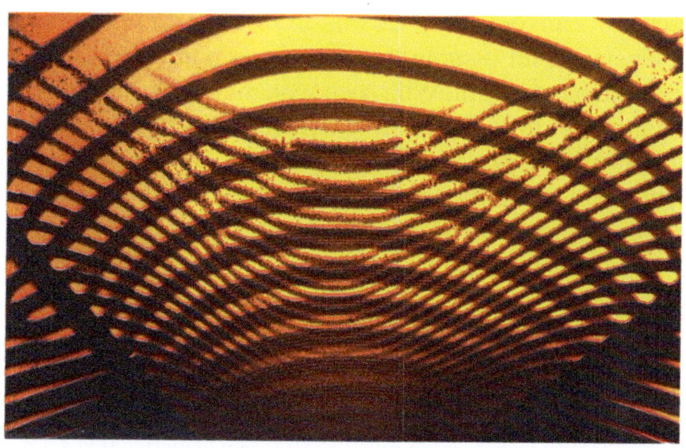

Figure 7.18 Two stacked PP patterns propagated in a wedge-like geometry in gel layers of different thickness. The layers are separated by a thin, nonpermeable polymer film. The curvatures of the patterns depend on gel thickness (see Example 7.2)

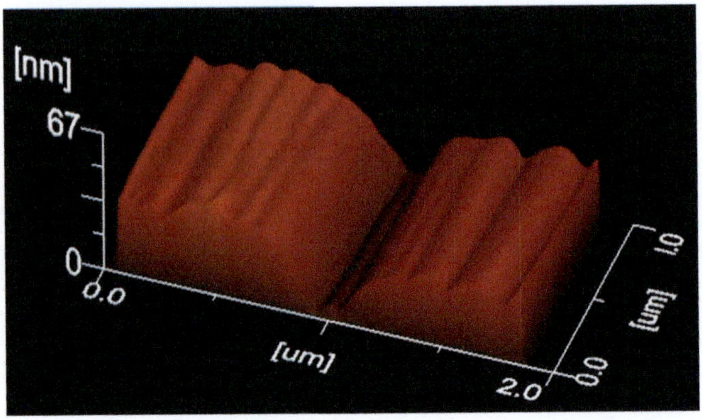

Figure 7.19 Atomic force microscope image of PP bands produced by reaction between $AgNO_3$ and $K_2Cr_2O_7$ propagated from two parallel lines separated by 2 μm. The smallest resolved bands have a width of 23 nm

thicknesses. Following the laws of thickness-dependent 'refraction' discussed in Example 7.2, the two sets of periodic bands have different curvatures giving the appearance of crossing rings. While these are only simple examples, the idea of spatial separation has tremendous potential, especially if one realizes how frequently nature – the ultimate microfabricator – employs such compartmentalization to deal with 'incompatible' reactions and build its intricate structures.[65–67]

The last issue deals with the smallest scale and the relevance of periodic precipitation for nanotechnology. We have seen earlier in this chapter that PP allows control of the heights of the periodic wrinkles with nanoscopic precision – what about the wrinkles' width? Can PP resolve bands that are of truly nanoscopic dimensions (<100 nm)? The atomic force microscopy image in Figure 7.19 demonstrates that it can, down to ∼25 nm. Before we pass on this result as mere technical achievement, let us note that the nanoscopic bands have only ∼3000 molecules per 1 nm band length. With such scant numbers, the entire formalism of PP based on continuous diffusive fields and RD equations simply should not apply. Yet, the bands form! Clearly, these reactions work in ways we still do not fully understand, and there are many new things we can learn about them. We will have more to say about nanoscale RD in later chapters.

REFERENCES

1. Grzybowski, B.A. and Campbell, C.J. (2007) Fabrication using 'programmed' reactions. *Mater. Today*, **10**, 38.
2. Liesegang, R.E. (1896) Ueber einige Eigenschaften von Gallerten [On a unique property of gelatin]. *Naturwiss. Wochenschr. [Scientific Weekly Review]*, **11**, 353.
3. Sultan, R.F. (2002) Propagating fronts in periodic precipitation systems with redissolution. *Phys. Chem. Chem. Phys.*, **4**, 1253.

4. Chopard, B., Droz, M., Magnin, J. et al. Liesegang patterns: effect of dissociation of the invading electrolyte. *J. Phys. Chem. A*, **103** 1432.
5. Heinsch, H.K. (1988) Liesegang rings, in *Crystals in Gels and Liesegang Rings*, Cambridge University Press, Cambridge. ch. 5.
6. Avnir, D. and Kagan, M. (1984) Spatial structures generated by chemical-reactions at interfaces. *Nature*, **307**, 717.
7. Hunt, P.A., Mitchell, P.B. and Paton, T.R. (1977) 'Laterite profiles' and 'lateritic ferricretes' on the Hawkesbury sandstone. *Geoderma*, **19**, 105.
8. Heaney, P.J. and Davis, A.M. (1995) Observation and origin of self-organized textures in agates. *Science*, **269**, 1562.
9. Levan, M.E. and Ross, J. (1987) Measurements and a hypothesis on periodic precipitation processes. *J. Phys. Chem.*, **91**, 6300.
10. Antal, T., Droz, M., Magnin, J. and Rácz, Z. (1999) Formation of Liesegang patterns: a spinodal decomposition scenario. *Phys. Rev. Lett.*, **83**, 2880.
11. Muller, S.C. and Ross, J. (2003) Spatial structure formation in precipitation reactions. *J. Phys. Chem. A*, **107**, 7997.
12. Jablczynski, M.C.K. (1923) La formation rythmique des précipités: les anneaux de Liesegang [Rhythmic formation of the precipitates: Liesegang rings]. *Bull. Soc. Chim. France*, **4**, 1592.
13. Matalon, R. and Packter, A. (1955) The Liesegang phenomenon: I. Sol protection and diffusion. *J. Colloid Sci.*, **10**, 46.
14. Kawano, T. and Imai, H. Characteristically shaped ZnO particles produced by periodic precipitation in organic gel media. *J. Cryst. Growth*, **283**, 490.
15. Stone, D.A. and Goldstein, R.E. (2004) Tubular precipitation and redox gradients on a bubbling template. *Proc. Natl Acad. Sci. USA*, **101**, 11537.
16. Muller, S.C., Kai, S. and Ross, J. (1982) Curiosities in periodic precipitation patterns. *Science*, **216**, 635.
17. Sultan, R.F., Al-Kassem, N.K., Sultan, A.A.H. and Salem, N.M. (2000) Periodic trends in precipitate patterning schemes involving two salts. *Phys. Chem. Chem. Phys.*, **2**, 3155.
18. Volford, A., Izsák, F., Ripszám, M. and Lagzi, I. (2007) Pattern formation and self-organization in a simple precipitation system. *Langmuir*, **23**, 961.
19. Narita, K., Matsumoto, K. and Igawa, M. (2006) Application of Liesegang ring formation on a gelatin film to the determination of sulfate concentration in individual rain droplets. *Anal. Sci.*, **22**, 1559.
20. Shreif, Z., Mandalian, L., Abi-Haydar, A. and Sultan, R. (2004) Taming ring morphology in 2D $Co(OH)_2$ Liesegang patterns. *Phys. Chem. Chem. Phys.*, **6**, 3461.
21. Msharrafieh, M. and Sultan, R. (2006) Dynamics of a complex diffusion-precipitation-re-dissolution Liesegang pattern. *Chem. Phys. Lett.*, **421**, 221.
22. Bensemann, I.T., Fiałkowski, M. and Grzybowski, B.A. (2005) Wet stamping of microscale periodic precipitation patterns. *J. Phys. Chem. B*, **109**, 2774.
23. Shreif, Z., Al-Ghoul, M. and Sultan, R. (2002) Effect of competitive complex formation on patterning and front propagation in periodic precipitation. *ChemPhysChem*, **3**, 592.
24. Sorensen, J.S. and Madsen, H.E.L. (2000) The influence of magnetism on precipitation of calcium phosphate. *J. Cryst. Growth*, **216**, 399.
25. George, J. and Varghese, G. (2005) Migrating triplet precipitation bands of calcium phosphates in gelatinous matrix. *J. Mater. Sci.*, **40**, 5557.
26. Hantz, P. (2000) Pattern formation in the $NaOH + CuCl_2$ reaction. *J. Phys. Chem. B*, **104**, 4266.

27. Zrinyi, M., Galfi, L., Smidroczki, E. *et al.* (1991) Direct observation of a crossover from heterogeneous traveling-wave to Liesegang pattern-formation. *J. Phys. Chem.*, **95**, 1618.
28. Büki, A. The Liesegang phenomenon, http://www.insilico.hu/liesegang, last update 19 April 2007 (25 February 2008).
29. Morse, H.W. and Pierce, G.W. (1903) Diffusion and supersaturation in gelatine. *Phys. Rev. Ser. I*, **17**, 129.
30. Jones, R.A.L. (2002) *Soft Condensed Matter*, Oxford University Press, Oxford.
31. Galfi, L. and Rácz, Z. (1988) Properties of the reaction front in an $A + B \rightarrow C$ type reaction–diffusion process. *Phys. Rev. A*, **38**, 3151.
32. Droz, M. (2000) Recent theoretical developments on the formation of Liesegang patterns. *J. Statistical Phys.*, **101**, 509.
33. Lifshitz, E.M. and Landau, L.D. (1984) *'Statistical Physics (Course of Theoretical Physics, Vol. 5)'*, 3rd edn, Butterworth-Heinemann, Oxford.
34. Henisch, H.K. (1991) *'Periodic Precipitation: A Microcomputer Analysis of Transport and Reaction Processes in Diffusion Media, With Software Development'*, Pergamon, New York.
35. Chopard, B., Luthi, P. and Droz, M. (1994) Reaction–diffusion cellular-automata model for the formation of Liesegang patterns. *Phys. Rev. Lett.*, **72**, 1384.
36. Fialkowski, M., Campbell, C.J., Bensemann, I.T. and Grzybowski, B.A. (2004) Absorption of water by thin, ionic films of gelatin. *Langmuir*, **20**, 3513.
37. Matsuzawa, S., Hondo, Y., Kawauchi, Y. *et al.* (1987) Crystal-growth in syndiotactic polyvinyl-alcohol hydrogels. *Colloid Polym. Sci.*, **265**, 810.
38. Karpatismidroczki, E., Buki, A. and Zrinyi, M. (1995) Pattern forming precipitation in gels due to coupling of chemical-reactions with diffusion. *Colloid Polym. Sci.*, **273**, 857.
39. Smoukov, S.K., Bitner, A., Campbell, C.J.et al. (2005) Nano- and microscopic surface wrinkles of linearly increasing heights prepared by periodic precipitation. *J. Am. Chem. Soc.*, **127**, 17803.
40. Paszewski, M., Smoukov, S.K., Klajn, R. and Grzybowski, B.A. (2007) Multilevel surface nano- and microstructuring via sequential photoswelling of dichromated gelatin. *Langmuir*, **23**, 5419.
41. Fialkowski, M., Bitner, A. and Grzybowski, B.A. (2005) Wave optics of Liesegang rings. *Phys. Rev. Lett.*, **94**, 018303.
42. Ferralis, N., Szmodis, A.W. and Diehl, R.D. (2004) Diffraction from one- and two-dimensional quasicrystalline gratings. *Am. J. Phys.*, **72**, 1241.
43. Dammann, H. and Gortler, K. (1971) High-efficiency in-line multiple imaging by means of multiple phase holograms. *Opt. Commun.*, **3**, 3.
44. Rogers, J.A., Schueller, O.J.A., Marzolin, C. and Whitesides, G.M. (1997) Wave-front engineering by use of transparent elastomeric optical elements. *Appl. Opt.*, **36**, 5792.
45. Smith, H.I. (1996) A proposal for maskless, zone-plate-array nanolithography. *J. Vac. Sci. Technol. B*, **14**, 4318.
46. Barrett, H. (1972) Fresnel zone plate imaging in nuclear medicine. *J. Nucl. Med.*, **13**, 382.
47. Black, D.N. and Wiltse, J.C. (1987) Millimeter-wave characteristics of phase-correcting Fresnel zone plates. *IEEE Trans. Microwave Theory Techn.*, **35**, 1122.
48. Hayasaki, Y., Itoh, M., Yatagai, T. and Nishida, N. (1999) Nonmechanical optical manipulation of microparticle using spatial light modulator. *Opt. Rev.*, **6**, 24.
49. Carter, D.J.D., Gil, D., Menon, R. *et al.* (1999) Maskless, parallel patterning with zone-plate array lithography. *J. Vac. Sci. Technol. B*, **17**, 3449.
50. Campbell, C.J., Baker, E., Fiałkowski, M. *et al.* (2005) Self-organization of planar microlenses by periodic precipitation. *J. Appl. Phys.*, **97**, 126102.

51. Nakajima, F., Hirakawa, Y., Kaneta, T. and Imasaka, T. (1999) Diffractive optical chemical sensor based on light absorption. *Anal. Chem.*, **71**, 2262.
52. Iizuka, K. (1985) *Engineering Optics (Springer Series in Optical Sciences, Vol. 35)*, 3rd edn, Springer-Verlag, Berlin/Heidelberg.
53. Droz, M. and Magnin, J. (1999) Liesegang patterns: studies on the width law. *J. Chem. Phys.*, **110**, 9618.
54. Antal, T., Droz, M., Magnin, J. *et al.* (1998) Derivation of the Matalon–Packter law for Liesegang patterns. *J. Chem. Phys.*, **109**, 9479.
55. Seemann, R., Brinkmann, M., Kramer, E.J. *et al.* (2005) Wetting morphologies at micro-structured surfaces. *Proc. Natl Acad. Sci. USA*, **102**, 1848.
56. Groves, J.T. (2007) Bending mechanics and molecular organization in biological membranes. *Annu. Rev. Phys. Chem.*, **58**, 697.
57. Margolis, L.B., Samoilov, V.I., Vasiliev, J.M. and Gelfand, I.M. (1975) Quantitative evaluation of cell orientation in culture. *J. Cell Sci.*, **17**, 1.
58. Dunn, G.A. and Brown, A.F. (1986) Alignment of fibroblasts on grooved surfaces described by a simple geometric transformation. *J. Cell Sci.*, **83**, 313.
59. Curtis, A. and Wilkinson, C. (1997) Topographical control of cells. *Biomaterials*, **18**, 1573.
60. Flemming, R.G., Murphy, C.J., Abrams, G.A. *et al.* (1999) Effects of synthetic micro- and nano-structured surfaces on cell behavior. *Biomaterials*, **20**, 573.
61. Curtis, A. and Wilkinson, C. (2001) Nanotechniques and approaches in biotechnology. *Trends Biotechnol.*, **19**, 97.
62. Andersson, H. and Berg, A.v.d. (2004) Microfabrication and microfluidics for tissue engineering: state of the art and future opportunities. *Lab on a Chip*, **4**, 98.
63. Grzybowski, B.A., Qin, D. and Whitesides, G.M. (1999) Beam redirection and frequency filtering with transparent elastomeric diffractive elements. *Appl. Opt.*, **38**, 2997.
64. Boisdé, G. and Harmer, A. (1996) *Chemical and Biochemical Sensing with Optical Fibers and Waveguides*, Artech House, Norwood, MA.
65. Warren, G. and Wickner, W. (1996) Organelle inheritance. *Cell*, **84**, 395.
66. Alberts, B., Johnson, A., Lewis, J. *et al.* (2002) *Molecular Biology of the Cell*, 4th edn, Garland Science, Abingdon, UK.
67. Weng, G., Bhalla, U.S. and Iyengar, R. (1999) Complexity in biological signaling systems. *Science*, **284**, 92.

8

Reaction–Diffusion at Interfaces: Structuring Solid Materials

The usefulness of reaction–diffusion (RD) is not limited to modifying gels and porous materials. Although RD processes cannot rapidly move chemicals inside of solids (but see Chapter 10 for some interesting nanoscale phenomena), they can be efficient in promoting interfacial reactions that either deposit solid materials or structure bulk solids. In this chapter, we will focus on two types of processes where diffusion inside and outside of micropatterned gels is coupled to interfacial reactions that either deposit or dissolve solids.

8.1 DEPOSITION OF METAL FOILS AT GEL INTERFACES

Microstructured metal foils are important components of several modern technologies, including magnetic disk drive heads,[1] NMR microcoils[2] and micro-/nanoelectronic devices.[3] While the majority of these applications use planar micropatterned foils, there has recently been a growing interest in preparing three-dimensional (3D) metallic microstructures which are interesting in the context of lightweight materials,[4,5] micro-waveguides[6] and structural and electrical components of microrobots.[7]

Typically, freestanding micropatterned 3D metal sheets are fabricated by vacuum deposition techniques (e.g., thermal evaporation, electron beam evaporation, sputtering, and metalorganic chemical vapor deposition, MOCVD) that are

Chemistry in Motion: Reaction–Diffusion Systems for Micro- and Nanotechnology Bartosz A. Grzybowski

not only cumbersome and cost-intensive but also limited by the 'line-of-sight' deposition, which makes metallization of vertical walls or features with negative inclines difficult. Alternative electroplating techniques can metallize diverse 3D topographies, but the detachment of freestanding foils from the conductive supports is challenging. RD combined with electroless plating offers an interesting alternative.[8]

Electroless plating, also known as chemical plating, is a nongalvanic plating method that deposits metal coatings from aqueous solution without the use of external electrical power. Instead, it uses catalytic chemical reactions to reduce metal salts to pure metal. Here, we will illustrate this process to plate copper films onto micropatterned agarose 'masters', such as those we have used earlier for wet stamping. We will see how the combination of electroless plating with RD allows for control of the two- and three-dimensional topographies of the foils and the locations over which they deposit.

Electroless deposition of copper is based on the reduction of Cu(II) salts to metallic Cu(0) catalyzed by colloidal palladium particles stabilized with tin. In a typical plating procedure (Figure 8.1), agarose gel masters are first soaked in an aqueous sensitizing solution (1L contains 5g $SnCl_2$ and 40 mL of 37% HCl assay, pH = 0.8) for $t_{sens} = 2$–48 h, washed in deionized water for $t_{wash} = 10$ min–24 h, and soaked in an aqueous activating solution (1L contains 2.5 mL of 2% w/v $PdCl_2$ solution, 1.0 mL of 37% HCl assay, pH = 1.7) for $t_{activ} = 5$ min–2 h. The masters are then transferred into a copper plating solution made by dissolving $C_4H_4KNaO_6 \cdot 4H_2O$ (potassium sodium tartrate, 25 g L^{-1}) and KOH (12.5 g L^{-1}),

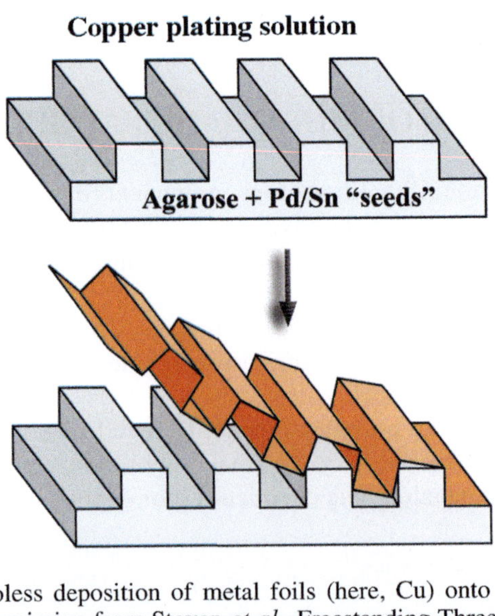

Figure 8.1 Electroless deposition of metal foils (here, Cu) onto micropatterned gels. Reproduced with permission from Stoyan *et al.*, Freestanding Three-dimensional Copper Foils Prepared by Electroless Deposition on Micropatterned Gels, Advanced Materials, 2005, 17, 6, 751–755, copyright Wiley-VCH

adding anhydrous $CuSO_4$ (5 g L^{-1}), and mixing in an aqueous solution of formaldehyde (or 25.0 mL of 37% solution).[9]

In the activation process, $SnCl_2$ reduces $PdCl_2$ to form colloidal particles of Pd(0) stabilized by tin cations. Subsequently, during the plating process, these palladium colloids catalyze electron transfer between Cu(II) (here, in the form of a tartrate complex) and formaldehyde according to the following redox reaction:

$$[Cu(C_4H_6O_4)_2]^{2-} + 2HCHO + 4OH^- \rightarrow Cu + 2HCOO^- + H_2 + 2H_2O + 2C_4H_4O_6{}^{2-}$$

Although this reaction is thermodynamically favorable, its rate constant is negligible unless a palladium catalyst is present.

In our system, the gel serves as a reservoir of palladium colloids that catalyze the reduction of metal cations at the gel/solvent interface, resulting in the deposition of a thin metallic foil. Importantly, the local magnitudes of diffusive fluxes of the plating solution towards the micropatterned gel control the topography of the deposited metal films while the sizes of the seed particles 'presented' at the gel surface determine the film roughness. To understand these phenomena, we will begin with a mathematical analysis, which illustrates how to deal with RD at interfaces and how to describe realistically the migration of colloidal particles through porous media.

8.1.1 RD in the Plating Solution: Film Topography

The speed of deposition and final thickness of the plated foil depend on the local rate of deposition of copper at the surface of the patterned stamp. At that boundary, mass conservation dictates that the rate/flux of Cu^{2+} ions delivered to the surface via diffusion equals the rate of consumption by the plating reaction. Mathematically, this condition is given by $D(\nabla c)_s \cdot \boldsymbol{n} = kc_s$, where D and c are the diffusion coefficient and concentration of Cu^{2+} ions, respectively, $\boldsymbol{n}$ is the outward surface normal, k is the first-order reaction constant and the subscript 's' denotes that this equation applies to the surface of the stamp. Away from the stamp, at a distance H, the concentration is homogeneous and equal to the bulk concentration, c_0 (note that in practice this condition is achieved by mixing the plating solution). Furthermore, we note that the boundary condition at the stamp's surface may be simplified considerably in the case of 'fast' plating reactions (i.e., for $kH/D \gg 1$) to give $c_s = 0$. In the case of copper plating described here,[10] $k \sim 10^{-3}$ cm s^{-1}, $D \sim 10^{-5}$ cm^2 s^{-1} and $H \sim 0.1$ cm, such that $kH/D \sim 10 \gg 1$. Under these circumstances, diffusion is rate limiting, and the rate of copper deposition is directly proportional to the diffusive flux at the stamp surface – i.e., to $D(\nabla c)_s \cdot \boldsymbol{n}$. This means that the metal should be plated most rapidly in places where the concentration gradients are steepest.

To determine the gradients and fluxes, it is necessary to solve the steady-state diffusion equation (using the pertinent boundary conditions) for the concentration of metal ions. To illustrate this procedure, consider a simple case of parallel lines of square cross-section ($L \times L$) and spaced by L (Figure 8.2(a), where $L_1 = L_2 = \frac{1}{2}L$ and $H_1 = L$), for which we wish to find steady-state concentration profiles of the Cu^{2+} cations. Although translational symmetry simplifies the problem to two dimensions, the boundary conditions on the micropattern remain nontrivial for an analytical solution. To simplify our task, we split the problem into three sub-problems, which we will later superimpose (i.e., 'add together'; see Section 2.4) to give the overall solution. Each sub-problem (labeled I-a, I-b and II in Figure 8.2(b)) requires the solution of the steady-state diffusion equation on a rectangular domain with exactly one nonhomogeneous boundary (Section 2.2.1). Unlike in previous examples, however, the spatial domains are different (i.e., I and II) and must be 'stitched' together such that concentrations and concentration gradients remain continuous across the boundaries of the neighboring domains.

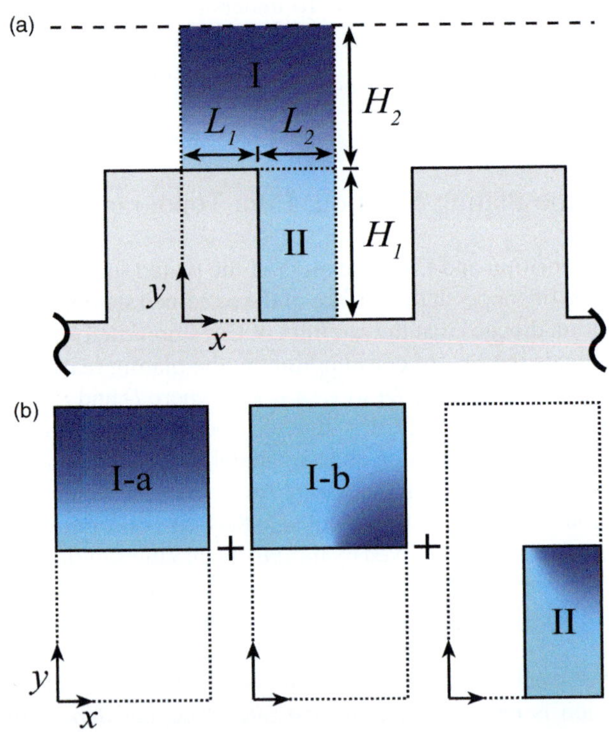

Figure 8.2 (a) Scheme of the parallel line geometry and relevant RD domain (blue) described in the text. (b) Graphical illustration of the three solution components, which may be superimposed to give the overall solution

1. RD in domain I ($0 \le x \le L_1 + L_2$, $H_1 \le y \le H_1 + H_2$). Diffusion in this domain is governed by the steady-state diffusion equation $\partial^2 c/\partial x^2 + \partial^2 c/\partial y^2 = 0$, where the Cartesian coordinates x and y are defined in Figure 8.2(a). Due to the symmetry of the stamp's features, we obtain 'no flux' (also called 'symmetry') boundary conditions at $x = 0$ and $x = L_1 + L_2$ – specifically, $(\partial c/\partial x)_{x=0} = (\partial c/\partial x)_{x=L_1+L_2} = 0$. The concentration away from the stamp at $y = H_1 + H_2$ is constant and given by $c(x, H_1 + H_2) = c_0$, and at $y = H_1$ the concentration is described by the following piece-wise function:

$$c(x, H_1) = \begin{cases} 0 & \text{for } 0 \le x \le L_1 \\ f(x) & \text{for } L_1 \le x \le L_1 + L_2 \end{cases} \tag{8.1}$$

where $c(x, H_1) = 0$ for $0 \le x \le L_1$ satisfies the required boundary condition at the stamp surface and $f(x)$ is an unknown function to be determined by matching concentrations and fluxes with domain II (discussed further below).

Using the separation of variables technique we learned in Section 2.2.1, we may solve the diffusion equation considering *only one* nonhomogeneous boundary condition at a time (note that there are two, at $y = H_1$ and at $y = H_1 + H_2$) and then superimpose the solutions. Let us first consider the nonhomogeneous condition at $y = H_1 + H_2$ and set the concentration at $y = H_1$ to zero, $c(x, H_1) = 0$. Denoting this problem as I-a, the solution is given by

$$c_{\text{I}-\text{a}}(x, y) = c_0(y - H_1)/H_2 \tag{8.2}$$

Similarly, if we consider the nonhomogeneous condition at $y = H_1$, setting $c(x, H_1 + H_2) = 0$, we obtain (the reader is encouraged to verify this result) the following solution denoted as I-b:

$$c_{\text{I-b}}(x, y) = A_0(H_1 + H_2 - y) + \sum_{n=1}^{\infty} A_n \cos(\kappa_n x) \sinh[\kappa_n(H_1 + H_2 - y)] \tag{8.3}$$

Here, $\kappa_n = n\pi/(L_1 + L_2)$ are the appropriate eigenvalues and A_n are unknown coefficients to be determined by the interdomain matching process. The overall solution for this domain is then given by adding the two solutions: $c_{\text{I}}(x, y) = c_{\text{I-a}}(x, y) + c_{\text{I-b}}(x, y)$.

2. RD in domain II ($L_1 \le x \le L_1 + L_2$, $0 \le y \le H_1$). In the second domain, the steady-state diffusion equation remains unchanged; however, the boundary conditions differ. At the stamp's surface, the concentration is fixed at zero due to the 'fast' plating reaction – i.e., $c(L_1, y) = 0$ and $c(x, 0) = 0$. At $x = L_1 + L_2$, we have the same 'symmetry' condition as described above, such that $(\partial c/\partial x)_{x=L_1+L_2} = 0$. Finally, at $y = H_1$ that concentration must be equal to that of the neighboring domain (i.e., I), such that $c(x, H_1) = f(x)$. Solving this equation via separation of variables, we obtain the following solution to problem II:

$$c_{\mathrm{II}}(x, y) = \sum_{m=0}^{\infty} B_m \sin[\alpha_m(x - L_1)]\sinh(\alpha_m y) \tag{8.4}$$

Again, $\alpha_m = (m + \frac{1}{2})\pi/L_2$ are the appropriate eigenvalues, and B_m are unknown coefficients to be determined by the matching process.

3. Matching the solutions by collocation. The unknown coefficients A_n and B_m may now be determined by equating the concentrations and the concentration gradients at the domain interface (i.e., at $y = H_1$ and $L_1 \leq x \leq L_1 + L_2$). Mathematically, these conditions are given by $c_{\mathrm{I}}(x, H_1) = c_{\mathrm{II}}(x, H_1)$ and $(\partial c_{\mathrm{I}}/\partial y)_{y=H_1} = (\partial c_{\mathrm{II}}/\partial y)_{y=H_1}$, or in terms of the explicit solutions:

$$A_0 H_2 + \sum_{n=1}^{\infty} A_n \cos(\kappa_n x)\sinh(\kappa_n H_2) = \sum_{m=0}^{\infty} B_m \sin[\alpha_m(x - L_1)]\sinh(\alpha_m H_1) \quad \text{or}$$

$$c_0/H_2 - A_0 - \sum_{n=1}^{\infty} A_n \kappa_n \cos(\kappa_n x)\cosh(\kappa_n H_2) = \sum_{m=0}^{\infty} B_m \alpha_m \sin[\alpha_m(x - L_1)] \cosh(\alpha_m H_1) \tag{8.5}$$

Unfortunately, because each equation contains two types of eigenfunctions, we cannot use their orthogonality properties to solve for A_n and B_m explicitly (as we might typically do in a single domain; Section 2.2). Nevertheless, we can solve for an arbitrarily large (but finite) set of coefficients using only linear algebra. In other words, we can reduce the problem of solving the two-dimensional diffusion equation on a complex geometry into a single matrix equation – this useful trick is worth remembering!

While there are several ways in which to set up this matrix equation for N coefficients, we will describe arguably the most intuitive approach, known as collocation. In this method, we equate the concentrations and fluxes (i.e., y components of the flux vector) at N different positions on $y = H_1$. Specifically, let $x_i = i(L_1 + L_2)/(N - 1)$, where $i = 0, 1, \ldots, N$. Along the stamp's surface, $x_i \leq L_1$, we require that $c_{\mathrm{I}}(x_i, H_1) = 0$ and along the boundary between regions I and II, $x_i > L_1$, we stipulate $c_{\mathrm{I}}(x_i, H_1) = c_{\mathrm{II}}(x_i, H_1)$. These conditions may be expressed explicitly as follows:

$$A_0 H_2 + \sum_{n=1}^{N-1} A_n \cos(\kappa_n x_i)\sinh(\kappa_n H_2) = 0 \quad \text{for} \quad x_i \leq L_1$$

$$A_0 H_2 + \sum_{n=1}^{N-1} A_n \cos(\kappa_n x_i)\sinh(\kappa_n H_2) - \sum_{m=0}^{N-1} B_m \sin[\alpha_m(x_i - L_1)]\sinh(\alpha_m H_1) = 0 \quad \text{for} \quad x_i > L_1 \tag{8.6}$$

This gives N equations for our $2N$ unknown coefficients. Another set of N equations is obtained by matching the fluxes on the domain $L_1 \leq x \leq L_1 + L_2$. This is done by introducing another set of equally spaced positions, $x_j = L_1 + jL_2/N$, where $j = 1, 2, \ldots N$, and equate the fluxes at these locations:

$$A_0 + \sum_{n=1}^{N-1} A_n \kappa_n \cos(\kappa_n x_j)\cosh(\kappa_n H_2) + \sum_{m=0}^{N-1} B_m \alpha_m \sin[\alpha_m(x_j - L_1)]\cosh(\alpha_m H_1)$$
$$= c_0/H_2 \qquad (8.7)$$

In this way, we obtain $2N$ linear equations for our $2N$ unknown coefficients, which may easily be solved numerically using common linear algebra routines from Matlab or Numerical Recipes.[11] As regards technical details, we note that in order to obtain accurate numerical solutions, it is advisable to solve for *scaled* coefficients, defined as $\bar{A}_n = A_n \sinh(\kappa_n H_2)$ and $\bar{B}_m = B_m \sinh(\alpha_m H_1)$. Otherwise, the hyperbolic functions may lead to poorly scaled matrices and large numerical errors.

We can now represent the overall solution graphically as in Figure 8.3. The model predicts that metal deposition is most rapid near the top corners of the pattern followed by the top surface and then the wells (with the lowest/inner corners metallizing last) where the flux of fresh Cu^{2+} is smallest. Generalizing to other types of patterns, we expect the metallization to proceed from 'top to bottom'. This effect is vividly illustrated in experimental optical micrographs in Figure 8.4 showing metallization of two-level cross 'pyramids', which begins at the uppermost face (Figure 8.4(b)) and proceeds through the middle level (from the outer corners inward; Figure 8.4(c)) before finally reaching the bottom surface (Figure 8.4(d)).

The useful aspect of such sequential metallization is that by controlling the time of plating, t_{plate}, it is possible to metallize only selected regions of the substrate and prepare perforated foils. For example, the foils in Figure 8.5(a) were plated on gels

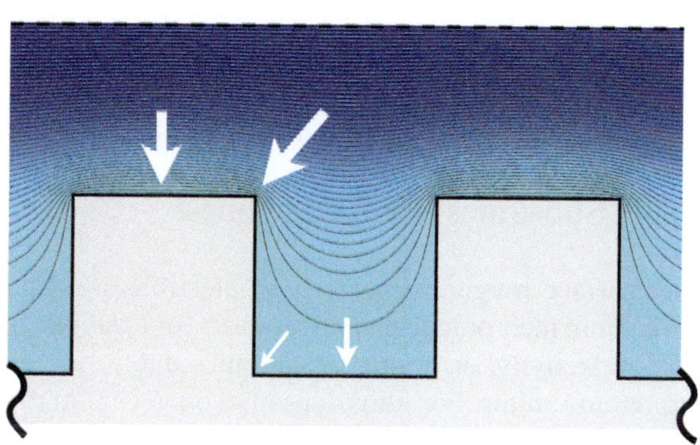

Figure 8.3 Contour plot of the two-dimensional concentration profile around an array of parallel lines. The arrows illustrate the magnitudes of the deposition fluxes at different locations

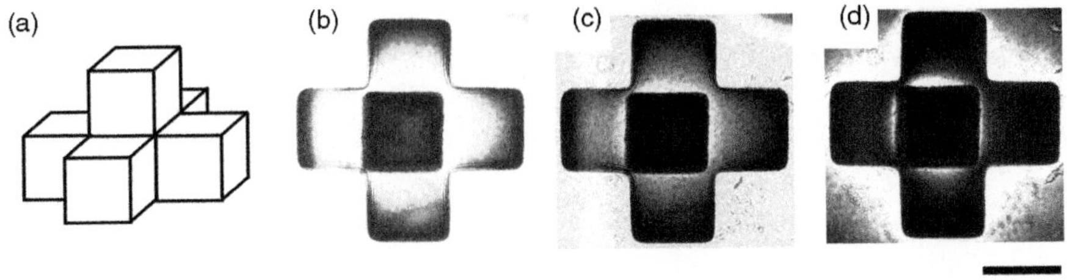

Figure 8.4 RD-controlled electroless deposition onto microscopic pyramids. Scale bar 150 μm. Reproduced with permission from Stoyan *et al.*, Freestanding Three-dimensional Copper Foils Prepared by Electroless Deposition on Micropatterned Gels, Advanced Materials, 2005, copyright Wiley-VCH

decorated with arrays of small hexagonal (left) and star-shaped (right) wells. After 30–60 min of plating, the surface of the gel gave a sturdy, several micrometers thick, detachable membranes metallized everywhere but in the well regions. With longer plating times ($t_{plate} \sim 2$ h), the top surface of the gel and the sidewalls of the microscopic wells were metallized to give perforated foils shown in Figure 8.5(b). Finally, with long plating times ($t_{plate} > 12$ h), metal was deposited over the entire micropattern to give continuous foils such as those shown in Figure 8.5(c,d).

In addition to the plating times, there are several other experimental parameters with which to control the topography of deposited films. For instance, increasing the aspect ratio of the features on the gel surface limits diffusion of reactants into the grooves of the pattern, and hence promotes metallization of its top portions. In contrast, placing the gel on a rotating support (akin to a rotating-disk electrode,[12] used in analytical chemistry) increases the flux of Cu^{2+} onto the surface, and facilitates metallization even in deep wells. Last but not least, one can combine the diffusive effects outside and inside of the stamp. One way to do so is by adjusting the concentration of reagents used to prepare catalytic Pd/Sn seeds required for plating. When these concentrations are low, transport of palladium or tin cations into the gel is severely limited in the grooves between the micropattern's features. As a result, catalytic seeds do not form in the grooves, but only at the tops of the features, which subsequently become the loci of selective copper deposition (even after days of plating).

8.1.2 RD in Gel Substrates: Film Roughness

A high degree of surface roughness/porosity of electrolessly deposited films is often a limiting factor in their practical applications – for example, in optics due to decreased optical reflectivity, or in microelectronics due to increased electrical resistance of the 'grainy' films. For films deposited on gels, film roughness (from the side facing the gel) can be controlled by an interesting RD process involving the catalytic Pd/Sn seeds present at the gel surface.

When the gel substrate is initially soaked in a sensitizing solution and subsequently washed in water, a gradient of $[Sn^{2+}]$ is established in the gel, with the

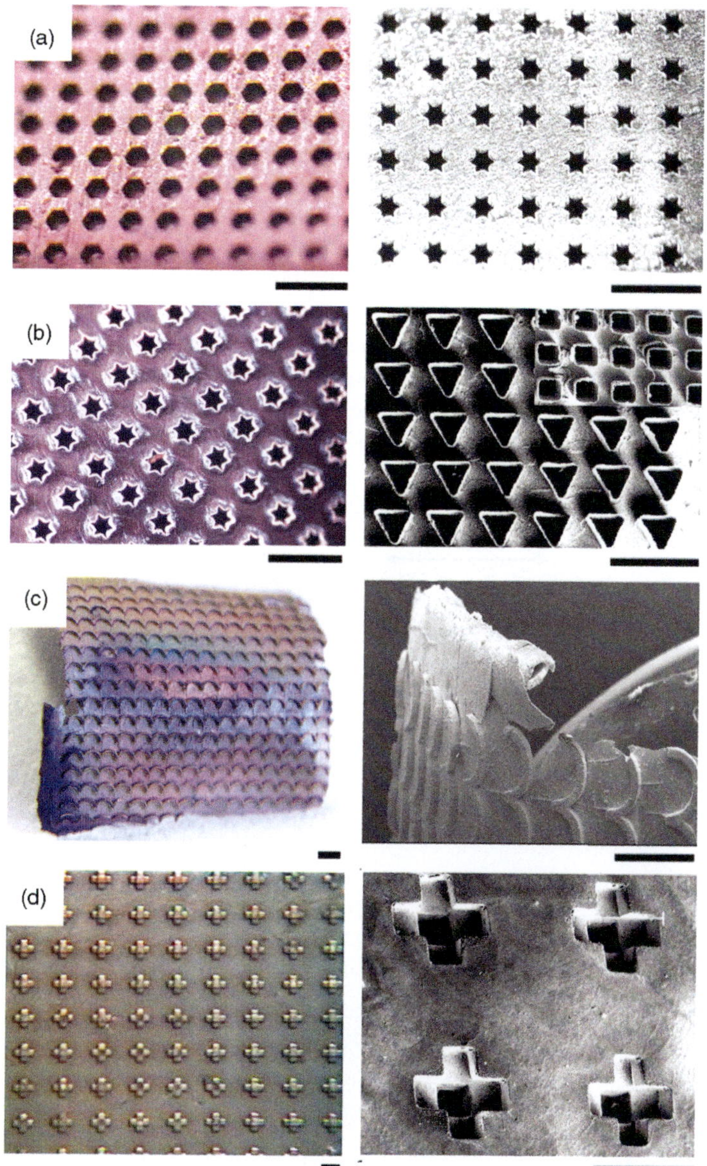

Figure 8.5 Copper foils prepared on micropatterned gels. (a) Flat membranes; (b) membranes with sidewalls of the wells metallized; (c, d) fully metallized foils. All scale bars are 500 μm. The inset in the right image in (b) shows a foil in which the metallized square features are 50 μm. The right image in (c) is a scanning electron microscopy zoom-out of the foil shown in the left image. The right image in (d) has the backside of the foil shown in the left image. Reproduced with permission from Stoyan *et al.*, Freestanding Three-dimensional Copper Foils Prepared by Electroless Deposition on Micropatterned Gels, Advanced Materials, 2005, copyright Wiley-VCH

concentration of tin cations increasing with distance from the gel surface (along the x coordinate in Figure 8.6(a)). Upon immersion in the activating solution, Pd^{2+} cations diffuse into the gel and react with Sn^{2+} in 1:3 stoichiometry to form colloidal complexes consisting of a metallic palladium core stabilized by strongly adhered tin cations.[13] It has been shown experimentally that these complexes are free to aggregate and form larger particles whose radii, R, are characterized by an exponential distribution, $p(R) = \exp(-R/R_{av})/R_{av}$, where $R_{av} \sim 1.5$ nm is the mean particle radius.[14] The diffusivities of Pd/Sn particles within the gel matrix depend on the particle sizes, with smaller particles migrating more rapidly than larger ones. Therefore, after soaking in the activating palladium solution, there is a distribution of particle sizes at the surface of the gel that is governed by the RD processes of both washing and activation. Because these surface particles 'seed' the plating of films, their sizes determine the surface roughness of the final foils. Our goal is then to describe these RD processes and predict the size distribution of

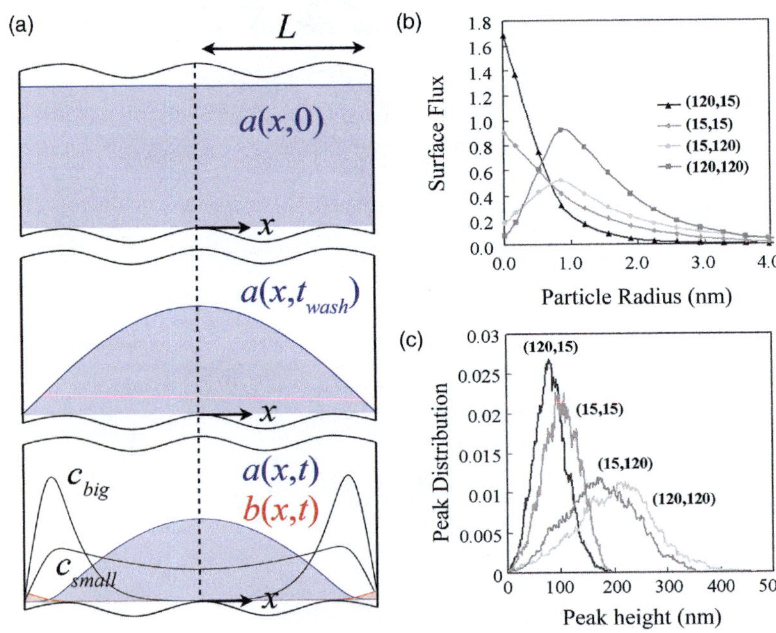

Figure 8.6 (a) Characteristic concentration profiles after (i) sensitizing, (ii) washing and (iii) activating (top to bottom). During activation, colloidal particles of different sizes (black curves) are generated at the reaction front and diffuse within the gel. (b) Calculated surface flux distributions of colloidal particles of different sizes for different values of (t_{wash}, t_{activ}). Gels soaked for shorter times have mostly small colloidal particles at their surfaces – consequently, the foils deposited on these films are smoother than those on gels soaked for longer times. (c) Normalized atomic force microscopy distributions of surface roughness for four copper foils prepared with different times of gel washing, t_{wash} (first number in parentheses; in minutes) and Pd^{2+} soaking, t_{activ} (second number)

colloidal seeds at the gel surface as a function of the experimentally controlled variables such as the washing and the activation times.

For simplicity, let us approximate the gel as a semi-infinite, unpatterned slab of thickness $2L$ and with interfaces at $x = \pm L$ (Figure 8.6(a)) exposed to (i) the sensitizing tin solution, (ii) the aqueous washing solution and finally (iii) the activating palladium solution. Let the concentrations of Sn^{2+} and Pd^{2+} in their respective solutions be denoted a_o and b_o, and those in the gel as $a(x, t)$ and $b(x, t)$, respectively. The gel is initially soaked in the sensitizing tin solution for a time, t_{sens}, which is much larger than the characteristic diffusion time, $t_{sens} \gg \tau_{diff} = L^2/D_a$, where $D_a \approx 5 \times 10^{-6}\,cm^2\,s^{-1}$ is the diffusion coefficient of Sn^{2+} in the gel (as estimated experimentally). This procedure results in a uniform concentration, a_o, of tin ions within the gel. The gel is then immersed in pure water for a time, t_{wash}, during which a concentration gradient develops and some of the Sn^{2+} species diffuse out of the gel. This process may be modeled by the one-dimensional diffusion equation $\partial a/\partial t = D_a \partial^2 a/\partial t^2$, where the initial condition is $a(x,0) = a_o$, and the boundary conditions are $a(L,t) = 0$ at the gel surface and $(\partial a/\partial x)_{x=0} = 0$ due to symmetry. The solution of this time-dependent diffusion problem may be solved by separation of variables (see Section 2.2.1) to give $a(x, t) = 2a_o \sum_{n=0}^{\infty} (-1)^n \cos(\kappa_n x)\exp(\kappa_n{}^2 D_a t)/\kappa_n$, where $\kappa_n = (n + \frac{1}{2})\pi/L$. For longer times (i.e., $t_{wash} \geq \tau_{diff}/2$), this series solution is very well approximated by its first term – thus, after washing, the concentration of Sn^{2+} in the gel is given by

$$a(x, t_{wash}) \approx (4a_o/\pi)\cos(\pi x/2L)\exp(-\pi^2 D_a t_{wash}/4L^2) \tag{8.8}$$

This concentration profile provides the necessary initial condition for Sn^{2+} in the gel directly prior to soaking in the activating palladium solution.

To describe the formation of the Sn/Pd colloids during the activation process, we introduce a second-order reaction term, $R(x,t) = ka(x,t)b(x,t)$. The choice of second-order kinetics is made for simplicity as the actual reaction kinetics is not known in detail. The products of this reaction are the colloidal seeds of size R drawn from a discretized version of the exponential distribution discussed previously, $p_i = C\exp\left[-(i-\frac{1}{2})\Delta R/R_{av}\right]$, where the indices $i = 1, 2, \ldots, \infty$ correspond to particle radii, $R_i = (i - \frac{1}{2})\Delta R$, and $C = [\exp(\Delta R/R_{avg}) - 1]/\exp(\Delta R/2R_{avg})$ is the necessary normalization constant. The local concentration of colloidal particles of size i is denoted $c_i(x, t)$ and evolves via diffusion with a size-dependent diffusion coefficient, $D_{c,i}$. To describe the size-dependent diffusivities of the particles within the gel, $D_{c,i}$, the well-known Stokes–Einstein relation (Example 8.1) needs to be corrected for the gel environment. While there is no exact formula to do so, various approximations have been developed – here, we use an expression based on the simulations of Clague and Phillips (summarized in Amsden[15]), in which the diffusivity scales as

$$D_{c,i} = D_i^o \left[1 + \frac{2\varphi}{3}\left(\frac{R_i + R_f}{R_f}\right)^2\right]^{-1} \exp\left(-\pi\varphi^{0.174\ln(59.6R_f/R_i)}\right) \tag{8.9}$$

where $D_i^o = k_B T/6\pi\eta R_i$ is the diffusivity in pure solution, φ is the volume fraction of gel (for agarose, this is essentially the same as its mass concentration[16] in g mL^{-1}), $R_f = 0.6$ nm is the fiber radius in the gel and R_i is the radius of the diffusing particles. The first bracketed term accounts for hydrodynamic drag effects imparted on the diffusing solute by the gel fibers, and the exponential term describes the 'sieving' of solute particles by the gel pores.

Example 8.1 Stokes–Einstein Equation

First derived by Einstein in 1905, the so-called Stokes–Einstein equation combines thermodynamic and fluid-mechanical considerations to approximate the diffusion coefficient of solute molecules in a liquid. The model Einstein developed applies to dilute solutions comprising solute molecules much larger than the surrounding solvent, which is treated as a continuous medium. At all times, the random motions of the solute molecules (treated as small spheres of radius R) are driven by forces due to local gradients of the chemical potential. For an ideal solution, the chemical potential, μ, is given by $\mu = \mu_o + kT \ln c$, where μ_o is the potential of some reference state, k is the Boltzmann constant, T is the temperature of the solution and c is the concentration of the solute molecules. For a solute molecule moving along direction x (or any other direction), the force generated by the gradient of μ is $F_\mu = (\kappa T/c)(\partial c/\partial x)$. This force is balanced by the force of viscous drag, F_D, exerted by the solvent on the solute molecule. The expression for drag was developed in 1851 by the famous fluid mechanician and mathematician Sir George Gabriel Stokes. Stokes' formula is $F_D = 6\pi\eta vR$, where η is the solvent viscosity and v is the particle steady-state (terminal) velocity. Equating these two forces and rearranging the terms gives the flux of the molecules, j_x (defined as the number of moles of molecules moving across a unit area per unit time) as

$$j_x = cv = -\frac{\kappa T}{6\pi\eta R}\frac{\partial c}{\partial x}$$

Since the molecules are jiggling randomly – that is, they are diffusing – the flux has to be equal to that prescribed by Fick's first law of diffusion (Section 2.1), $j_x = -D(\partial c/\partial x)$. By comparing the two expressions for j_x, we then obtain the Stokes–Einstein formula: $D = kT/6\pi\eta R$. Simplistic as the model might appear, it has proven surprisingly accurate and has found widespread use in modeling dilute liquid systems.[17]

With these assumptions, the partial differential equations describing reaction and diffusion in the agarose slab during activation read as follows:

$$\frac{\partial \bar{a}}{\partial \bar{t}} = \frac{\partial^2 \bar{a}}{\partial \bar{x}} + \upsilon_a \mathrm{Da} \cdot \bar{a}\bar{b}; \quad \frac{\partial \bar{b}}{\partial \bar{t}} = \frac{D_b}{D_a}\frac{\partial^2 \bar{b}}{\partial \bar{x}} + \upsilon_b \mathrm{Da} \cdot \bar{a}\bar{b}; \quad \frac{\partial \bar{c}_i}{\partial \bar{t}} = \frac{D_{c,i}}{D_a}\frac{\partial^2 \bar{c}_i}{\partial \bar{x}} + \upsilon_c p_i \mathrm{Da} \cdot \bar{a}\bar{b} \tag{8.10}$$

where we have introduced dimensionless variables $\bar{a} = a/a_0, \bar{b} = b/a_0, \bar{c} = c/a_0, \bar{t} = tD_a/L^2, \bar{x} = x/L$, and the Damkohler number (Section 4.2) $\mathrm{Da} = kL^2 a_o/D_a$. The υ's are the stoichiometric coefficients, such that $\upsilon_a = -3$, $\upsilon_b = -1$ and $\upsilon_c = \nu_{Sn_3Pd}/V_{avg}$, where $\nu_{Sn_3Pd} \approx 96\,\mathring{A}^3$ is the volume of the 3:1 Sn/Pd complex and $V_{avg} = 8\pi R^2_{avg}$ is the average volume of the colloidal seeds; these estimates give $\upsilon_c \approx 0.001$.

Before solving these governing RD equations, we need to specify the appropriate initial and boundary conditions. Prior to soaking in the activating solution, there are neither palladium ions nor Pd/Sn colloids present in the gel, such that $\bar{b}(\bar{x}, 0) = 0$ and $\bar{c}_i(\bar{x}, 0) = 0$. As described by equation (8.8), the concentration of Sn^{2+} ions in the gel directly after washing is given by $a(\bar{x}, 0) = (4/\pi)\exp(-\pi^2 \bar{t}_{wash}/4)\ \sin(\pi\bar{x}/2)$. Boundary conditions at the gel surface ($\bar{x} = 1$) are $\bar{a}(0, \bar{t}) = 0$, $\bar{b}(0, \bar{t}) = b_0/a_0$, $\bar{c}_i(0, \bar{t}) = 0$ and approximate the plating solution as a perfect sink for the colloidal seeds that diffuse out of the gel. Finally, by symmetry, the fluxes at the center of the slab ($\bar{x} = 0$) are all zero: $(\partial\bar{a}/\partial\bar{x})_{x=0} = (\partial\bar{b}/\partial\bar{x})_{x=0} = (\partial\bar{c}_i/\partial\bar{x})_{x=0} = 0$. With these assumptions the RD equations can be integrated numerically using, for example, the forward time centered space (FTCS) finite difference integration scheme (Section 4.4).

The time-dependent solutions of these equations for the distribution of the Pd/Sn particles in the gel are bell-shaped curves widening towards the gel/solvent interfaces (Figure 8.6(a), bottom plot). Curves corresponding to smaller particles widen more rapidly than those corresponding to larger ones. This means that for short activation times (e.g., $t_{activ} = 15$ min), it is mostly the small particles that reach the gel surface; for longer times (e.g., $t_{activ} = 120$ min), the small particles have already diffused out of the gel, and it is mostly the larger ones that are found at the gel/solution interface. These effects are quantified in Figure 8.6(b), which shows size distributions of particles crossing the gel boundary calculated for two different washing and activation times.

Importantly, the sizes of the catalytic seeds presented at the surface are proportional to the degree of surface roughness. This trend is illustrated in Figure 8.6(c) showing statistics of the heights of surface deformations (measured by atomic force microscopy) for four copper foils prepared under different experimental conditions. The foils formed on stamps that were soaked for t_{activ} 15 min were substantially smoother than those prepared on stamps soaked for $t_{activ} = 120$ min.

Figures 8.6b and 8.6c indicate that the washing time – though less important than the activation time – also plays a role in controlling films roughness. For long washing times, Sn^{2+} becomes increasingly depleted from the gel, and the Pd/Sn particles are generated deeper in the gel bulk. Consequently, during the plating, it takes longer for the smaller particles to diffuse out of the gel completely, and the films are less rough than those obtained for shorter washing times (e.g., compare $t_{wash} = 120$ min, $t_{activ} = 15$ min with $t_{wash} = 15$ min, $t_{activ} = 15$ min).

8.2 CUTTING INTO HARD SOLIDS WITH SOFT GELS

Let us now examine a RD process in which a gel substrate is used not as a mere substrate for the deposition, but as a 'tool' for cutting into solids. Can a piece of jelly cut into metal, or glass, or semiconductor? Surprisingly, it can, provided it is soaked in a substance that dissolves/etches the solid. In this respect, agarose is a very useful 'jelly' as it can withstand a variety of highly potent etchants (e.g., HF for glass or $HCl/FeCl_3$ for glass, see Chapter 5). This chemical robustness of micropatterned agarose stamps allows for printing useful micro- and nanoarchitectures into solid materials.[18–20]

Consider a familiar wet stamping arrangement shown in Figure 8.7, in which a micropatterned agarose stamp is soaked with a solution of water-based etchant and placed on a substrate to be etched. If the substrate's surface is hydrophilic (e.g., indium–tin oxide or glass) such that it promotes spreading of the etchant, the entire assembly is placed under mineral oil containing a small amount of surfactant. The surfactant helps the oil to penetrate in between the features of the stamp, thus preventing any spilling or lateral spreading of the etchant and limiting etching to the areas of contact between the stamp's microfeatures and the substrate (in addition, the oil minimizes etchant evaporation and stamp drying).

When the tops of the stamp's features come into conformal contact with the substrate, etching of the solid commences. In this process, the stamp's bulk acts as a two-way chemical 'pump', simultaneously supplying fresh etchant to (orange arrows in Figure 8.7(a)) and removing reaction products from (violet arrows) the gel/substrate interface. The net result of these two processes is that the stamp 'sinks' into the substrate. In doing so, it causes only minimal horizontal etching of the substrate since the sidewalls of the stamp's features lose physical contact with the walls of the wells they create – in other words, etching continues only at the interface between the tops of the features and the bottoms of the indentations in the substrate. Ultimately, the stamp imprints its surface micropattern into the substrate material.

Before we turn to the practical aspects of this method, let us have a brief look at the RD equations of this system.

8.2.1 Etching Equations

As a result of the etching reaction at the stamp/solid interface, a gradient of etchant concentration is established throughout the stamp. Mass conservation requires that the rate of etchant consumption at the interface be exactly balanced by the diffusive flux of etchant onto that surface, $D(\nabla c)_s \cdot \boldsymbol{n} = kc_s$, where we have assumed a

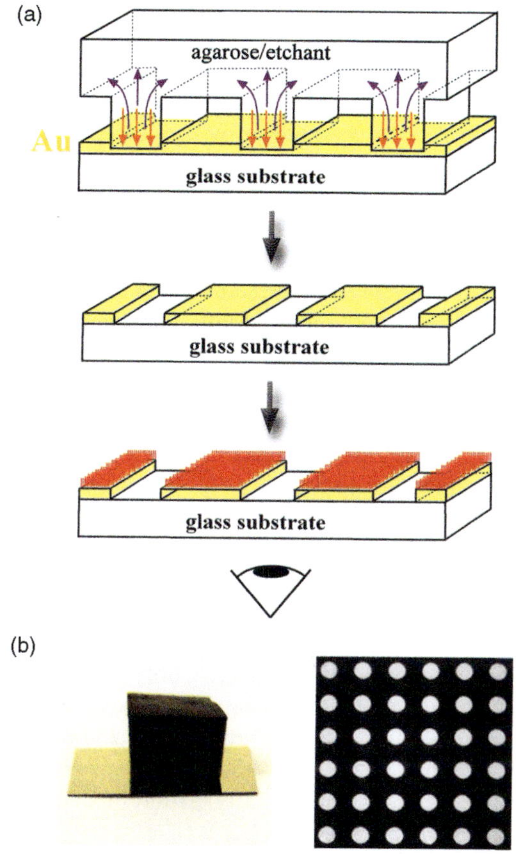

Figure 8.7 Reaction–diffusion microetching. In this example, an agarose stamp cuts into a thin layer of gold supported on glass. (a) Scheme of the process. (b) Left panel shows the actual experiment (the stamp soaked with etchant is dark brown). The right panel shows an array of etched circles, 40 μm in diameter. Reproduced from reference 26 with permission, copyright 2007, The Royal Society of Chemistry

first-order etching reaction (cf. Section 8.1.1). Let us approximate the area being etched as a planar surface in contact with a semi-infinite stamp/reservoir, which initially contains a uniform concentration, c_o, of etchant. The concentration profile within the stamp, $c(x, t)$, is governed by the one-dimensional time-dependent diffusion equation, $\partial c/\partial t = D\partial^2 c/\partial x^2$, with initial condition $c(x,0) = c_o$, boundary condition far from the interface $c(\infty,t) = c_o$, and boundary condition describing the etching reaction $kc(0,t) = D(\partial c/\partial x)_{x=0}$. In terms of dimensionless variables $\bar{c} = c/c_o$, $\bar{x} = kx/D$ and $\bar{t} = k^2 t/D$, the solution to this problem is given by $\bar{c}(\bar{x},\bar{t}) = \mathrm{erf}(\bar{x}/2\sqrt{\bar{t}}) + \exp(\bar{x}+\bar{t})\mathrm{erfc}(\bar{x}/2\sqrt{\bar{t}}+\sqrt{\bar{t}})$, where erf( ) and erfc( ) are the error and complementary error functions (Example 2.1 provides a hint as to how to verify this result). Using this equation and noting that the rate of etching (depth, H, per unit time) can be expressed as $R(\bar{t}) = \mathrm{d}H(\bar{t})/\mathrm{d}\bar{t} = \nu_s k c_o \bar{c}(0,\bar{t})$ or

$R(\bar{t}) = \nu_s kc_o(\partial\bar{c}/\partial\bar{x})_{x=0}$, where ν_s is the molar volume of the solid, we obtain $R(\bar{t}) = \nu_s kc_o \exp(\bar{t})\text{erfc}(\sqrt{\bar{t}})$.

Having solved for the general case, we may now investigate the extreme cases where the etching rate is controlled either by the delivery of fresh etchant from the stamp's bulk or by the speed of the interfacial reaction; these asymptotics will be useful when treating specific experimental systems. For times much longer that the characteristic time, $t \gg D/k^2$ or $\bar{t} \gg 1$, the etchant concentration at the surface is maintained close (but not equal!) to zero by the etching reaction, which is 'fast' relative to etchant diffusion (Section 8.1.1). In this case, it is the diffusion of fresh etchant that limits/controls the etching rate, which can be approximated as $R(\bar{t}) = \nu_s Dc_o/\sqrt{\pi\bar{t}}$. This approximation may be derived either by (i) solving the diffusion equation above with the boundary condition $c(0,t) = 0$ at the surface, or (ii) from the general etching rate expression by noting that $\text{erfc}(\sqrt{\bar{t}}) \approx \exp(-\bar{t})/\sqrt{\pi\bar{t}}$ for $\bar{t} \gg 1$. Similarly, for times much smaller that the characteristic time, $t \ll D/k^2$ or $\bar{t} \ll 1$, the etching process is limited by the etching reaction while the concentration profile of the etchant in the stamp is approximately uniform – i.e., there are almost no concentration gradients. In this case, the etching rate is approximated as $R(\bar{t}) = \nu_s kc_o$. Again, the approximation may be derived (i) by solving the diffusion equation with the boundary condition $(\partial c/\partial x)_{x=0} = 0$ at the surface, or (ii) by Taylor-expanding the general etching rate expression to zero order (next order correction is $\sim 0.1\bar{t}$). It is important to note that these limiting cases are derived in the case of a semi-infinite domain with no inherent length scale L. For finite sized systems/stamps, characterized by a length L, the conditions for 'diffusion-limited' and 'reaction-limited' etching are slightly different and given by $kL/D \gg 1$ and $kL/D \ll 1$, respectively.

8.2.1.1 Gold etching

To see how these methods are used in practice, consider the case of a gold substrate etched using an aqueous mixture of iodide and triiodide as an etchant. Here, the etching reaction is given by $2\text{Au(s)} + \text{I}^- + \text{I}_3^- \rightarrow 2[\text{AuI}_2]^-$, where I_3^- is the limiting reagent, and the kinetics may be approximated as $r = k[\text{I}_3^-]$ with $k \sim 7 \times 10^{-5}\ \text{cm s}^{-1}$. The diffusion coefficient of the etchant within the gel stamp is given by $D \sim 10^{-5}\ \text{cm}^2\ \text{s}^{-1}$, such that $D/k^2 \sim 1$ h, which is comparable to the experimental etching times of gold films tens of micrometers thick. Therefore, gold etching is neither diffusion-limited nor reaction-limited, and we must use the general method described above, in which the boundary condition at the surface is given by $kc(0,t) = D(\partial c/\partial x)_{x=0}$ and the etching rate is given by $R(t) = \nu_{\text{Au}} kc_o \exp(k^2t/D)\text{erfc}(\sqrt{k^2t/D})$. Integrating this expression, we find that the etching depth is given by

$$H(t) = (\nu_{\text{Au}} Dc_o/k)[2\sqrt{k^2t/\pi D} + \exp(k^2t/D)\text{erfc}(\sqrt{k^2t/D}) - 1] \quad (8.11)$$

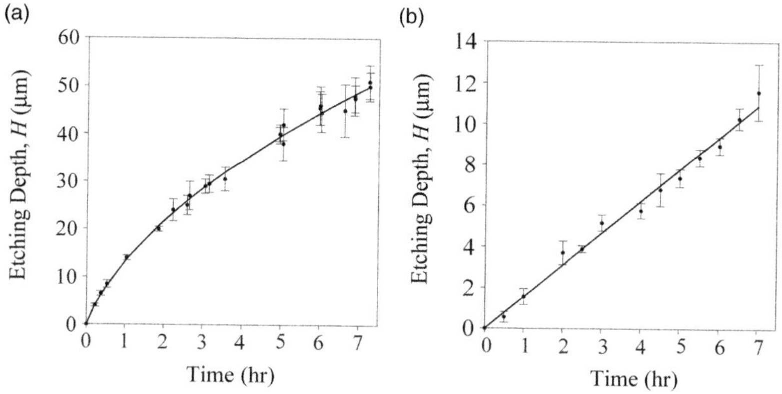

Figure 8.8 Depth of etched gold as a function of time for (a) gold substrate (25% aqueous solution of TFA etchant, Transene Company) and (b) glass substrate (0.6 M HF etchant). Data points correspond to experimental results; the line gives a theoretical fit. Reproduced with permission from Stoyan *et al.*, Freestanding Three-dimensional Copper Foils Prepared by Electroless Deposition on Micropatterned Gels, Advanced Materials, 2005, copyright Wiley-VCH

Figure 8.8(a) compares the experimental and theoretical etching depth of a gold film as a function of time; the agreement is excellent!

8.2.1.2 Glass and silicon etching

Glass and silicon surfaces can be etched by hydrofluoric acid (HF), for which the first-order etching rate is considerably smaller than that for metals – e.g., $k \sim 10^{-8}\,\mathrm{cm\,s^{-1}}$ for silicon dioxide.[21] Therefore, the characteristic time scale, $D/k^2 \sim 10^{11}\,\mathrm{s}$, is much larger than any experimentally relevant time scale, and reaction is the rate-limiting process. Following the arguments in Section 8.2.1., the etching rate is then given by $R(t) = \nu_s k c_o$, which may be integrated to give $H(t) = \nu_s k c_o t$. In other words, etching proceeds at a constant rate, with the etching depth increasing as a linear function of time (Figure 8.8(b)). Furthermore, because the etching process is entirely reaction-controlled, the geometry of the stamp/substrate interface has no effect on the etching process – wherever the stamp contacts the substrate, the dissolution reaction proceeds at a constant rate, and the topography of the etched relief is identical to that of the micropattern in the stamp. This will be important in the context of etching curvilinear and multilevel structures discussed later in this chapter.

With the model having been covered, we can now talk some useful microfabrication.

8.2.2 Structuring Metal Films

There are many good reasons to microstructure metal films: thin metal foils with holes can be used as membranes (Figure 8.9(a)) for separating small particles, disjoint metal

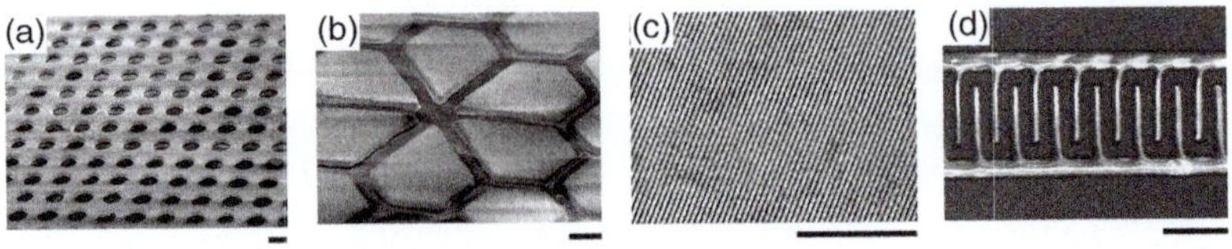

Figure 8.9 Etching micropatterns into metals. (a) Array of circular holes etched through a freestanding, 100 μm thick copper foil; the holes correspond to disjoint, raised features in the gel stamp. (b) Disjoint metal plates obtained from the same foil; the lines along which the foil is 'cut' correspond to the network of connected, raised lines in the stamp. (c) Array of 1.5 μm lines spaced by 2.5 μm etched in a 100 nm thick layer of nickel on glass. (d) Interdigitated electrodes etched in a 150 nm thick layer of gold on polystyrene. All scale bars = 50 μm. Reproduced with permission from Stoyan *et al.*, Freestanding Three-dimensional Copper Foils Prepared by Electroless Deposition on Micropatterned Gels, Advanced Materials, 2005, copyright Wiley-VCH

plates (Figure 8.9(b)) are building blocks of many micromechanical devices, arrays of very thin metallic lines (Figure 8.9(c)) can polarize electromagnetic radiation, and metallic micropatterns on dielectrics (Figure 8.9(d)) are essential components of electronic circuits. In making such structures, our RD method offers speed and simplicity without the need to protect the unetched regions or to use high-power lasers. The structures shown in Figure 8.9 were all fabricated on a bench-top by a straightforward application of a gel stamp soaked in appropriate commercial etchants (for gold, 25% water solution of TFA etchant (Transene Company, Danvers, MA); for copper and nickel, 35% water solution of $FeCl_3$-based PCB RadioShack etchant, Cat. 276-1535; for iron, 5–10% HNO_3; in all cases stamps were soaked for 2 h).

One other exciting use of RD microetching is its combination with self-assembled monolayers (SAMs) for biological applications.[22] It has been known for several decades that molecules known as alkane thiols form extremely regular, one-molecule-thick layers on metals such as gold, silver, copper, palladium and platinum[23] (Figure 8.10).

These layers not only mask the underlying metal, but also modulate effective surface properties depending on the thiol terminal functionality (X in Figure 8.10(a)). For instance, when X is a hydrophilic group such as OH or COOH, the surface

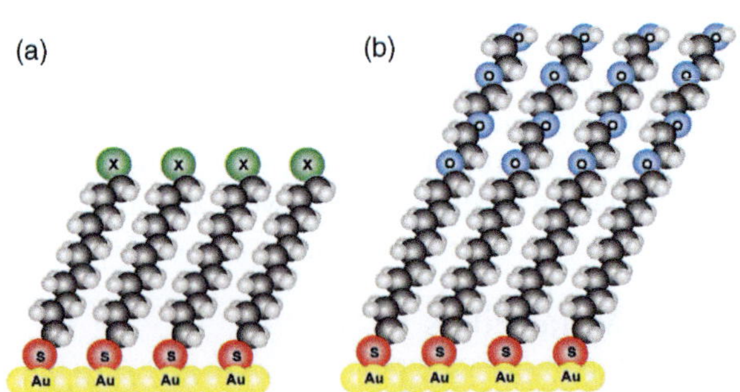

Figure 8.10 Schemes of self-assembled monolayers (SAMs) of alkane thiols on gold. The surface properties of the SAM depend on the terminal group denoted X in (a). The SAM in (b) presents EG_3 functionalities known to resist adsorption of proteins and/or cells

becomes hydrophilic and water spreads on it readily; when X bears an electric charge (e.g., $N(CH_3)_3^+$ or SO_3^-), the surface captures oppositely charged particles; when X is a biospecific group such as biotin, the surface can capture protein constructs containing avidin subunit.[23] With a judicious choice of X, one can engineer almost any imaginable surface property! In the 1990s, the groups of George Whitesides and Don Ingber at Harvard discovered that when X is an oligo(ethylene glycol) ($(OCH_2CH_2)_n$, abbreviated as EG_n; Figure 8.10(b)) the SAM surface prevents adsorption of proteins and/or cells.[24,25] Combination of this property with surface patterning creates a powerful tool for cell biology. To see why, consider a metal film (usually gold supported on glass) whose surface has been covered with EG_n thiols everywhere except small, disjoint 'islands'. When cells are applied onto such a surface, they avoid the adhesion-resistant EG_n SAMs and localize exclusively onto the islands. Once there, the cells spread until they hit the island edges and ultimately assume the island shapes. If the islands are circular, the cells have circular shapes; if the islands are triangular, the cells are also triangles (Figure 8.11).

Importantly, the imposition of such unnatural geometries is equivalent to imposing mechanical stresses on the cells. These stresses, in turn, can cause the reorganization of cell inner components and can alter cell function. Over the last decade, cell micropatterning has been used extensively to study and control a wide range of cellular functions (see Ref. 26 and references therein) including organization of cytoskeleton (Example 8.2), apoptosis (i.e., cell 'suicide'), differentiation, cell polarization, propagation of viruses in micropatterned assemblies of cells as well as mitotic spindle orientation and cell adhesion.

Despite these spectacular achievements, the SAM-on-gold approach suffers from one serious limitation – it can be used to analyze only fixed (i.e. 'dead') cells and is not suitable for real-time and 3D analyses of cellular and intracellular processes. For technical reasons, high-resolution optical imaging of cells (usually

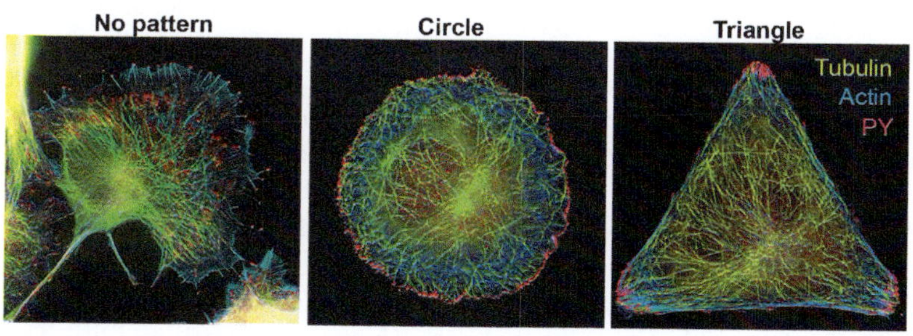

Figure 8.11 Cancer B16F1 cells cultured on substrates without a pattern and on substrates patterned with circular and triangular microislands. Immunofluorescence staining of fixed, micropatterned cells reveals spatially separated components of the cytoskeleton: focal adhesions (pY, red) which are the 'sticky pads' through which cells adhere to the surrounding world; microtubules (tubulin, green), which are constantly growing and shrinking fibers loosely termed cell 'bones'; and actin filaments (blue), which are cell tensile elements ('tendons'). The sticky pad–bone–tendon terminology should not be used in the presence of serious biologists. Reproduced from reference 26 with permission. Copyright 2007, The Royal Society of Chemistry

by means of fluorescence spectroscopy) has to be done through the flat substrate/cell interface which does not distort/deflect the probing beam. Unfortunately, in the gold/SAM systems, even ultrathin gold films supporting the SAMs attenuate fluorescence and prevent high-quality imaging of fluorescently labeled cellular components in *living* micropatterned cells. The best one can do is to fix the cells and then stain them with copious amounts of biospecific dyes to obtain 'static' images. Clearly, imaging cell cadavers is not very exciting.

The solution is to somehow make the islands transparent. One way is to pattern the EG_n SAM first and then etch the unprotected metal over the island regions. The EG_n thiols, however, are too chemically sensitive to withstand harsh etching conditions and they lose their cell-repellant properties in the process. A much more efficient approach is to use our RD microetching method, which selectively removes gold from the island regions without affecting the rest of the metal film. Once the 'windows' into the gold layer are etched (see Fig. 8.7b), the remaining portion of the surface can be modified with a cell-resistant EG_n SAM.

Transparency of the islands opens some quite unprecedented possibilities. Not only do the islands now create arrays of morphologically identical 'designer' cells whose internal components can be manipulated by the imposed geometries, but they permit *real-time* imaging of the *dynamic* events taking place *inside* such cells. One instance in which these capabilities combine to answer an important biological question related to cancer is discussed in Example 8.2. Other geometries and more biology can be found in reference 26.

Example 8.2 RD Microetching for Cell Biology: Imaging Cytoskeletal Dynamics in 'Designer' Cells

One of the major uses of the micropatterned cells is in manipulating cytoskeletal components. The cytoskeleton is a maze of protein fibers of various types that endow cells with mechanical robustness and, at the same time, allow them to migrate via a series of extension–contraction cycles (Fig 1). Cell migration/motility is, in turn, of paramount importance in cancer metastasis, where it enables invasive cells to detach from primary tumors and travel to other body parts to start new loci of disease – usually with fatal consequences. Understanding how cytoskeleton components interact with one another in space and time could thus bring us closer to understanding – and later eliminating – cell migration in metastatic cancer.

Somewhat paradoxically, the ability to immobilize live cells on micropatterned islands provides a unique tool with which to study cell motility. This is because the patterned cells continue to operate their motility machinery, and the cells behave as if placed on a treadmill (by analogy to a human running on a treadmill). At the same time, the imposed shape directs the otherwise spatially entangled cytoskeletal components to specific locations within an island (Figure 8.12 in the main text), which greatly simplifies their analysis.

In this example, we will illustrate how placing cells on transparent, triangular microislands etched by RD into gold can help answer an important (and outstanding) cell biological question: namely how the highly dynamic cystoskeletal fibers called microtubules (MTs) find/target cell's focal adhesions (FAs; see Fig. 1). The importance of this targeting process lies in the fact that when the MTs growing from the cell's centrosome find the FAs at the rear end of a moving cell, they cause FA disassembly and thus allow a motile cell to detach from the substrate and crawl forward. What we wish to find out is whether the MTs explore all possible directions randomly before finally localizing onto the FAs, or are they somehow guided towards them? If the first scenario were true, one could slow down metastatic cells by preventing the growth of microtubules; if, however, the MTs are somehow guided toward the FAs, we should probably try to disrupt the guiding mechanism, not the MT growth itself.

In unpatterned cells, the FAs are spaced densely and distinguishing random from guided MT growth is not possible. In sharp contrast, in triangular cells, all focal adhesions localize to the vertices of the triangle (Figure 8.12 in the main text). Given this 'discrete' localization, the mechanism of MT/FA targeting can be readily established by monitoring the trajectories of the growing MTs. The figure 2 (and the corresponding movies available at http://www.dysa.northwestern.edu/Research/Celldynamics.dwt) quantifies these trajectories by the angles, θ, between instantaneous growth directions of MTs and the lines connecting MT ends at a given instant of time with the nearest vertex of the triangle (Fig. 2c). This metric easily distinguishes between random and guided growth modes – the fact that the $f(\theta)$ versus θ probability distribution plots over all annular regions Fig. 2 peak around $\theta = 0°$ indicates that MTs are indeed guided towards the FAs, and that the degree of guidance increases as the MT ends move away from the cell center (compare distributions for different annular regions). Although the cause of guidance remains unknown, the microetched islands allowed us to solve at least part of the puzzle.

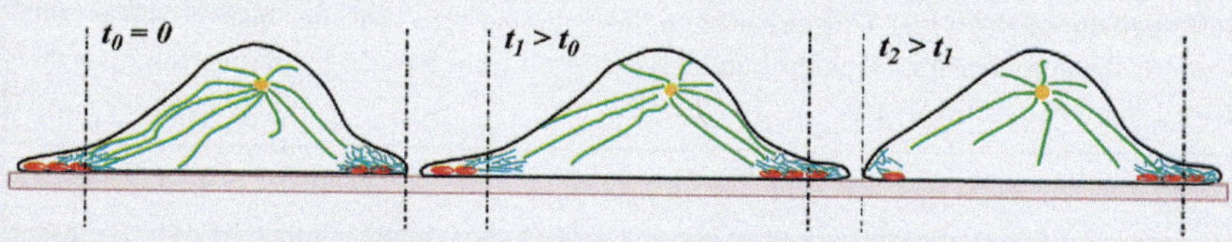

Figure 1 Cell motility – a prime. After the cell breaks its spatial symmetry (i.e., polarizes) to define its 'leading' and 'rear' regions, actin filaments (blue) near the leading edge push the cell membrane to form a protrusion called lamellipodium that adheres to the substrate via 'sticky' focal adhesions (red). These events are accompanied by the disassembly of focal adhesions in the rear in a process that is facilitated by transient interactions of the adhesions with the ends of the microtubules (green) growing from the cell centrosome (yellow). Overall, the cell 'crawls' forward via a series of push-and-pull motions caused by constant reorganization of the cell cytoskeletal machinery

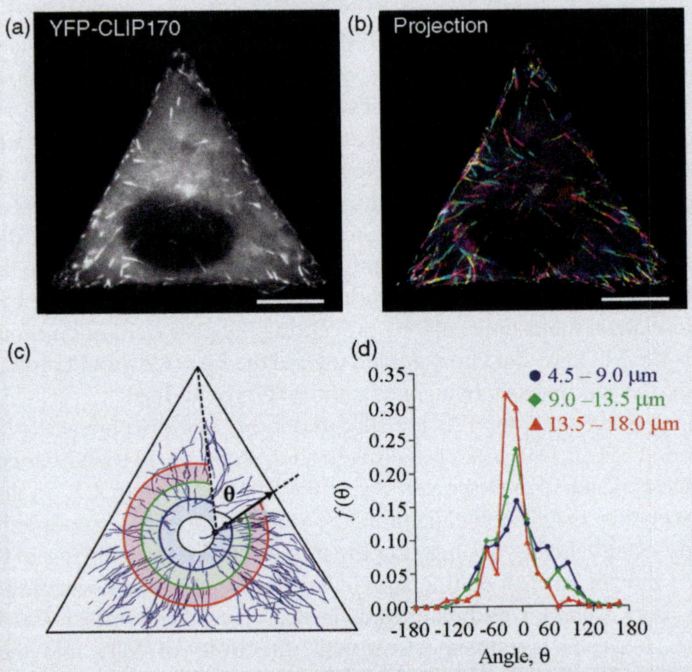

Figure 2 above shows live-cell molecular dynamics imaging in cells immobilized onto microetched, triangular islands. (a) A single frame in time-lapse series of a B16F1 cell having the end of the growing microtubules labeled with fluorescent YFP proteins (bright spots). Scale bar, 10 m. (b) Superposition of multiple consecutive frames (here, 12) gives representative MT growth trajectories. The color scheme corresponds to the elapsed time (from blue through yellow to red). Scale bar, 10 m. (c) Blue curves show MT growth trajectories. The arrow gives instantaneous direction of growth of one MT end – that is, a vector parallel to the line joining the ends positions (colored dots) at two consecutive times. (d) Probability distributions of growth directions within annular shells around a centrosome (cf. shaded regions in (c)). Clustering of angles, , around zero indicates bias of MT growth towards the corners/FAs. (Figures 1 and 2 reproduced from Ref. 26 by permission from the Royal Society of Chemistry)

8.2.3 Microetching Transparent Conductive Oxides, Semiconductors and Crystals[19]

Etching micropatterns in transparent conducting oxides such as indium–tin oxide (ITO) or zinc oxide (ZnO) and in semiconductors (e.g., GaAs) is of great importance for the fabrication of optoelectronic devices (ITO electrodes), sensors and on-chip UV lasers (ZnO), as well as integrated circuits, solar cells and optical switches (GaAs). Since all of these applications rely on the ability to define pertinent microscopic architectures, a variety of methods have been developed to micropattern these materials. Serial methods based on laser scanning or focused ion beam (FIB) techniques through protective "masks" offer direct maskless and resistless patterning, but require expensive equipment and are relatively slow, especially at high resolutions. Parallel etching techniques through protective "masks" (wet etching, reactive ion etching) are usually more rapid and less expensive but give rise to pattern underetching, and the solvents used to remove the masking material place severe compatibility restrictions on the possible plastic/ flexible substrate materials.

The familiar question is: can RD offer any advantages? It depends. On the one hand, it is utterly unrealistic to expect that agarose stamps will somehow rival the automated and efficient infrastructure of the microelectronics industry. On the other hand, RD can be a useful method of bench-top prototyping available to scientists with even rudimentary experimental facilities. Let us have a look at some optimized RD microetching formulations.

1. **ITO.** Microetching of polycrystalline ITO films is usually not an easy affair, requiring relatively harsh conditions. Since ITO is degraded mostly by undissociated halogen acids, procedures using HCl require high etchant concentrations and elevated temperatures, causing agarose to lose its structural integrity and even dissolve. A mixture of 2 M HCl (17% of assay)/0.2 M $FeCl_3$ minimizes these complications and allows for etching at room temperature while maintaining the stamp structural rigidity for several hours. Because this formulation is compatible with glass substrates (i.e., it does not degrade them), it can be used to pattern thin (100–200 nm) ITO/glass substrates at etch rates of about 30 nm h^{-1}. For ITO supported on polymeric substrates (e.g., poly(ethylene terephthalate), PET) it is possible to use more powerful HF etchants, such as 2:1:100 HF:H_2O_2: H_2O, which give etch rates of about 600 nm h^{-1}. This procedure, however, requires extreme caution as HF is able to painlessly penetrate skin, extract calcium from bones and destroy soft tissues. Working with HF requires – and we mean always! – special Norfoil gloves, antidote calcium gluconate next to the hood and familiarity with exposure symptoms and emergency response procedures.

Since the ITO surface is hydrophilic, both types of etchants spread on it thus reducing the spatial resolution of etching to ~50 μm. As mentioned at the beginning of Section 8.2, placing the substrate and the stamp under light mineral oil containing 1:1000 v/v Triton X-100 surfactant limits lateral spreading and improves resolution to several micrometers, adequate for rapid (tens of minutes to hours) prototyping of microelectrode systems (Figure 8.12(a)), RFID tags or low-end displays.

2. **ZnO.** In contrast to ITO, zinc oxide is relatively easy to etch with agarose stamps soaked in a 1:1:300 (v/v/v) mixture of acetic acid, phosphoric acid and water. Since agarose tolerates this dilute-acid etchant well, the stamps can be

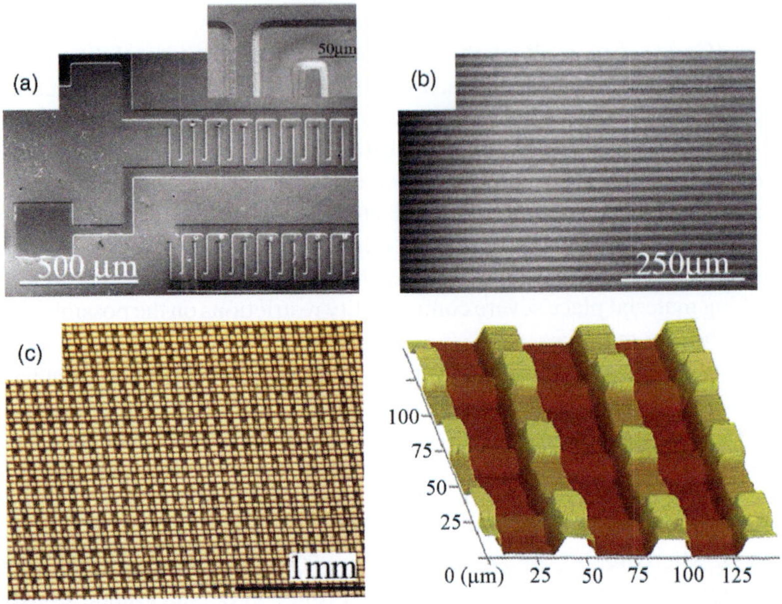

Figure 8.12 (a) Scanning electron microscope (SEM) image of an array of interdigitated microelectrodes etched into ITO on glass. (b) SEM image of an array of 10 μm lines etched in ZnO supported on aluminum. (c) Optical (left) and atomic force microscope (right) images of a multilevel structure etched into GaAs by sequential application of arrays of parallel lines along perpendicular directions. Reprinted from reference 19, copyright (2006), American Chemical Society

reused multiple times (>20 and even after prolonged soaking in the etchant solution) without noticeable reduction in patterning quality. In addition, the etchant does not wet the ZnO surface, so that stamps do not need to be immersed in oil but simply placed onto the substrate to fabricate patterns (Figure 8.12(b)) with resolution down to ~200 nm (below this limit, effects associated with the porosity of the agarose matrix become dominant, and the features are ill-defined). Commercially available ZnO films (100–300 nm thick) are etched completely and uniformly within tens of seconds over areas of several square centimeters.

3. **GaAs.** Finally, a ‘dream-case’ substrate that is etched at ~30 $\mu m\, h^{-1}$, with spatial resolution down to ~200 nm and over areas of several square centimeters using stamps soaked in 1:1:20 v/v/v mixture of H_2SO_4:H_2O_2:H_2O. Since the etchant does not wet GaAs, no immersion in oil is necessary, and even deep patterns are etched within minutes by simply placing the stamp on the substrate. With the ability to control etch depth, sequential application of stamps can give multilevel structures such as that shown in Figure 8.12(c).

With proper optimization of the etchant/stamp composition, RD is able to deal with other substrates for which water-based etchants are known[20] (e.g., InGaAsP, InP, Si, SiO_2). While going over the myriad of possible etchant/substrate combinations is not the objective of our discussion, we will discuss one more, very important system in detail.

8.2.4 Imprinting Functional Architectures into Glass

Optical transparency combined with mechanical and chemical robustness make glass a material of choice for a range of applications in optics and microfluidics, often in the form of curvilinear (e.g., lenses) or multilayer (e.g., fluidic mixers) architectures. In 8.2.1.2 we showed that microetching of glass initiated from agarose stamps is entirely reaction-controlled indicating that the geometry of the stamp/substrate interface should have no effect on the etching process, and the microstructure etched in glass should reproduce the topography of the agarose stamp faithfully. This observation creates some exciting opportunities for fabrication. Recall from Chapter 6 that gels and soft materials can be decorated with curvilinear and/or multilevel surface reliefs using either RD or multistep photolithography. Now, with the RD microetching technique, we could use these easy-to-make substrates to first make appropriate agarose stamps and then simply imprint their topographies into glass.[20] This imprinting procedure based on HF etchant is illustrated in Figure 8.13 and is performed under oil to minimize the etchant spreading over the hydrophilic glass surface. Note that the optimal arrangement for this microetching system is one in which the stamp is placed upside down and the glass substrate is placed onto it, often with small additional weight ($\sim$50 g) to ensure conformal contact between the two materials and prevent sliding of the agarose stamp on the surface.

The concentration of HF around 0.6 M does not markedly affect the stamp's mechanical integrity for up to several days and yields constant etching rates of about 2 $\mu m\, h^{-1}$, adequate for he fabrication of both microoptical components and microfluidic systems. For example, Figure 8.14(a) shows arrays of both convex and concave microlenses imprinted into glass by agarose stamps replicated from commercially available epoxy masters. The etched substrates reproduce the topography of the epoxy originals down to less than 500 nm and focus light as efficiently as the masters themselves. Another example is illustrated in Figure 8.14(b), where an agarose stamp imprints a passive microfluidic mixer that utilizes 'herring-bone' ridges at the bottom of the channel[4] to mix laminarly flowing liquids by chaotic advection (see Section 6.6 and references therein). When the top portion of this structure is sealed, and two streams are flowed through it, the device achieves complete mixing within $\sim$1 cm of the channel's length (compared to $\sim$20 cm for purely diffusive mixing). Finally, Figure 8.14(c) shows a complex multilevel structure, in which fluids delivered by sets of perpendicular micro-channels encounter alternating hurdles at intersections. As the streams 'collide', they not only turn by 90° but also partly mix at each intersection. As a result, the outlet channels are characterized by a continuous concentration gradient, and the system acts as a gradient diffuser.

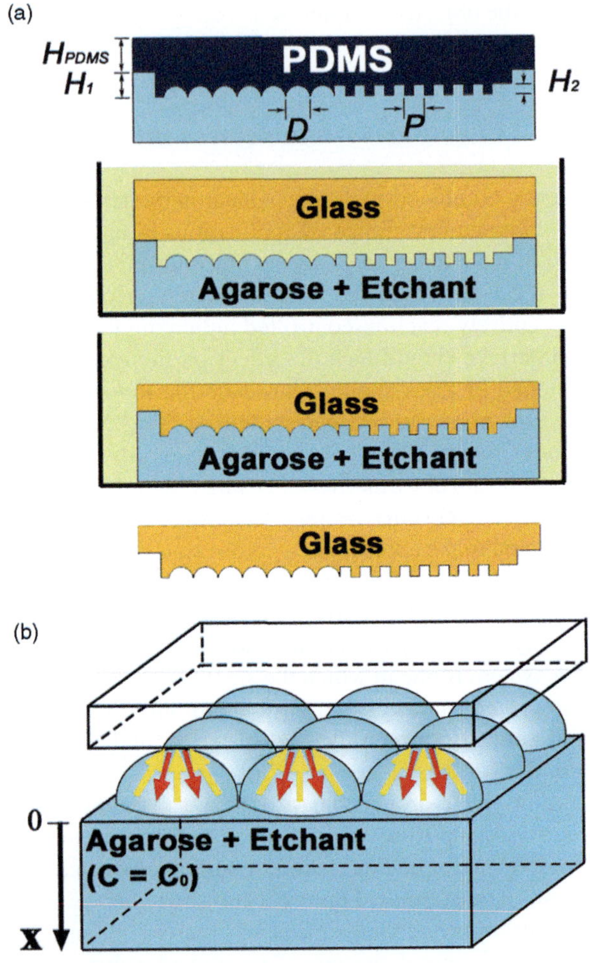

Figure 8.13 (a) Etching of curvilinear or multilevel features in glass. An agarose stamp (10% w/w high gel strength agarose) is made by casting against a curvilinear or multilevel master (typical dimensions: $H_1 = 200$ nm–150 µm; $H_2 = 0.1$–50 µm; $D = 100$–250 µm; $P = 1$–200 µm), soaked in an HF etchant/surfactant solution, inverted and immersed in light mineral oil. The substrate is then placed directly on top of the stamp. When etching is complete, the substrate is removed, rinsed with distilled water and cleaned in Piranha solution to remove oil residues. (b) During etching (here, of curvilinear microlenses), agarose supplies the HF etchant (yellow arrows) to dissolve the substrate while removing the etching products (SiF_6^{2-}; red arrows) into the stamp's bulk. Reprinted from reference 20 with permission from Wiley-VCH

In all of these demonstrations, the highest lateral resolution achieved for typical glass substrates (e.g., cover slides) is ~500 nm and the degree of surface roughness (RMS) is around 33.0 ± 13.40 nm over a 25 µm^2 area, which compares favorably with much more sophisticated methods such as reactive ion etching. The depth of

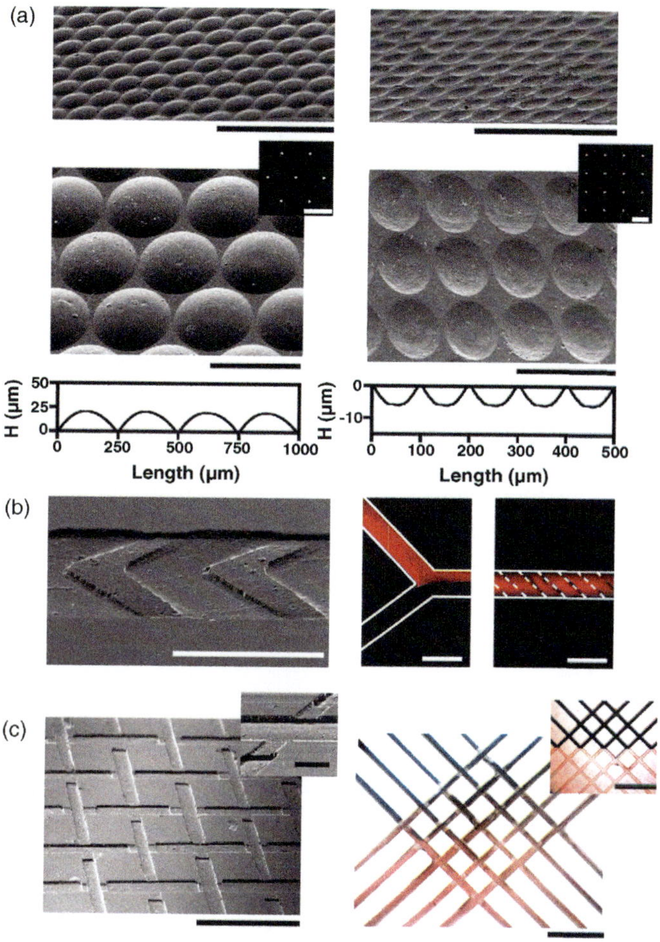

Figure 8.14 Microscale etching of curvilinear and multilevel architectures in glass. (a) Convex (left) and concave (right) microlenses etched in glass over large areas (top; left scale bar = 1 mm, right scale bar = 500 μm). Close-up views (middle) and profilometric scans (bottom) show faithful reproductions of the masters. Insets illustrate focusing of light by the lens arrays. Scale bars are 300 μm on the left and 200 μm on the right. (b) Left: SEM image of a staggered-herringbone-mixer architecture (top scale bar = 1 mm, bottom scale bar = 200 μm) printed into glass. Right: fluorescence images of Rhodamine B solution flowing in a herringbone-mixer device microetched in glass (channel depth = 75 μm, herringbone height = 20 μm, channel width = 200 μm). This system achieves complete mixing within 1 cm of the channel length (scale bar = 200 μm). (c) Left: multilevel gradient-diffuser array etched in glass and composed of intersecting microchannels (channel depth = 20 μm, channel width = 100 μm) with alternating hurdles at intersections (hurdle height = 8 μm; scale bar = 1 mm; inset scale bar = 100 μm). Right: when aqueous solutions of Coomassie Blue and Rhodamine are flowed in at 1 $mL\,h^{-1}$ from the upper and lower left, respectively, they undergo a 90° turn while mixing at the hurdles. As a result of this local mixing, the fluids in the outlet channels have continuous concentration (and color) gradients – blue-to-violet in the upper-right outlets and pink-to-violet in lower-left outlets. In channels without the hurdles (right inset), the fluids undergo only a 90° turn without mixing. (Scale bar = 1 mm; inset scale bar = 1 mm). Reprinted from reference 20, with permission from Wiley-VCH

the etched features can be up to several hundreds of micrometers deep and is limited predominantly by the rigidity/integrity of agarose, and not by the nature of the RD process.

8.3 THE TAKE-HOME MESSAGE

The message from this chapter is that RD can make a soft gel into a razor-sharp tool for cutting into various types of solids with micrometer and even submicrometer precision. RD microetching, which combines the ease and flexibility of replicating soft materials with the durability and other desirable properties of solids, should prove particularly useful to scientists and engineers wishing to rapidly prototype and test various small-scale structures in hard substrates for uses in microfluidics, microoptics and microelectronics. Of course, the method can no doubt be optimized further by identifying porous materials (gels, polymers) that are mechanically more rugged than agarose (to increase spatial resolution) and support more concentrated etchants (to allow for higher etch rates).

REFERENCES

1. Alodan, M.A. and Stover, L.E. (1999) Selective plating without masking. *Electrochim. Acta*, **44**, 3721.
2. Coutrot, A.L., Dufour-Gergam, E., Quemper, J.M. *et al.* (2002) Copper micromoulding process for NMR microinductors realization. *Sens. Actuators A*, **99**, 49.
3. Osaka, T., Takana, N. and Yokoshima, T. (2003) Microfabrication of electro- and electroless-deposition and its application in the electronic field. *Surf. Coat. Technol.*, **169**, 1.
4. Zaumseil, J., Meitl, M.A., Hsu, J.W.P. *et al.* (2003) Three-dimensional and multilayer nanostructures formed by nanotransfer printing. *Nano Lett.*, **3**, 1223.
5. Brittain, S.T., Sugimura, Y., Schueller, O.J.A. *et al.* (2001) Fabrication and mechanical performance of a mesoscale space-filling truss system. *J. Microelectromech. Syst.*, **10**, 113.
6. Ebbesen, T.W., Lezec, H.J., Ghaemi, H.F. *et al.* (1998) Extraordinary optical transmission through sub-wavelength hole arrays. *Nature*, **391**, 667.
7. Grayson, A.C.R., Shawgo, R.S., Johnson, A.M. *et al.* (2004) A BioMEMS review: MEMS technology for physiologically integrated devices. *Proc. IEEE*, **92**, 6.
8. Smoukov, S., Bishop, K.M., Campbell, C.J. and Grzybowski, B.A. (2005) Freestanding three-dimensional copper foils prepared by electroless deposition on micropatterned gels. *Adv. Mater.*, **17**, 751.
9. Mishra, K.G. and Paramguru, R.K. (1996) Kinetics and mechanism of electroless deposition of copper. *J. Electrochem. Soc.*, **143**, 510.
10. Schlesinger, M. and Paunovic, M. (2000) *Modern Electroplating*, John Wiley & Sons, Ltd, New York.
11. Press, W.H., Teukolsky, S.A., Vetterling, W.T. and Flannery, B.P. (1992) *Numerical Recipes in C: The Art of Scientific Computing*, Cambridge University Press, New York.
12. Bard, A.J. and Faulkner, L.R. (2001) *Electrochemical Methods: Fundamentals and Applications*, 2nd edn, John Wiley & Sons, Ltd, New York.

13. Cohen, R.L. and West, K.W. (1973) Generative and stabilizing processes in tin-palladium sols and palladium sol sensitizers. *J. Electrochem. Soc.*, **120**, 502.
14. Cohen, R.L. and West, K.W. (1972) Characterization of tin–palladium solutions. *Chem. Phys. Lett.*, **16**, 128.
15. Amsden, B. (1998) Solute diffusion within hydrogels. Mechanisms and models. *Macromolecules*, **31**, 8382.
16. Johnson, E.M., Berk, D.A., Jain, R.K. and Deen, W.M. (1995) Diffusion and partitioning of proteins in charged agarose gels. *Biophys. J.*, **68**, 1561.
17. Deen, W.M. (1998) *Analysis of Transport Phenomena*, Oxford University Press, New York.
18. Smoukov, S.K., Campbell, C.J., Klajn, R. *et al.* (2005) Cutting into solids with micropatterned gels. *Adv. Mater.*, **17**, 1361.
19. Smoukov, S.K. and Grzybowski, B.A. (2006) Maskless microetching of transparent conductive oxides (ITO and ZnO) and semiconductors (GaAs) based on reaction–diffusion. *Chem. Mater.*, **18**, 4722.
20. Campbell, C.J., Smoukov, S.K., Bishop, K.J.M. *et al.* (2006) Direct printing of 3D and curvilinear micrometer-sized architectures into solid substrates with sub-micrometer resolution. *Adv. Mater.*, **18**, 2004.
21. Judge, J.S. (1971) Study of dissolution of SiO_2 in acidic fluoride solutions. *J. Electrochem. Soc.*, **119**, 1772.
22. Whitesides, G.M., Ostuni, E., Takayama, S. *et al.* (2001) Soft lithography in biology and biochemistry. *Annu. Rev. Biomed. Eng.*, **3**, 335.
23. Witt, D., Klajn, R., Barski, P. and Grzybowski, B.A. (2004) Applications, properties and synthesis of omega-functionalized n-alkanethiols and disulfides: the building blocks of self-assembled monolayers. *Curr. Org. Chem.*, **8**, 1763.
24. Chen, C.S., Mrksich, M., Huang, S. *et al.* (1997) Geometric control of cell life and death. *Science*, **276**, 1425.
25. Singhvi, R., Kumar, A., Lopez, G.P. *et al.* (1994) Engineering cell-shape and function. *Science*, **264**, 696.
26. Kandere-Grzybowska, K., Campbell, C.J., Mahmud, G. *et al.* (2007) Cell motility on micropatterned treadmills and tracks. *Soft Matter*, **3**, 672.

9

Micro-chameleons: Reaction–Diffusion for Amplification and Sensing

After three chapters on forcing reaction–diffusion (RD) to make well-defined structures, we are now going to give these systems some freedom of choice. In return for this benevolence, we will expect RD to report back to us – in the form of visual patterns – the values of experimental parameters (e.g., thickness of a gel film supporting a process, equilibrium constants of reactions involved, etc.) that dictate specific choices our systems make. In doing so, we hope to capitalize on the inherent nonlinearity of the RD equations, which can potentially translate small parameter changes into large differences in the RD patterns that ultimately emerge. To put it succinctly, we will use RD processes as chemical 'amplifiers' and 'sensors'.

Although modern technology, especially optoelectronics, has used nonlinear amplification as a basis for such important devices as lasers or power and frequency amplifiers, it has not been able to apply it as broadly and flexibly as biological systems do. Indeed, in biology the coupling between inherently nonlinear (bio) chemical kinetics and the transport of chemicals makes nonlinear amplification phenomena ubiquitous at virtually all length scales. On the level of macromolecules, various ultrasensitive protein/gene regulatory cascades[1] play the role of developmental 'programs' and amplify molecular events into spatial and/or temporal patterns up to cellular[2] or even organism[3,4] scales. In humoral immune response, B lymphocytes recognize and respond to new antigens by amplifying the production of antibodies that ultimately help destroy the foreign invader.[5] In collections of microorganisms, cAMP signaling between the individual members of an ensemble translates/amplifies into their collective behaviors and visual

Chemistry in Motion: Reaction–Diffusion Systems for Micro- and Nanotechnology Bartosz A. Grzybowski

appearance.[6] Finally, among large organisms, nonlinear predator–prey dynamics can cause amplification (or extinction) of the entire species.[7]

The systems we describe in this chapter draw inspiration from a particular mode of biological amplification, in which molecular or cellular events are transformed into color and/or pattern changes (Figure 9.1). For example, chameleons change

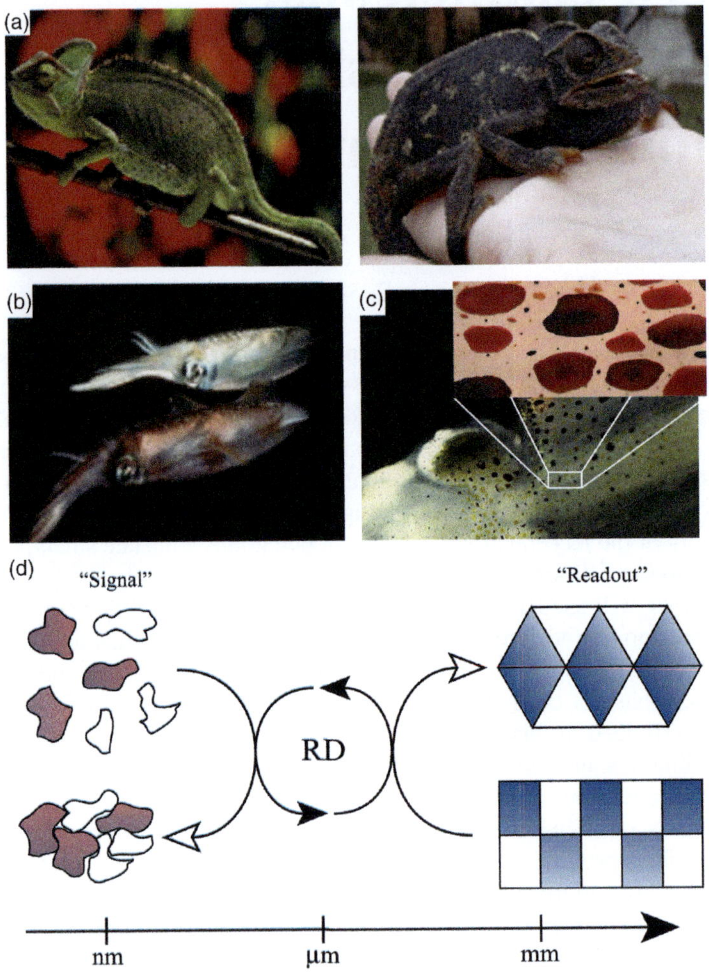

Figure 9.1 (a) Chameleons, such as this *Chamaeleo chameleon*, are green when calm (left) but turn dark brown when threatened (right). (b) Squids display different colors depending on their 'mood' and environment. (c) Close up of the skin of the squid *Lolliguncula brevis*; note the collection of light-reflecting iridophores. (d) Schematic illustration of RD amplification of a nanometer-scale 'signal' into a macroscopic, visual 'readout'. Reprinted, with permission, from *Chaos*, Microchameleons: Nonlinear chemical microsystems for amplification and sensing, Bishop *et al.* Copyright (2006), American Institute of Physics

their skin color (e.g., to indicate reproductive desires, to provoke combat or submit) as a result of endocrinal processes[8,9] leading to microscopic movements of light-reflecting cells called iridophores and the crystals within these cells. Similarly, squids respond to feelings of fear by becoming iridescent. In this case, stimulation mediated by acetylcholine[10] causes either a gel–sol phase transition in platelets contained in iridophores or a change in the platelet viscosity and thickness, with thinner platelets scattering light of lower wavelengths.[11]

Our objective is to develop artificial systems based on RD that would, in a very primitive sense, behave like chameleons or squids and would change their *macroscopic* appearance in response to underlying *microscopic/molecular* properties or processes ('signals'). To make these systems 'sense' and amplify rapidly, their dimensions should be as small as possible since the characteristic times governing diffusive migration of chemicals (and, therefore, of pattern formation) scale with the square of the characteristic length, $\tau = L^2/D$. For typical diffusion coefficients of small molecules in gels, $D \sim 10^{-5}$–$10^{-6}\,cm^2\,s^{-1}$, pattern formation can occur within seconds provided that these patterns have dimensions in tens to hundreds of micrometers. While this size range is perfectly suited for wet stamping (WETS), the challenge remains to somehow teach the WETS-imposed patterns to make choices and, like chameleons, change in response to external 'signals'.

9.1 AMPLIFICATION OF MATERIAL PROPERTIES BY RD MICRONETWORKS

Our first 'chameleon' changes its appearance when placed on films differing in thickness by only few micrometers.[12] Consider an agarose stamp micropatterned in bas relief and soaked in an aqueous solution of a metal salt such as iron(III) chloride ($FeCl_3$ 'ink'). When the stamp is placed onto a dry, absorptive substrate (e.g., dry gelatin 'paper') uniformly doped with potassium hexacyanoferrate, $K_4Fe(CN)_6$ (Figure 9.2), the water it contains rapidly wets the 'paper' surface by capillarity (at a rate of several micrometers per second) and slowly diffuses into its bulk ($D_w \sim 10^{-7}\,cm^2\,s^{-1}$; see Example 6.1). At the same time, the Fe^{3+} cations from the stamp are delivered diffusively into an already wetted, thin layer (~10 μm) of gelatin near the surface, where they react with hexacyanoferrate anions to give a deep-blue precipitate known as Prussian blue. As the cations diffuse away from the stamped features, they precipitate all $[Fe(CN)_6]^{4-}$ anions they encounter. Consequently, the unconsumed hexacyanoferrate anions between the features experience a concentration gradient and diffuse toward the incoming reaction fronts (Figure 9.2(a), top). Because the production of the precipitate selectively reduces the mobility of the large $[Fe(CN)_6]^{4-}$ ions (but not of the smaller Fe^{3+}), the color fronts propagating from the nearby features prevent the premature 'escape' of hexacyanoferrate.[13] By synchronizing the outflow of $[Fe(CN)_6]^{4-}$ versus the inflow of Fe^{3+}, the color fronts are able to come very close

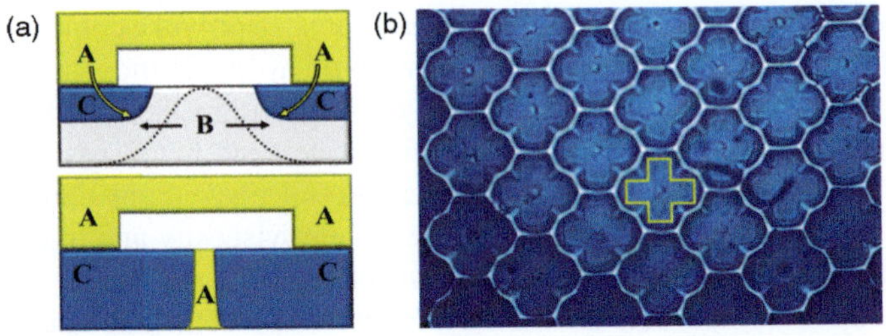

Figure 9.2 (a) Schematic illustration of the WETS of A = $FeCl_3$ onto a dry gelatin film containing B = $K_4Fe(CN)_6$. A diffuses from the stamp into the gelatin layer, where it precipitates B to give a colored C (here, Prussian blue). Unprecipitated B experiences a concentration gradient (dotted line) that causes it to diffuse towards the incoming A. Because the diffusion coefficient of B depends nonlinearly on the amount of precipitate produced at location x, $D_B(x) = D_B^0 \exp[-\lambda C(x)]$, the counterpropagating color fronts are sharp. When all B is consumed, these fronts come to a halt very close (down to 100 nm) to one another, leaving a clear line in between. Mathematical details of this phenomenon are described in detail in Campbell *et al.*[13] (b) Wet stamping of an array of cross-shaped microfeatures (outlined by yellow lines) creates a curvilinear Voronoi tiling via the Prussian blue reaction. The field of view is about 1 mm × 0.6 mm

to one another and – when all hexacyanoferrate ions are ultimately consumed – leave a sharp, clear region between themselves[13] (Figure 9.2(a), bottom). The clear lines thus developed delineate a new pattern that is a transform of the original geometry of the stamp's microfeatures. When the features are disconnected from one another (as in arrays of posts of different shapes), this transform is always the so-called Voronoi tessellation of the stamped pattern. Figure 9.2(b) shows one example, in which the stamped array of crosses evolves into a plane-filling polygonal tiling.

In contrast, when the features form a connected micronetwork, the same stamp can produce dramatically different patterns depending on the properties of the 'paper' onto which it is applied. Figure 9.3 illustrates such systems that 'respond' to different paper thickness by choosing between the so-called tile-centering (TC; white lines bisect the angles between nearby features) or dual-lattice (DL; white lines form perpendicular to the features) patterns.

This bimodality is due to the fact that the $FeCl_3$ 'ink' redistributes within this network prior to entering the gel 'paper'. To understand the origin of this effect, let us first examine the gradients of water content, ρ, that form within the network during WETS (Figure 9.4).

1. As water flows out from the network's features into dry gelatin, a gradient of ρ is established in the vertical, z, direction such that there is less water near the gelatin surface ($z \approx 0$) than at the bases of the features ($z \approx h$).
2. Along a cross-section of each stamped feature (i.e., in the yz plane direction), the edge regions ($y \approx \pm d/2$) lose water more rapidly than the

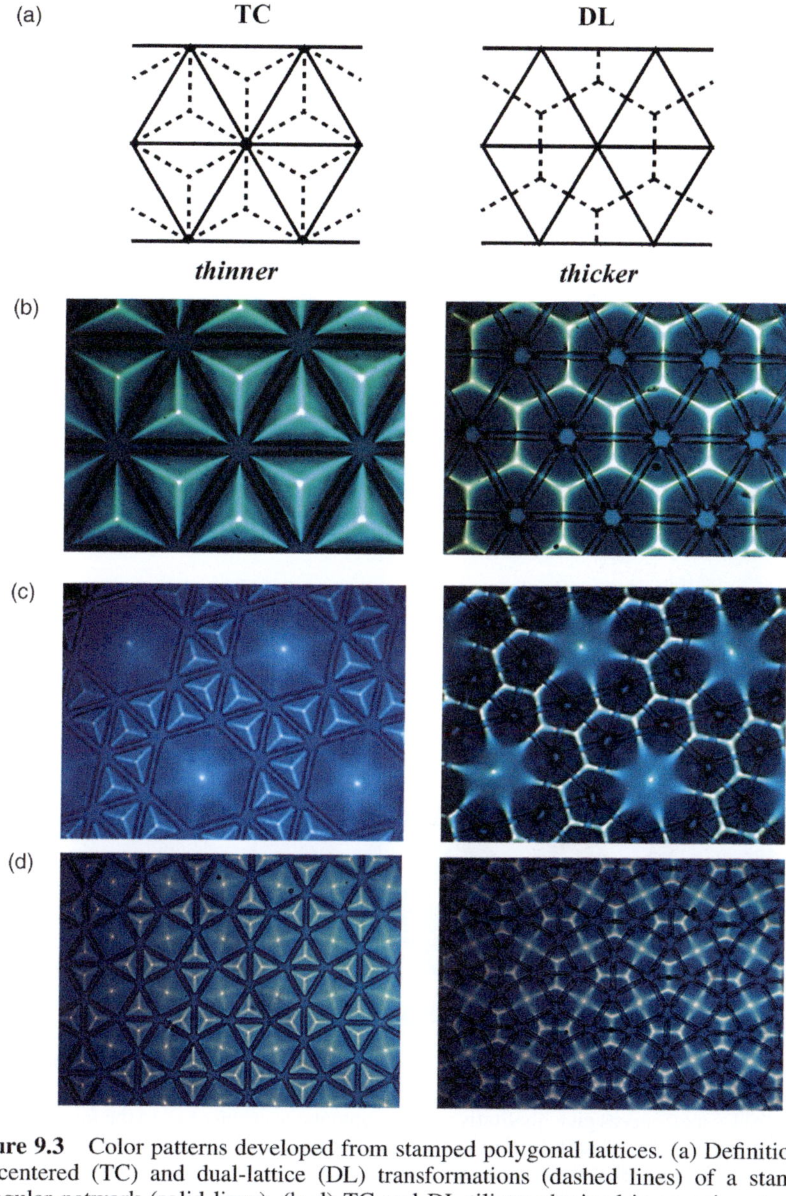

Figure 9.3 Color patterns developed from stamped polygonal lattices. (a) Definition of tile-centered (TC) and dual-lattice (DL) transformations (dashed lines) of a stamped triangular network (solid lines). (b–d) TC and DL tilings obtained in experiments with networks of different geometries and/or dimensions. (b) Left: TC tiling emerges from a network of equilateral triangles stamped onto a thin ($H = 15\,\mu m$) gel 'paper' containing potassium hexacyanoferrate; right: the same stamp gives a very different DL pattern on a thicker paper (here, $\sim 40\,\mu m$). (c, d) Identical micronetworks give TC solutions when applied on thinner gels ($H = 10\,\mu m$; left column) and DL solutions on thicker gels ($H = 30\,\mu m$; right column). Reproduced, with permission, from reference 37, copyright (2005), American Chemical Society

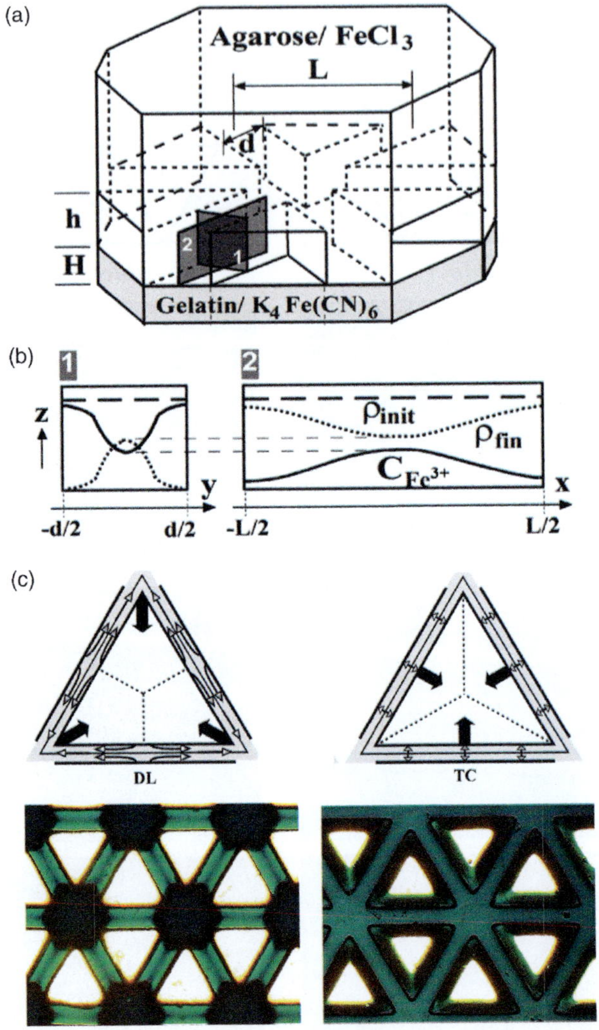

Figure 9.4 Bimodal RD patterning with micronetworks. (a) The scheme illustrates WETS of a network of connected features (here, triangular lattice) and defines pertinent dimensions. (b) Dashed curves give qualitative water contents, $\rho(y)$ and $\rho(x)$, in the features. Solid lines correspond to Fe^{3+} concentration profiles. The left panel is the *yz* cross-section of the feature; the right panel is the *xz* cross-section along the feature. The numbers correspond to the shaded planes in (a). (c) The diagrams illustrate the directions of flow of Fe^{3+} in agarose (thin arrows) and directions of the propagation of RD fronts in gelatin (thick arrows) for TC (left) and DL (right) solutions. Experimental images of the stamped micronetwork developing a DL tiling when applied onto a thicker gel 'paper' and a TC tiling when applied onto a thinner paper. The node-to-node distance in the network is about 300 μm. Reproduced, with permission, from reference 37, copyright (2005), American Chemical Society

feature's center (Figure 9.4(b), left). The evolution of the water content ρ within the agarose gel can be approximated by a diffusion-like equation, $(\partial/\partial t)\rho(y,t) = D(\partial^2/\partial y^2)\rho(y,t)$, which is formally appropriate when the times scales of water migration are significantly larger than those characterizing the relaxation of the polymer chains within the gel (even when this condition fails, Fickian diffusion often provides a good, first-order model[14,15]). This equation can be applied to describe the outflow of water from a stamped feature whose yz cross-section is approximated as a one-dimensional gel slab bounded at $y = \pm d/2$, and characterized by spatially uniform diffusivity, D, uniform initial content of water, $\rho(y,t=0) = \rho_{init}$, and constant outflow rate at the edges, $(\partial\rho/\rho y)_{d=\pm y/2} =$ constant. Solving by one of the methods we learned in Chapter 2 gives water content profiles that are bell-shaped and symmetric at $y = 0$.

3. Along coordinate x connecting the nearby nodes of the network, ρ is lowest near the center of the feature ($x = 0$) and increases towards the nodes ($x \approx \pm L/2$) (Figure 9.4(b), right). This can be rationalized by noting that water flows out of the stamp at an approximately constant rate and the direction of the water flux, $\boldsymbol{j} = -D\nabla\rho$, is everywhere perpendicular to the edges of the stamped features. Since the amount of water transferred from a given element of the feature per unit time is proportional to the density of the flux lines, and because this density is the lowest for the elements located near the network's nodes, the nodes lose less water than the centers of the edges.

The key observation now is that the differences in the water content described in (1)–(3) above translate into the gradients of Fe^{3+} concentration and mobility within the printed networks. On the one hand, the Fe^{3+} ink wants to diffuse from the drier regions (where $[Fe^{3+}]$ is high) to the wetter ones ($[Fe^{3+}]$ low). Such diffusion promotes transport of the ink towards the network's nodes (which lose the least amount of water; cf. (3) above) and formation of DL tilings (Figure 9.4(c), left). On the other hand, continual drying of the features during water outflow decreases lateral diffusivity of Fe^{3+} cations and prevents their 'escape' to the network's nodes. Under such circumstances, the cations enter gelatin from all locations of the stamped micronetwork and ultimately develop TC tilings (Figure 9.4(c), right). Which of the two tendencies wins out depends on the relative speeds of gel drying versus Fe^{3+} diffusion towards the nodes. Importantly, both of these effects depend on the geometry/dimensions of the network (for details, see Ref. 12) and on the thickness of the gel 'paper', H, which we will now discuss in some detail.

To see how the pattern type depends on H, let us first define parameter ρ^* along the center of a feature (i.e., along coordinate x at $y = 0$; cf. Figure 9.4(b)) such that if the water content ρ therein drops below ρ^*, the Fe^{3+} cations cannot escape to the nodes. Next, recall from our previous discussion that water drained from the stamped network wets the gelatin's surface by capillarity and also diffuses into the gelatin's bulk. Consequently, the wetting front that propagates away from the network features drags behind itself a layer of water traveling under the gel's surface. Because on thick gels water penetrates deeper than on thin ones (Section 7.4.2), the speed of the wetting front, v_{wet}, should decrease with H, as indeed verified experimentally.[12] During the wetting, water content $\rho(x, y=0)$ decreases approximately linearly with time, t (cf. point (2) above), and the amount of water lost by the features is proportional to the area of the wetted surface. It follows that $\rho_{init} - \rho(x, y=0,\tau) \propto v_{wet}(H)t$. Therefore, the concentration of water at the center of a feature drops down to ρ^* after some characteristic time $\tau_{closure}(H) \sim (\rho_{init} - \rho^*)/v_{wet}(H)$ which increases with increasing H. On the other hand, the characteristic time needed for the Fe^{3+} cations to migrate/ 'escape' towards the network's nodes is independent of H and is roughly inversely proportional to the characteristic concentration gradient along the feature: $\tau_{esc} \sim L/(\rho_{init} - \rho^*)$. The TC tiling develops when $\tau_{esc} > \tau_{closure}$ and Fe^{3+} cations are trapped in the features. Conversely, if $\tau_{esc} < \tau_{closure}$, the cations 'escape' to the nodes before the features dry out. The critical thickness, H^*, of the gel layer corresponding to the TC–DL transition is determined by $\tau_{closure}(H^*) = \tau_{esc}$. Because $v_{wet}(H)$ is a decreasing function of H, TC solutions develop on thinner gels (Figure 9.3, left) and DL solutions on thicker ones (Figure 9.3, right). A striking manifestation of this effect is shown in Figure 9.5 in which the gel thickness increases continuously ($\sim$5 μm mm^{-1}) from top left to bottom right, resulting in a crossover from the TC to the DL mode. The transition zone is fairly narrow (about two unit cells of the network; $\sim$500 μm) confirming that the system is very sensitive to the changes in H and in the absorptivity of the patterned substrates. The reader might find it interesting that one start-up company is currently working on applying a similar RD system in dental diagnostics where the idea is to amplify the absorptivity/porosity of the teeth enamel into RD patterns initiated from micronetwork stamps (TC = good, nonadsorptive and nonporous enamel; DL = bad enamel).

Figure 9.5 When the thickness of the gel 'paper' varies continuously, the two types of tilings are observed. In the image, the gel is $\sim$10 μm thick in the top left corner (TC tiling) and $\sim$35 μm thick in the bottom right corner (DL tiling). Reprinted, with permission, from reference 37, copyright (2005), American Chemical Society

9.2 AMPLIFYING MACROMOLECULAR CHANGES USING LOW-SYMMETRY NETWORKS

As already mentioned, the principle of micronetwork sensing is that the changes in a substrate's material properties should translate into different modes of ink delivery. Importantly, the more sensitive the delivery modes to the property changes, the better the 'sensor'. In this section, we will work on increasing the sensitivity of our networks by introducing an additional, 'asymmetric' mode of ink delivery through only some of the network's nodes. The objective of this exercise is to detect a macromolecular phase transition and amplify it into macroscopic RD patterns.

To study phase transitions occurring at molecular scales, it is first necessary to translate the changes that accompany them into a form perceptible to our senses. While some transitions (e.g., those involving separation of phases,[16] changes in order parameter[17] or changes in system symmetry[18]) manifest themselves by pronounced visual changes detectable even by the naked eye, others (e.g., conformational transitions in macromolecules[19,20]) alter only the physical/material properties of a substance without visual effects and can usually be studied only with high-end equipment. The behavior of familiar dry gelatin films upon temperature increase provides one illustrative example. At low temperatures, gelatin is composed of partly folded triple helices of amino acid chains; however, when the temperature is increased above $T_c \sim 39\,^{\circ}$C, these helices unwind/denature into so-called random coils. Although the visual appearance of the film does not change, the denatured random coils have less of a propensity to coordinate water molecules than the helices. Consequently, substrates at temperatures $T > T_c$ are less absorptive – by a few per cent – than those at $T < T_c$.[21] The question we wish to ask is whether RD can amplify these small changes into patterns that would switch abruptly and cleanly at T_c thus reporting the occurrence of the helix-to-coil phase transition.

Networks having all nodes of the same type cannot differentiate between gelatin films below and above T_c and give the same types of tiling at all temperatures (TC on thinner and DL on thicker gels). If, however, the network comprises nodes of different types that lose water at different rates, migration of the ink within the stamped network is very sensitive to the relative speeds with which these nodes dehydrate. Consider a network shown in Figure 9.6 and comprising three-fold ($\underline{3}$) and four-fold ($\underline{4}$) nodes.[21] As before, this network delivers Fe^{3+} cations to dry gelatin doped with $[Fe(CN)_6]^{4-}$. Unlike in previous examples, however, the amount of water flowing out of a node per unit time depends on and decreases with the node's degree – that is, the 3-fold ($\underline{3}$) nodes lose water more rapidly than the 4-fold ($\underline{4}$) nodes. This can be explained by the flux argument developed in the previous section or by intuitive reasoning that in order to wet the same area of the gel 'paper', the three features around a $\underline{3}$ node must spill water at a higher rate than the four features around a $\underline{4}$ node. The key thing is that this differential draining of the nodes creates asymmetric water profiles along the features connecting nodes $\underline{3}$ and $\underline{4}$.

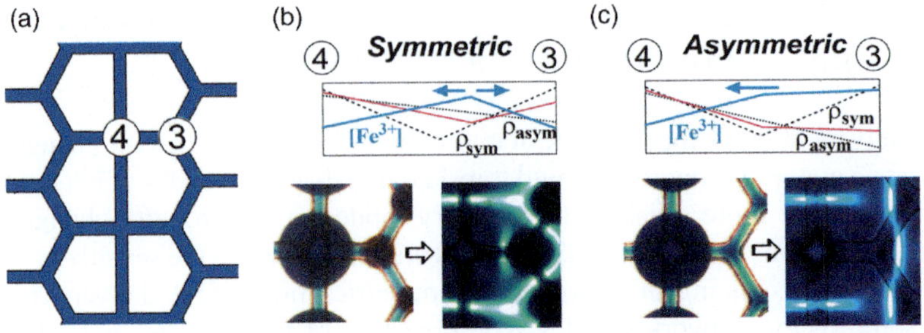

Figure 9.6 (a) Scheme of an asymmetric network comprising three-fold and four-fold nodes. Typical distances between the nodes are ~300 μm and the features are ~50 μm wide. (b, c) The schemes in the top row illustrate instantaneous concentrations of water (red lines) and iron cations (blue lines) along a line joining nodes 3 and 4. Dotted lines give symmetric and asymmetric components of $\rho(x, t)$. The situation in (b) corresponds to slow drainage of water from the network (small α) allowing migration of Fe^{3+} (blue arrows) to both nodes and formation of DL tiling. When water is drained rapidly, Fe^{3+} migrates only to the 4 nodes and gives an 'asymmetric' RD pattern. The experimental images illustrating both of these scenarios are shown in the bottom row. Reprinted, with permission, from reference 37, copyright (2005), American Chemical Society

The red lines in Figure 9.6(b,c) show qualitative water content profiles, $\rho(x, t)$, along a feature connecting $\underline{3}$ and $\underline{4}$ nodes of length L, and with the water contents at the nodes obeying $\rho_3(t) < \rho_4(t)$. At this point it is convenient to approximate these profile as a sum of two contributions: (i) asymmetric, $\rho_{asym}(x, t)$, due to the difference in the water contents between the nodes, $\rho_{asym}(x,t) \approx \rho_4(t) - [\rho_4(t) - \rho_3(t)]x/L$; and (ii) symmetric, triangle-like correction, ρ_{sym}, accounting for the fact that nodes lose less water than the edges (see Section 9.1), $\rho_{sym}(x, t) \approx -f(t)x/L$ for $x < L/2$ and $f(t)(x - L)/L$ for $x \geq L/2$, where $f(t)$ is a monotonically increasing function of time. Because, as discussed previously, water is drained from the network at an approximately constant rate (at least in the early stages of the process), we can also write $f(t) \approx \gamma t$ and $\rho_{3,4}(t) \approx \rho_{init}(1 - \sigma_{3,4}\alpha t)$, where $\sigma_{3,4}$ are some positive constants ($\sigma_3 > \sigma_4$), and α and γ are parameters describing the rate of water loss from the nodes and the features, respectively. The type of pattern that emerges is then determined by the magnitude of α. For low values of α, the overall water profile (red line in Figure 9.6(b)) exhibits a minimum between the nodes, and the Fe^{3+} cations diffuse along the concentration gradients (blue line) towards both types of nodes to give a 'symmetric' DL pattern. On the other hand, when α is large, the steep asymmetric term dominates the overall water profile and the cations migrate only towards the 4 nodes, where they initiate 'asymmetric' patterns in the gelatin layer (Figure 9.6(c)). For a given geometry of the network, the crossover between the two types of patterns occurs at some critical value, α^*, corresponding to the gradient of water content, $d\rho(x)/dx$, at the $\underline{3}$ node being equal to zero – that is, when the profile switches from V-shaped to monotonically decreasing (from $\underline{3}$ to $\underline{4}$). This condition leads to the following approximate expression for α^*: $(\sigma_3 - \sigma_4)\alpha^* \approx \gamma$, with the asymmetric RD pattern appearing when $\alpha > \alpha^*$. In other words, substrates absorbing water rapidly give rise to asymmetric RD patterns, while those absorbing water slowly give rise to symmetric DL ones. Note that in networks having only one type of nodes, the $\sigma_3 - \sigma_4$ term would be zero and no crossover could be expected. It is this term that gives our RD system additional choices beyond the simple TC–DL transition and, as it turns out, allows for the amplification of the helix-to-coil phase transition which otherwise is accompanied by only a small drop of α at T_c.

This is illustrated in Figure 9.7(a) which shows gels at different temperatures. For $T < T_c$, the patterns that emerge are almost exclusively of the 'asymmetric' type expected for more absorptive helical conformation of the gelatin fibers. At around T_c, these patterns switch to symmetric DL mode characteristic of the less absorptive films comprising random-coil protein chains. Remarkably, pattern crossover is virtually binary and much more manifest than the small absolute changes in the film absorptivity (few per cent; Figure 9.7(b)). The RD amplification is also more pronounced than a small peak resolved in traditional differential scanning microcalorimetry (DSC; Figure 9.7(c)).

9.3 DETECTING MOLECULAR MONOLAYERS

In the previous two examples, amplification was mediated by the changes in the *bulk* absorptivity of the gel films. With the periodic precipitation (PP) reactions we studied in Chapter 7, it is possible to amplify directly from the level of one-molecule-thick substrates into macroscopic patterns.

Although many of the mechanistic aspect of PP remain obscure, it is generally believed that aggregation of the precipitate into discrete, periodic bands involves charged colloidal complexes.[22–24] This property could provide a basis for sensing surface reactions, especially on substrates whose redox potentials are close to those of the inorganic species involved in periodic precipitation. Let us apply this idea to detect and 'amplify' deformation of the so-called self-assembled monolayers (SAMs)[25,26] of thiolates on gold (see Section 8.2.2) and amplify this process into PP patterns. Our chemical amplifier will be based on the reaction between $AgNO_3$ delivered from WETS and $K_3Fe^{II}(CN)_6$ immobilized in a thin layer of a gel film (Figure 9.8). When this reaction is initiated in a film supported by a dielectric/nonconductive material (glass, silicon), it gives rise to uniform precipitation and no banding. However, when the gel layer rests on a gold surface, the same process produces multiple and distinct PP bands. The explanation of this difference is that when in contact with the gel, the metallic gold is reduced (i.e., $Au^0 + e^- \rightarrow Au^{-I}$; electrochemical potential $E = -2.15$ V), while iron in the cyanoferrate complex is oxidized ($Fe^{II} \rightarrow Fe^{III} + e^-$; $E = 0.36$ V). This scenario is consistent with additional obser- vations that (i) PP rings do appear on glass substrates when the gel is initially doped with Fe^{III} instead of Fe^{II} complex (i.e., $K_3Fe(CN)_6$ instead of

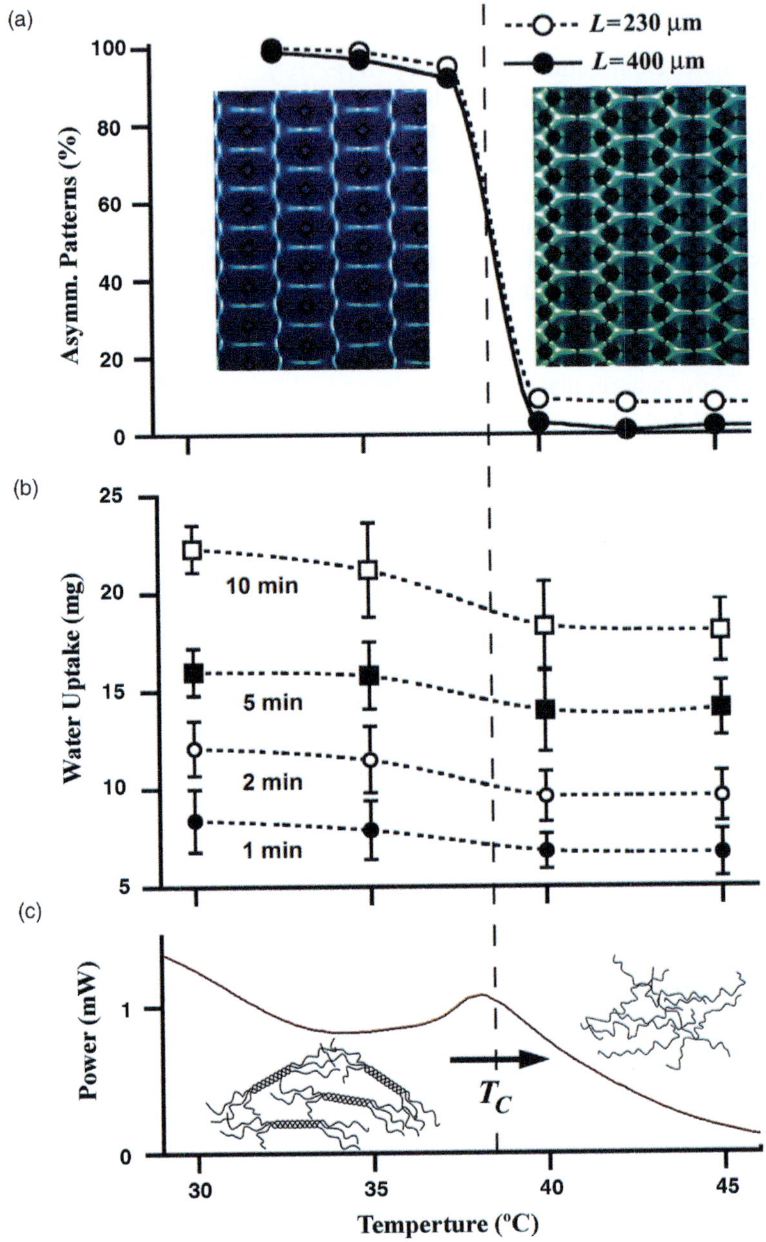

Figure 9.7 (a) Percent of asymmetric solutions/tilings as a function of T (for two networks differing in the distance between the $\underline{3}$ and $\underline{4}$ nodes, L). At around T_c patterns switch almost completely from asymmetric to symmetric ones. (b) Water uptake of gelatin films at different temperatures. Difference between uptakes at $T < T_c$ and $T > T_c$ is only few percent. (c) DSC scan of a gelatin/hexacyanoferrate solution showing helix-to-coil transition at temperature T_c equal to that of the crossover between RD patterns. Reprinted, with permission, from 37, copyright (2005), American Chemical Society

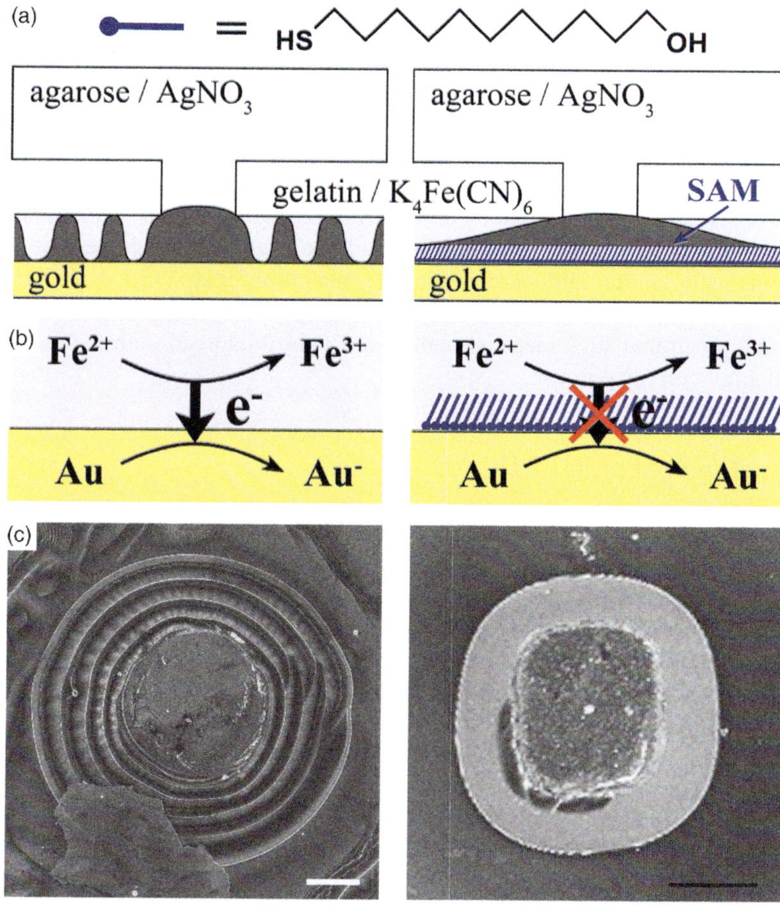

Figure 9.8 Is SAM there? Detection of SAMs of alkane thiolates on gold. (a) When $AgNO_3$ is delivered to a dry gelatin layer doped with $K_4Fe(CN)_6$ and supported on a clean gold substrate, PP gives distinct bands of $Ag_3Fe(CN)_6$ (dark gray). If, however, the gold surface is covered with a SAM of 11-mercaptoundecanol (shown above the diagrams), the precipitation of $Ag_4Fe(CN)_6$ proceeds continuously with no visible banding. (b) In the absence of the electron-insulating SAM, oxidation of Fe^{2+} to Fe^{3+} sets up the appropriate PP chemistry. This redox reaction is eliminated by the presence of the SAM, which prevents transfer of electrons. (c) Electron microscope images of (left) PP patterns forming on a bare gold substrate and (right) continuous and uniform precipitation zone forming in the presence of an insulating SAM. Scale bars = 10 μm. Reprinted, with permission, from *Chaos*, Microchameleons: Nonlinear chemical microsystems for amplification and sensing, Bishop *et al.* Copyright (2006), American Institute of Physics

$K_4Fe(CN)_6$); and (ii) no rings are observed on metallic silver supports (since silver does not have an oxidation state of −I).

For the purpose of chemical amplification, it is important to note that the appearance of PP patterns requires electron transfer from the gel layer into the metal; therefore, the presence of an insulating layer – even the one-molecule-thin SAM – should prevent pattern formation. Indeed, when experiments are carried out on gold surfaces covered with well-organized SAMs of alkyl thiolates, continuous precipitation is observed (cf. Figure 6(b)). Interestingly, with SAMs composed of thiols that are either poorly packed and/or can tunnel electrons[27] (e.g., phenyl thiol), PP bands do form but their quality is poorer than bands on unprotected gold. These observations suggest that by correlating the quality of a PP readout pattern (e.g., number of bands that form, spacing coefficient, etc.), it should be possible to develop this RD system into a sensitive analytical tool for studying defects in and electron transport through SAMs. Such development should be all the more welcome given that such measurements[28] are nontrivial even with the elaborate instruments currently in use.

9.4 SENSING CHEMICAL 'FOOD'

It is now time to make our amplifiers a bit livelier. The system[29] we are about to discuss in this section can create truly dynamic patterns and is capable of regulating its own behavior depending on the nature of the participating chemicals. Although more difficult to control and model (the reader might want to skip some theoretical details on the first reading), it is more 'biomimetic' than the micronetwork or PP amplifiers, and involves elements of primitive 'metabolism' and through-space signaling loosely analogous to that observed in certain types of microorganisms. It is also based on a very different (and more complex) set of chemical reactions involving autocatalytic and inhibitory loops (see Sections 3.4 and 3.5) that act in unison to induce chemical oscillations and traveling chemical waves rather than static RD patterns. As we will see shortly, coupling of these complex chemical kinetics to the system geometry imposed by WETS can create a 'starfish' companion to our chemical chameleons, which – depending on the chemical 'food' it is given – emits different types of colorful waves (Figure 9.9).

The system we use is a modified version of the famous Belousov–Zhabotinsky (BZ) oscillator,[30] in which chemical 'food' ($T = CH_2O$ (formaldehyde) or CH_3OH (methanol)) is delivered (Figure 9.9(a)) from a star-shaped agarose stamp into an agarose film soaked with the so-called Winfree formulation of the BZ reagent comprising several substrates and reaction intermediates (Example 9.1). Upon delivery, the 'food' triggers a complex network of reactions (Figure 9.10), involving: $A = BrO_3^-$; $B = Br_2$; $H = H^+$; $M = CH_2(COOH)_2$; $M' = BrCH(COOH)_2$; $O =$ oxidation products (i.e., $CH_2O = COOH = CO_2$); $P = HOBr$; $X = HBrO_2$; $Y = Br^-$; $Z = Fe(phen)_3^{3+}$ (oxidized ferroin indicator colored blue); and Z', $Fe(phen)_3^{2+}$ (reduced ferroin, pale red). These reactions manifest themselves in rhythmic color oscillations and propagating chemical waves. Remarkably, despite that fact that the food is delivered uniformly from the stamp, the waves are emitted only from discrete locations: When the star delivers formaldehyde 'food', the waves propagate only from the star's tips (Figure 9.9(b)); when the food

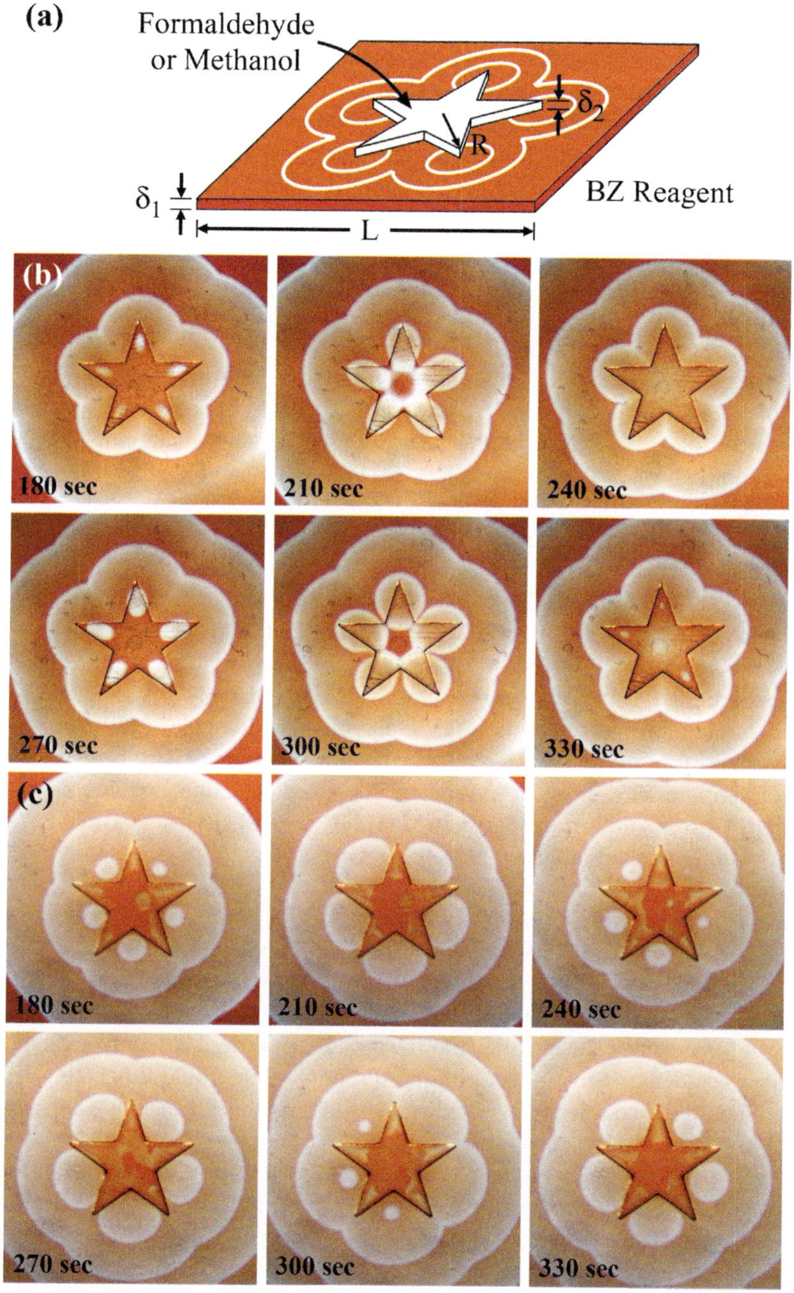

Figure 9.9 Emit what you eat – BZ 'starfish' sends different types of waves depending on the WETS-applied 'diet'. (a) Scheme of the experimental arrangement ($\delta_1 = \delta_2 = 400\,\mu m$, $R = 1$–$6\,mm$, $L = 5\,cm$). Experimental images obtained after stamping (b) formaldehyde and (c) methanol 'foods'. Reprinted from reference 19, with permission, copyright (2006), American Physical Society

is methanol, the waves originate exclusively from the vertices between the star's arms (Figure 9.9(c)). It therefore appears that these qualitatively different behaviors are a consequence of different ways in which the system 'metabolizes' the two types of 'foods'. To understand the origin of these phenomena, we will first consider the underlying chemical kinetics in a homogeneous, well-stirred system and then couple these kinetics to the diffusion of chemicals within the geometric boundaries imposed by WETS (beware, lots of math ahead!).

Example 9.1 Patterning an Excitable BZ Medium using WETS

An excitable medium is a nonlinear system that can propagate waves. In chemistry, such waves correspond to space-propagating concentration variations that can be made visible/colored by the use of appropriate indicator molecules. In this example, we will describe how to implement and pattern the classic Belousov–Zhabotinskii (BZ) system developed in the 1960s by Russian chemists Belousov and Zhabotinskii, and modified in the 1970s by A.T. Winfree.[30] First, the so-called Winfree solution is made from the following stock solutions: (1) 2.86% v/v H_2SO_4 and 7.14% w/v $NaBrO_3$ in deionized H_2O; (2) 10% w/v NaBr in deionized water; (3) 10% w/v malonic acid, $CH_2(COOH)_2$, in deionized water. From these ingredients, 8.5 mL batches of BZ reagent are made as needed by first adding 0.5 mL of bromide solution (2) and 1 mL of MA solution (3) to 6 mL of acidic bromate solution (1). Once the bromine color (yellow/orange) vanishes, 1 mL of 25 mM (standard) ferroin solution (1,10-phenanthroline ferrous sulfate serving as color indicator) is added. The final concentrations of the red Winfree solution are then 0.36 M H_2SO_4, 0.33 M $NaBrO_3$, 0.057 M NaBr, 0.10 M malonic acid and 0.0029 M ferroin.

The BZ medium to be patterned is typically a 3% w/w high-gel-strength agarose in the form of thin films (~400 μm thick). Immediately prior to use, these films are soaked in the Winfree solution for about 10 minutes. At the end of this period, a so-called 'oxidative excursion'[34] is induced by dipping a silver wire into the solution and stirring until the gel changes color from red (reduced state of ferroin indicator) to blue (oxidized state); this 'desensitized' gel is stable against accidental wave-generating perturbations that could otherwise occur when handling an active BZ system. After soaking, the gel sheet is placed on a glass slide and blotted dry with tissue paper to remove excess BZ reagent. The sheet is then covered with a Petri dish and allowed to rest for 2 minutes to equilibrate any hydration gradients that may have developed during drying.

The stamps from which the waves are to be patterned (e.g., pentagonal stars described in the main text) are made from 8% w/w solution of high-gel-strength agarose as in all other WETS procedures (see Chapter 5). Prior to use, the stamps are soaked for 30 minutes in a solution of a desired wave-generating 'triggering' reagent (either 0.13 M aqueous solution of formaldehyde or 1.0 M aqueous solution of methanol), blotted dry with filter paper and allowed to equilibrate on a glass plate for 2 minutes.

To initiate chemical waves, the stamp is applied onto the BZ film and covered with a Petri dish to prevent gel drying. Images of the resulting chemical waves are best captured from below the film. For classroom demonstrations, placing the system on a transparency projector works quite well (but do not overheat with too much light!). Most importantly, the waves that appear usually keep students awed (and quiet) for some time.

9.4.1 Oscillatory Kinetics

The starting point of our kinetic analysis of BZ oscillations is the so-called KFN model developed in 1974 by Fields, Korös and Noyes.[31,32] It can be shown[29] that with appropriate approximations of rate-determining steps, this model simplifies to the set of reactions (R2′) through (J) summarized in Figure 9.10(a). In addition, to describe the effects of the stamped methanol/formaldehyde 'foods', T, the model needs to be augmented to take into account these substances triggering (i) production of $HBrO_2$ via reaction with $HBrO_3$ (reaction (P1) in Figure 9.10 (a)) and (ii) production of Br^- by reaction with Br_2 (reaction (P2)). Together, these equations constitute the system 'metabolic' network, in which various intermediates are connected as in Figure 9.10(b).

Because the concentrations of A, H, M and T do not change significantly on the time scale of the experiments, their consumption by chemical reaction can be

(a)

	Reaction	Rate
(R2′)	$3H + X + 3Y \rightarrow 2B$	$r_2 = k_2[H][X][Y]$
(R3)	$2H + A + Y \rightarrow X + P$	$r_3 = k_3[H]^2[A][Y]$
(R4)	$2X \rightarrow A + P + H$	$r_4 = k_4[X]^2$
(R5′)	$3H + A + X + 2Z' \rightarrow 2X + 2Z$	$r_5 = k_5[H][A][X]$
(R8)	$B + M \rightarrow Y + H + M'$	$r_8 = k_8[M][B]$
(J)	$Z + M \rightarrow \frac{1}{2}fY + Z' + O$	$r_j = k_j[M][Z]$
(P1)	$T + H + A \rightarrow X + O$	$r_{P1} = k_{P1}[T][H][A]$
(P2)	$T + B \rightarrow 2Y + O$	$r_{P2} = k_{P2}[T][B]$

(b)

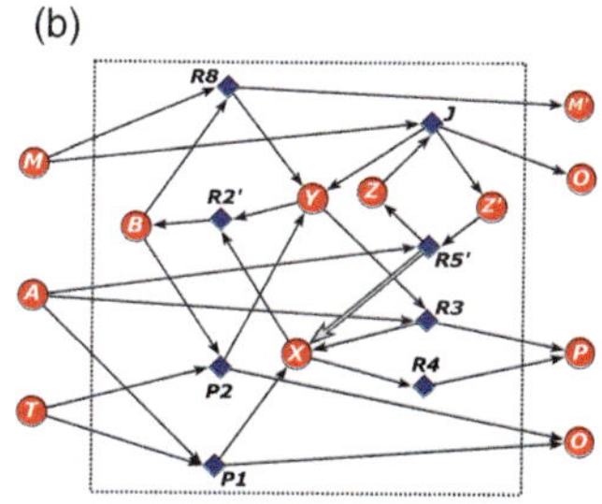

Figure 9.10 (a) Kinetic model accounting for perturbations due to formaldehyde and methanol. The reactions (left) and rates (right) governing the system's kinetics involve $A = BrO_3^-$; $B = Br_2$; $H = H^+$; $M = CH_2(COOH)_2$; $M' = BrCH(COOH)_2$; O = oxidation products (i.e., $CH_2O = COOH = CO_2$); P = HOBr; T = triggering reagent (i.e., CH_2O or CH_3OH); $X = HBrO_2$; $Y = Br^-$; $Z = Fe(phen)_3^{3+}$; $Z' = Fe(phen)_3^{2+}$; and f is a stoichiometric factor. (b) Graphical representation of the reactions listed in (a). Here, the red circles represent chemical species (for simplicity, H^+ is omitted), and the blue diamonds represent chemical reactions. The species on the left 'feed' the dynamic system of 'metabolic' reactions in the outlined region; waste products are emitted to the right. The highlighted gray arrow corresponds to the autocatalytic production of X by reaction (R5′). Such autocatalysis is necessary (but not sufficient) for oscillatory and/or wave-emitting behavior

neglected, and the steady-state approximation methods discussed in Chapter 3 (see Examples 3.1 and 3.2) can be used to reduce the problem to only four kinetic equations involving reactive intermediates X, Y, Z and B (for details, see Ref. 29):

$$
\begin{aligned}
d[X]/dt &= k_3[H]^2[A][Y] - k_2[H][X][Y] + k_5[H][A][X] - 2k_4[X]^2 + k_{P1}[H][A][T] \\
d[Y]/dt &= -k_3[H]^2[A][Y] - 3k_2[H][X][Y] + hk_j[M][Z] + k_8[M][B] + 2k_{P2}[T][B] \\
d[B]/dt &= 2k_2[H][X][Y] - k_8[M][B] - k_{P2}[T][B] \\
d[Z]/dt &= 2k_5[H][A][X] - k_j[M][Z]
\end{aligned}
\tag{9.1}
$$

The reader is encouraged to verify that using the so-called Tyson scaling[32] ($x = (2k_4/k_5[H][A])[X]$, $y = (k_2/k_5[A])[Y]$, $b = (2k_4k_8[M]/(k_5[H][A])^2)$ [B], $z = (k_4k_j[M]/(k_5[H][A])^2)[Z]$, $\tau = (k_j[M])t$, $f = 2h$, $\varepsilon_1 = k_j[M]/k_5[H][A]$, $\varepsilon_2 = 2k_4k_j[M]/k_2k_5[H]^2[A]$, $\varepsilon_3 = k_j/k_8$, $q = 2k_3k_4/k_2k_5$, $\alpha = 2k_4k_{P1}[T]/\ k_5^2[H][A]$ and $\beta = k_{P2}[T]/k_8[M]$) these equations can simplified to the following nondimensional form:

$$
\begin{aligned}
\varepsilon_1 dx/d\tau &= qy - xy + x(1-x) + \alpha \\
\varepsilon_2 dy/d\tau &= -qy - 3xy + fz + (1+2\beta)b \\
\varepsilon_3 db/d\tau &= 2xy - (1+\beta)b \\
dz/d\tau &= x - z
\end{aligned}
\tag{9.2}
$$

In the absence of the 'food'/triggering reagent, $\alpha = 0$ and $\beta = 0$, and the dynamics of this system are characterized by the activating species X, which autocatalyzes its own production, and inhibitors Y, Z and B, which prevent the autocatalysis of X as long as their concentrations remain above some critical levels. In this case, autocatalytic production of X proceeds only if $\omega = [fz + (1 + 2\beta)b]/3(x + \alpha) < 1$;[32] otherwise, the autocatalytic process is inhibited and no oscillations occur. At this point, it is important to note that both methanol and formaldehyde 'foods' are involved in the production of the activating species (since T is present in reaction (P1) and parameter α enters the kinetic equation for $dx/d\tau$) as well as the inhibitory ones (since T is present in reaction (P2) and β appears in the expressions for $dy/d\tau$ and $db/d\tau$). We will see later in the section that it is the balance between these processes that determines the mode of wave emission.

9.4.2 Diffusive Coupling

In the wet stamped star, the oscillation-controlling parameter ω may vary as the result of both chemical reactions and diffusive fluxes that modify local concentrations of the activating (X) or inhibiting (Y, Z, B) species. These effects are accounted for by combining the kinetic equations with the diffusion of all species (i.e., X, Y, Z, B and T) to give a familiar system of RD equations, $dC/dt = D_C\nabla^2 C + R_C$, where $C = [X], [Y], [Z], [B]$ and $[T]$, D_C is the diffusion coefficient of species C and R_C is the appropriate kinetic expression from Equation (9.2) above. For the 3% w/w agarose gels used, all diffusion coefficients are of the order of[33] $10^{-5}\,cm^2\,s^{-1}$ with the exception of ferroin whose larger size results in smaller diffusivity, $\sim 5.0 \times 10^{-6}\,cm^2\,s^{-1}$ (cf. Example 8.1). While these equations are too complicated to attempt any analytic solution, they can be integrated numerically – using methods we learned in Chapter 4 – on a two-dimensional grid and with no-flux boundary conditions for all species. Initially, the concentrations of the intermediates (X, Y, Z and B) are set to their steady-state values in the absence of T, and the effect of stamping is approximated by imposing a step function over the star-shaped stamped area ($[T] = [T]_0$ inside the star and $[T] = 0$ outside at $t = 0$).

9.4.3 Wave Emission and Mode Switching

Numerical simulations using experimental concentrations and rate constants reproduce the 'focused' wave emission faithfully (Figure 9.11), but they shed relatively little light on the very nature of this process. It is therefore instructive to use the numerical results to explain, at least qualitatively, the physical reality behind the phenomena we observe.

Here, we first note that in order to satisfy the $\omega < 1$ condition necessary to initiate oscillations in the patterned gel, the concentration of the 'food', [T], has to decrease slightly from its 'nominal value', $[T]_0$, delivered from the stamped star (Figure 9.12(a)). Diffusion facilitates such decrease by transporting T outwards from the star's contour – as a result, only the perimeter region I (Figure 9.12(b,c)) is oscillatory and capable of generating chemical waves. In this region, the concentration of T is such that an isolated system would oscillate. In our patterned geometry, however, region I is not isolated but coupled to nonoscillatory regions both inside (region II) and outside (region III) of the star. Therefore, diffusive fluxes of activating and/or inhibiting species into region I have the potential to dramatically affect its dynamic behavior (e.g., speed up, slow down, or even inhibit oscillations). Analysis of the calculated concentration gradients indicates that the most important fluxes are those of the activator X and inhibitors Z and B. These fluxes are comparable in magnitude and directed from region II toward region III.

Now, the crucial observation is that although both methanol and formaldehyde affect the production of both the activating and the inhibiting species (through parameters α and β, respectively), they do so to different extents. The relative importance of inhibition and activation in the 'metabolism' of these foods is then captured by a dimensionless parameter $R = \beta/\alpha$, which for the experimental conditions used is $R \sim 15$ for formaldehyde and $R \sim 14$ for methanol.

Formaldehyde food (Figure 9.12(b)). In the case of formaldehyde triggering ($R \sim 15$), the balance between the activating perturbation (see reaction (P1) in

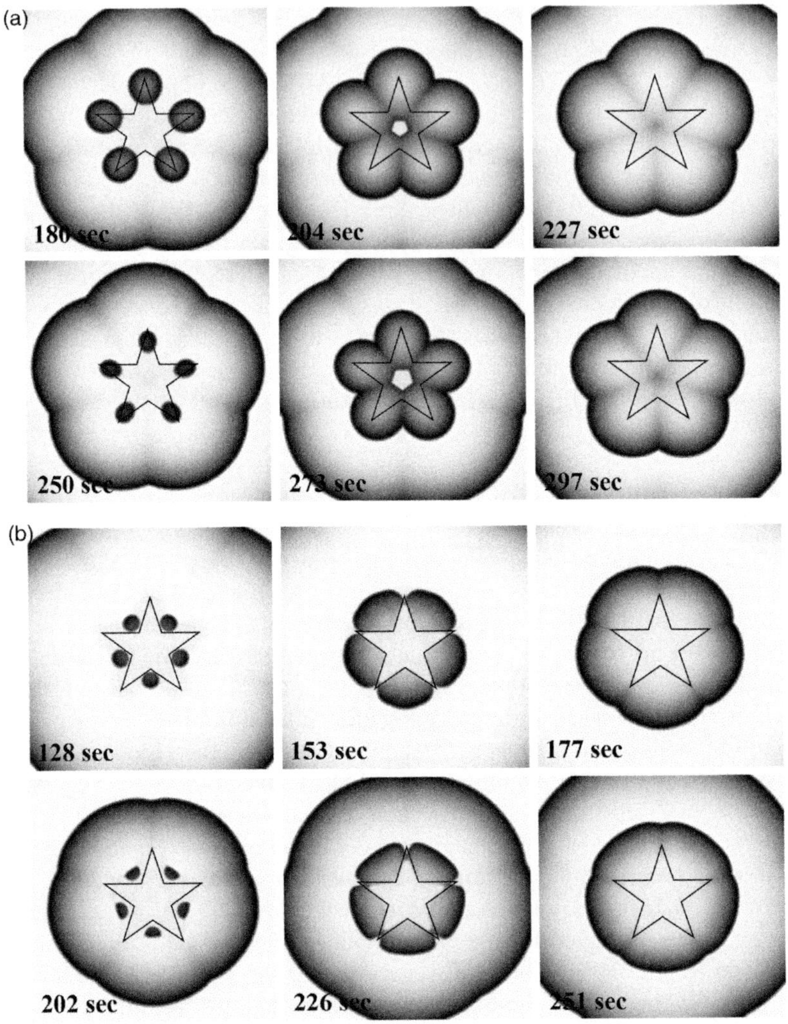

Figure 9.11 (a) Images of wave patterns from RD simulations for formaldehyde triggering ($\alpha = 0.0189$, $\beta = 0.284$ and $R = 15$). The star-shaped, stamped region is outlined for reference. (b) Images from RD simulations for methanol ($\alpha = 0.0680$, $\beta_0 = 0.952$ and $R = 14$). Reprinted from reference 29, with permission. Copyright (2006), American Physical Society

Figure 9.10) and the inhibiting perturbation (reaction (P2)) is shifted in favor of inhibition, resulting in inhibitor-controlled pattern formation. Examining the criterion for autocatalysis and oscillations, $[fz + (1 + 2\beta)b]/3(x + \alpha) < 1$, we note that an increase in R – either by decreasing α or by increasing β – enhances the inhibitory effects of Z and B. Therefore, for larger values of R, the autocatalytic process is controlled predominately by the diffusive delivery/removal of inhibitor

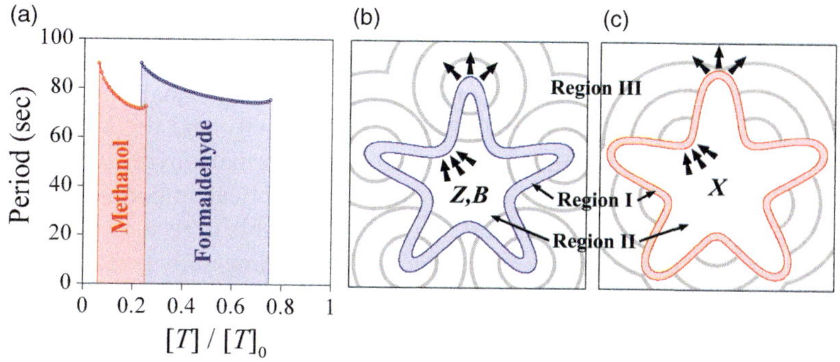

Figure 9.12 (a) Oscillation period as a function of scaled concentration of the triggering reagent for perturbation parameters (α and β) corresponding to formaldehyde (blue) and methanol (red). For concentrations other than those within the shaded region, the system does not oscillate. Note that for both formaldehyde and methanol, the oscillatory regimes require the concentrations of these 'foods' to be lower than those delivered from the stamp (i.e., $T/T_0 < 1$). Schematic illustrations of (b) inhibitor-controlled pattern formation (formaldehyde 'food') and (c) activator-controlled pattern formation (methanol 'food'). Shaded regions correspond to oscillatory regimes where 'food' concentration falls within the limits specified in (a). Reprinted from reference 29, with permission. Copyright (2006), American Physical Society

species into/from the oscillatory region. Due to the curvature of this region, the delivery of inhibiting species is enhanced between the star's arms (from region II to region I), while their removal is enhanced at the tips (from region I to region III). The criterion for autocatalysis is thus achieved only at the star's tips, creating focused oscillations, which, in turn, initiate waves into the surrounding excitable medium.

Methanol food (Figure 9.12(c)). The situation is reversed in the case of methanol triggering for which $R = 14$ and the activating perturbation is dominant. Notice that because our system is highly nonlinear, only a slight decrease in R (from 15 to 14) is sufficient[29] to cause the transition to the activator-controlled mode of pattern formation. Here, the activator is removed most effectively from the tips of the star and is 'flooding' the space between the arms, thereby initiating oscillations only from the latter regions. These oscillations couple to the surrounding excitable medium to initiate traveling waves.

9.5 EXTENSIONS: NEW CHEMISTRIES, APPLICATIONS AND MEASUREMENTS

Having explained the origin of wave propagation in the BZ star, let us consider the phenomena involved in a wider context. On the one hand, engineering of bimodal (or even multimodal) systems by coupling nonlinear chemical kinetics and system geometry appears a very general and powerful concept for chemical sensing. On the other hand, the complexity of this coupling makes rational design of such RD systems difficult. In other words, while the effects of specific chemical 'foods' can be rationalized *a posteriori*, predicting these effects *a priori* is no easy undertaking requiring both chemical intuition and careful theoretical analysis. A second observation we make concerns the structural complexity and the usefulness of the chemical foods – admittedly, methanol and formaldehyde we used are somewhat trivial, and their detection is of little practical importance. For the chemical amplifiers to be truly useful, they should be able to recognize important molecules such as enzymes, natural products or environmentally toxic substances. Constructing such systems will most certainly require chemistries other than that of BZ – fortunately, the number of chemical oscillators described in the literature is constantly growing and the chemical networks these systems rely on are by now fairly well understood. The curious reader is referred to excellent monographs,[35,36] which discuss various classes/families of oscillating and wave-generating systems. With the chemical diversity these systems offer, one should be able, at least in principle, to incorporate desired analytes into their kinetics. The challenge is then to establish which RD chemistries can be used to amplify which types of small-scale phenomena – such categorization appears a necessary step to extend the scope of this approach beyond the narrow confines of demonstrative experiments.

A separate, but probably even more important question is whether the RD amplification can lead to real-world applications. While it is idealistic to expect that chemical amplifiers will replace analogous electronic devices, the low cost and disposability of gel stamps suggest some interesting niche uses. Our personal favorites are the use of micronetwork stamps in dentistry (mentioned at the end of Section 9.1) and in dermatology as a noninvasive alternative to biopsy proce- dures to test for malignancy of moles (e.g., by applying a stamp resembling a transdermal patch and containing fluorescently labeled Mdm2, an E3 ubiquitin ligase, which can bind to an apoptotic p53 protein). In a world where so many things need to be sensed and amplified, it will certainly be possible to find more such applications.

There is also a challenge of transforming binary RD sensors into measuring devices. Can RD not merely detect the presence/absence of chemical analytes but also quantify their properties/concentrations or even interactions between them? It can.

One relatively simple example is illustrated in Figure 9.13 where the so-called Briggs–Rauscher (BR) oscillations and waves[37,38] are induced by Wet Stamping hydrogen peroxide (H_2O_2) onto a polyacrylamide (PAAm) gel containing a mixture of potassium iodate, malonic and sulfuric acids, manganese sulfate, and thyodene indicator. The BR system is sensitive to the antioxidants (e.g., vitamin C,[39] polyphenol compounds,[40,41] substituted diphenols[42]) whose delivery into the gel eliminates the micropatterned oscillators – significantly, the antioxidant concentration can be estimated by the number of eliminated oscillators.

The second example is much more practical and provides the icing on the cake to conclude this chapter: a RD system that quantifies the strength of molecular interactions! Although not really bimodal, this system has a truly extraordinary sensitivity and can measure binding constants between arbitrary proteins and

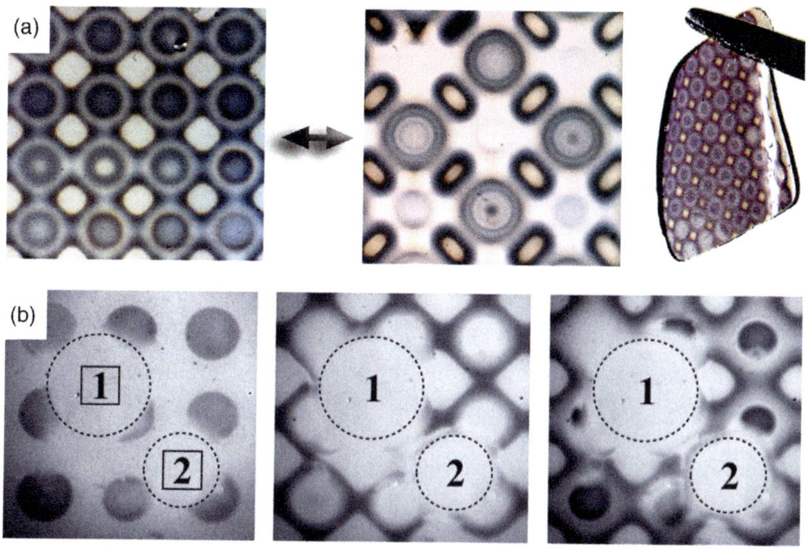

Figure 9.13 (a) Chemical oscillations and waves in a BR system patterned using WETS delivering H_2O_2 from 500 μm circular posts into a polyacrylamide (PAAm) gel.[37] The rightmost image shows a freestanding PAAm gel with over 50 active oscillators. (b) Spatially distributed sensing of antioxidants. An antioxidant (2,6-dihydroxybenzoic acid) was delivered to the PAAm film from 1 mm agarose cubes *before* introducing H_2O_2. The concentration of antioxidant was 0.01 M in cube 1 and 0.005 M in cube 2 – both cubes were applied for 20 s. Because antioxidants scavenge free radials necessary for BR oscillations, there are no waves in the regions influenced by the cubes. Note that region 1 is approximately two times larger than region 2, corresponding to the differences in antioxidant concentration. Reprinted, with permission, from reference 37, copyright (2005), American Chemical Society

small-molecule ligands. Measurements of such constants are all-important in the pharmaceutical industry (to quantify drug potency and selectivity), and are typically not an easy affair, as different proteins require different experimental protocols (called 'assays'). For many proteins, such protocols are not even known especially when the protein–ligand binding is not accompanied by some characteristic 'signal' such as change in fluorescence, UV, NMR or mass spectra, optical rotation or electrochemical response.[43] Let us see how RD solves this important problem in quite a straightforward way.

The system comprises a thin film of a PAAm gel modified with ligand molecules that bind to proteins of interest (Figure 9.14). Since the ligands are attached to the gel backbone covalently, they are immobile. In contrast, the proteins delivered to the gel from a micropatterned WETS stamp can diffuse freely. As they migrate through the gel, they are captured by the ligands, much like in standard affinity chromatography.[44] Our objective, however, is not to perform any chromatographic separation but to relate the degree of protein migration through the gel to the strength of protein–ligand binding. Intuitively, the stronger this binding, the more

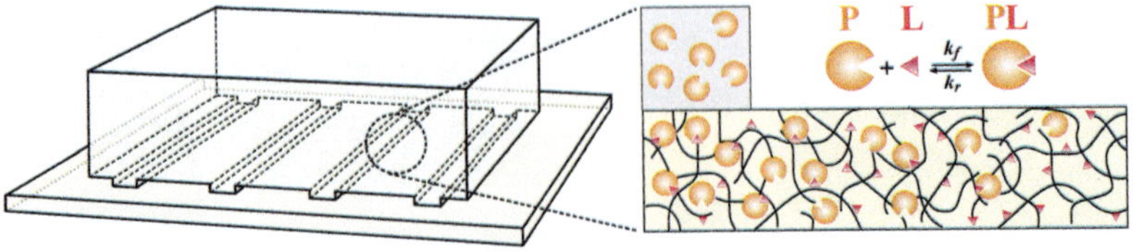

Figure 9.14 Scheme of the stamping arrangement and a zoom-in of the cross-sectional area near the edge of one of the stamped lines. Agarose stamp (gray) delivers proteins P (orange circles) to a PAAm gel (yellow) modified with protein-specific ligands (red triangles). The proteins diffuse through a gel and also bind to ligands reversibly with binding constant $K_b = k_f/k_r$

effective the gel should be in 'capturing' the proteins, and the slower the spreading of proteins away from the stamped features (Figure 9.15). To quantify the competition between protein migration and ligand binding, let us first write out the system of RD equations describing these phenomena. Denoting the concentration of free proteins as [P], that of the proteins bound to the ligands as [PL] and the concentration of immobile ligands as [L], we have

$$\begin{aligned} \partial[\mathrm{P}]/\partial t &= D_\mathrm{P}\nabla^2[\mathrm{P}] - k_\mathrm{f}[\mathrm{P}][\mathrm{L}] + k_\mathrm{r}[\mathrm{PL}] \\ \partial[\mathrm{L}]/\partial t &= -k_\mathrm{f}[\mathrm{P}][\mathrm{L}] + k_\mathrm{r}[\mathrm{PL}] \\ \partial[\mathrm{PL}]/\partial t &= k_\mathrm{f}[\mathrm{P}][\mathrm{L}] - k_\mathrm{r}[\mathrm{PL}] \end{aligned} \tag{9.3}$$

where all concentrations depend on spatial coordinates and on time, D_P is the diffusion coefficient of the free protein, k_f is the 'forward' rate of protein–ligand

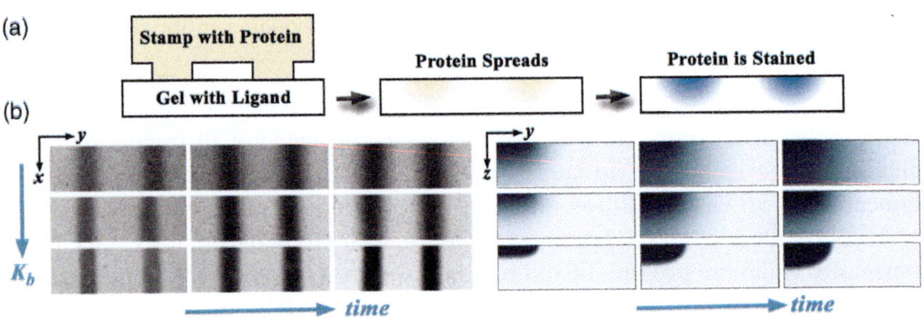

Figure 9.15 (a) Scheme of the stamping/staining procedure for the determination of K_b. The stamp is placed on a ligand-modified gel for a predetermined period of time, after which the spread protein is 'visualized' by staining. (b) The three left panels show the top views (xy plane) of lines stamped for different times (20, 60 and 100 min) and then stained with coomassie blue. Note the widening of the lines with time. The right three panels have the corresponding modeled side-views (yz plane) of protein concentration profiles in the gel. The top row corresponds to no ligand in the gel ([L] = 0; $K_b = 0$) and no binding. In this case, protein spreading is most pronounced. The middle row corresponds to a relatively weak protein–ligand binding ($K_b \sim 10^4\,\mathrm{M}^{-1}$; gel contains a weak ligand of human carbonic anhydrase, HCA). The bottom row illustrates stronger protein–ligand binding ($K_b \sim 10^6\,\mathrm{M}^{-1}$ HCA ligand). In this case, spreading is the least pronounced

binding and k_r is the 'reverse' rate for the dissociation of the protein–ligand complex. Our task is to find the protein–ligand binding constant given by the ratio of the forward and the reverse rates, $K_b = k_f/k_r$.

To do this, we must know how the concentration of proteins ([P] and [PL]) in the gel changes as a function of spatial coordinates and of time. Experimentally, such information is obtained by Wet Stamping the protein from an array of long, parallel lines (this geometry effectively reduces the problem to two dimensions, x and z) for different times, t, and then staining the patterned gels with coomassie blue (or other protein-staining reagent[45]) which translates/amplifies protein concentration profiles into easy-to-quantify color gradients (Figure 9.15).

These experimental profiles are targets for RD modeling and optimization. Here, we can first simplify the full kinetic equations (9.3) by noting that the protein–ligand association kinetics are fast ($k_f \sim O(10^5)\ \mathrm{M^{-1}\,s^{-1}}$) compared with protein diffusion ($D_P \sim O(10^{-5})\ \mathrm{cm^2\,s^{-1}}$). Therefore, the RD dynamics can be treated as a repeating sequence of two substeps occurring at different time scales. In the first, slow substep, [P] is allowed to diffuse; in the second, fast substep, the concentrations of [P] and [PL] are updated 'instantaneously' to account for the protein–ligand binding reaction. Mathematically, these two events for a two-dimensional system can be written in a discrete form as

$$\begin{aligned}
&(i) \quad [\mathrm{P}] \to [\mathrm{P}] + \Delta[\mathrm{P}] \quad \text{with} \quad \Delta[\mathrm{P}] = D_\mathrm{P}\left(\frac{\partial^2[\mathrm{P}]}{\partial x^2} + \frac{\partial^2[\mathrm{P}]}{\partial z^2}\right)\Delta t \\
&(\mathrm{ii}) \quad [\mathrm{PL}] \to [\mathrm{PL}] + \delta, \quad [\mathrm{L}] \to [\mathrm{L}] - \delta \quad \text{and} \quad [\mathrm{P}] \to [\mathrm{P}] - \delta \quad \text{with} \\
&\qquad K_\mathrm{b} = \frac{[\mathrm{PL}] + \delta}{([\mathrm{P}] - \delta)([\mathrm{L}] - \delta)}
\end{aligned} \tag{9.4}$$

where $[\mathrm{L}]_0$ is the initial ligand concentration. When integrated numerically, these equations yield the RD dynamics in terms of only two free parameters, D_P and K_b (as opposed to the three parameters D_P, k_f and k_r in Equations (9.3)), and two initial concentrations (that of the protein in the stamp, $[\mathrm{P}]_0$, and of the ligand in the gel, $[\mathrm{L}]_0$) which are known *a priori*. The key idea now is to vary the values of D_P and K_b such as to minimize the difference between the calculated and the experimental protein concentration profiles (see Example 9.2 for numerical details). The minimal difference pair, $\{D_P, K_b\}$, is then the end result of our protein binding assay.

Example 9.2 Calculating Binding Constants from RD Profiles

As discussed in the main text, the degree of protein spreading from the patterned lines can be visualized by staining. This procedure yields top-down images like the ones shown in the left-hand part of Figure 9.15(b). Importantly, since the optical absorbance of the staining reagent is proportional [50,51] to the local total protein concentration (i.e., [P] + [PL]) at a given location, color intensities can be translated into total concentration of protein integrated over gel depth (i.e., down the z direction; see right-hand part of Figure 9.15(b)). To estimate the values of D_P and K_b, the RD equations are first integrated with some 'guess' values of these parameters, and the modeled 'images' (cf. right-hand part of Figure 9.15(b)) are compared to the experimental ones. The parameters are then updated and the procedure is repeated until the best fit to experimental data is found.

(i) **Integration of RD equations.** Since the characteristic diffusion time is much longer than the reaction time, the integration of the RD equations can be separated into two separate steps. The step describing diffusive motion is treated efficiently by one of the finite difference methods we studied in Chapter 4 – for instance, by the alternating-direction implicit (ADI) version of the Crank–Nicholson algorithm (Sections 4.4.1 and 4.4.2). The ADI method works well on a square lattice, which also allows the nodes to be centered along all of the boundaries of the system (although such centering is not required, it simplifies coding of the algorithm).

After each diffusive move, the 'fast-reaction' step is performed, and [P], [L] and [PL] concentrations are updated on all nodes to satisfy chemical equilibria therein: $K_b = k_f/k_r = [PL]/[P][L]$. Mathematically, concentration updates (from time t to $t + \Delta t$) for each node can be written in terms of reaction extent, δ, as $[P]_t - \delta = [P]_{t+\Delta}$, $[L]_t - \delta = [L]_{t+\Delta}$ and $[PL]_t + \delta = [PL]_{t+\Delta}$. Knowing the concentrations at time t, the value of δ is found by solving $K_b = [PL]_{t+\Delta}/[P]_{t+\Delta}[L]_{t+\Delta}$, an equation which is quadratic in δ and gives

$$\delta = \frac{([P]_t + [L]_t + K_b^{-1}) \pm \sqrt{([P]_t + [L]_t + K_b^{-1})^2 - 4([P]_t[L]_t - [PL]_t K_b^{-1})}}{2}$$

Of course, only $[P]_{t+\Delta} \geq 0$, $[L]_0 \geq [L]_{t+\Delta} \geq 0$ and $[L]_0 \geq [PL]_{t+\Delta} \geq 0$ are allowed. Finally, the initial conditions at $t = 0$ are: $[P] = [P]_0$ and $[L] = [PL] = 0$ in the stamp, and $[L] = [L]_0$ and $[P] = [PL] = 0$ in the gel. For the boundary conditions, the concentration of the protein at the top of the stamp is held constant, and other boundaries are specified as no flux (von Neumann conditions, $\partial P/\partial y = 0$ and $\partial P/\partial z = 0$ using the coordinates indicated in the right-hand part of Figure 9.16(b)).

(ii) **Fitting D_P and K_b.** Each set of D_P and K_b parameters tested gives different protein concentration profiles for different stamping times. When the concentrations are normalized to between 0 and 1 (where $I = 0$ corresponds to [P] + [PL] = 0 at a given location; $I = 1$ corresponds to the maximal value of [P] + [PL] in the profile), the modeled and experimental images can be compared pixel by pixel. The difference between these two sets is then quantified by the sum of squared differences over all pixels (i, j) and times, t:

$$R = \sum_{i,j,t} (I(i,j,t)_{\text{model}} - I(i,j,t)_{\text{exp}})^2$$

As illustrated in the figure below, the $R(D_P, K_b)$ function is relatively smooth, and has one global minimum corresponding to the optimal set $\{D_P, K_b\}$. This minimum can be identified by various minimization methods such as the conjugate gradient algorithm described in detail elsewhere.[52] Note that one stamping experiment allows for several such optimizations, one for each stamped line. In other words, one stamping can create statistics of the binding constants.

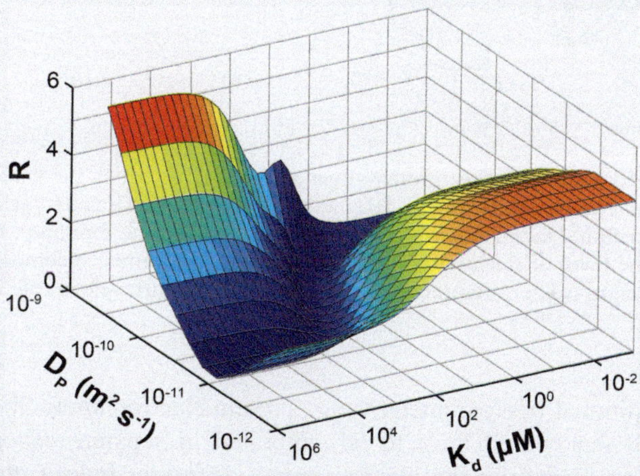

In the figure above, the error, R, is shown as a function of both D_P and K_b optimized against experimental data describing binding of m-aminobenzamidine ligand to bovine trypsin. The global minimum, $R \sim 5 \times 10^{-3}$, for this particular sample corresponds to $D_P = 1.66 \times 10^{-10}\,\text{m}^2\,\text{s}^{-1}$ and $K_b = 17.4\,\mu\text{M}$.

The precision with which this method can determine K_b values for various protein–ligand pairs is quite remarkable, as illustrated for three systems listed in Figure 9.16.[46–49] The lower limit of the method's sensitivity is sub-picomolar and it places no specific demands on the proteins/ligands it uses. Curiously, the key to its success is the diffusive nature of protein transport. Although external fields could certainly be added to the system to move proteins around more rapidly, they would

Protein	Ligand	K_d^{our}(μM)	K_d^{ref}(μM)	Ref
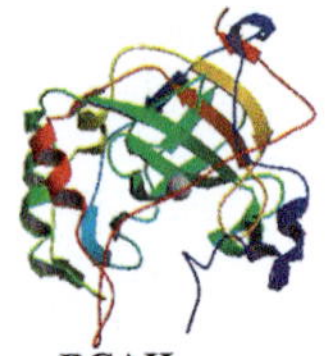 **BCAII**	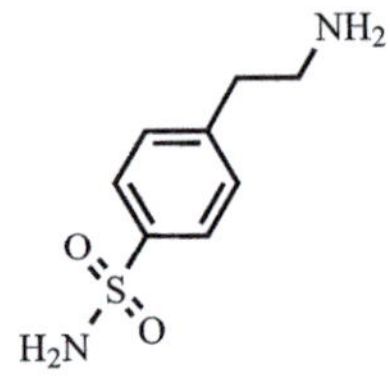	**2.90**	**1.69**	**[21]**
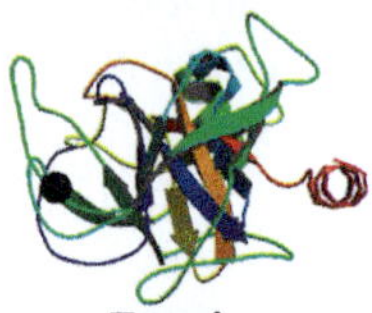 **Trypsin**	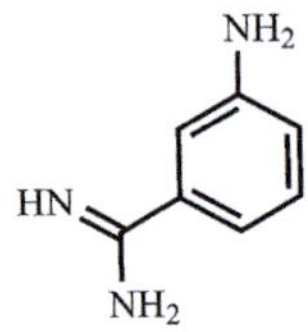	**14.8**	**26**	**[22]**
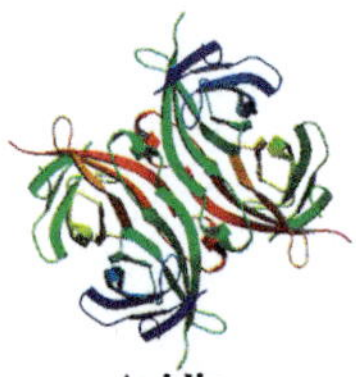 **Avidin**	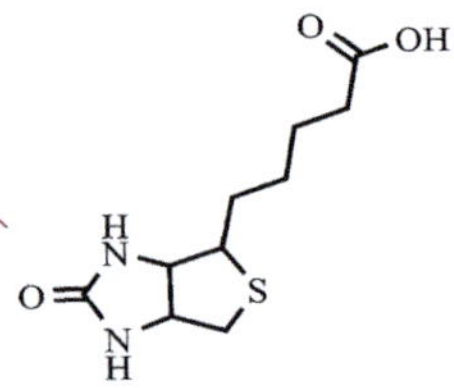	**<10^{-6}**	**~10^{-9}**	**[23,24]**

Figure 9.16 Examples of three protein–ligand systems, for which the values of binding constants, K_b, were measured using the RD method. For both BCAII and trypsin, the binding constants are within a factor of two of the values reported in the literature. Biotin–avidin binding is at the level of sensitivity of the RD approach (K_b in the picomolar range); the experimental value is about femtomolar. (Protein images courtesy of Protein database.)

introduce additional free parameters (e.g., protein electrophoretic mobility in an electric field) that would have to be calibrated in separate experiments and somehow decoupled from the always-present diffusion. Indeed, the beauty of RD is largely in its simplicity.

REFERENCES

1. Hooshangi, S., Thiberge, S. and Weiss, R. (2005) Ultrasensitivity and noise propagation in a synthetic transcriptional cascade. *Proc. Natl Acad. Sci. USA*, **102**, 3581.
2. Chu, S., DeRisi, J., Eisen, M. *et al.* (1998) The transcriptional program of sporulation in budding yeast. *Science*, **282**, 699.
3. Arnone, M.I. and Davidson, E.H. (1997) The hardwiring of development: organization and function of genomic regulatory systems. *Development*, **124**, 1851.
4. Davidson, E.H., Rast, J.P., Oliveri, P. *et al.* (2002) A genomic regulatory network for development. *Science*, **295**, 1669.
5. Alberts, B., Johnson, A., Lewis, J. *et al.* (2002) *Molecular Biology of the Cell*, Garland Science, New York.

6. Dormann, D., Vasiev, B. and Weijer, C.J. (2002) Becoming multicellular by aggregation: the morphogenesis of the social amoebae, *Dicyostelium discoideum. J. Biol. Phys.*, **28**, 765.
7. Renshaw, E. (1991) *Modeling Biological Populations in Space and Time*, Cambridge University Press, New York.
8. Rohrlich, S.T. and Porter, K.R. (1972) Fine-structural observations relating to production of color by iridophores of a lizard, *Anolis carolinensis. J. Cell Biol.*, **53**, 38.
9. Oshima, N. (2001) Direct reception of light by chromatophores of lower vertebrates. *Pigm. Cell. Res.*, **14**, 312.
10. Hanlon, R.T., Cooper, K.M., Budelmann, B.U. and Pappas, T.C. (1990) Physiological color change in squid iridophores. *Cell Tiss. Res.*, **259**, 3.
11. Cooper, K.M., Hanlon, R.T. and Budelmann, B.U. (1990) Physiological color change in squid iridophores. *Cell Tiss. Res.*, **259**, 15.
12. Bitner, A., Fialkowski, M. and Grzybowski, B.A. (2004) Color micropatterning with reconfigurable stamps. *J. Phys. Chem. B*, **108**, 19904.
13. Campbell, C.J., Fialkowski, M., Klajn, R. *et al.* (2004) Color micro- and nanopatterning with counter-propagating reaction–diffusion fronts. *Adv. Mater.*, **16**, 1912.
14. Alfrey, T., Gurnee, E.F. and Lloyd, W.G. (1966) Diffusion in glassy polymers. *J. Polym. Sci. C*, **12**, 249.
15. Gehrke, S.H., Biren, D. and Hopkins, J.J. (1994) Evidence for Fickian water transport in initially glassy poly(2-hydroxyethyl methacrylate). *J. Biomater. Sci. Polym. Ed.*, **6**, 375.
16. Idziak, I., Avoce, D., Lessard, D. *et al.* (1999) Thermosensitivity of aqueous solutions of poly(N,N-diethylacrylamide). *Macromolecules*, **32**, 1260.
17. Wittmann, J.C., Meyer, S., Damman, P. *et al.* (1998) Preparation and characterization of side-chain liquid crystalline polymer thin films aligned on PTFE friction-transferred layers. *Polymer*, **39**, 3545.
18. Huck, N.P.M., Jager, W.F., deLange, B. and Feringa, B.L. (1996) Dynamic control and amplification of molecular chirality by circular polarized light. *Science*, **273**, 686.
19. Helfand, E. (1971) Theory of kinetics of conformational transitions in polymers. *J. Chem. Phys.*, **54**, 4651.
20. Arrondo, J.L.R., Muga, A., Castresana, J. and Goni, F.M. (1993) Quantitative studies of the structure of proteins in solution by Fourier-transform infrared spectroscopy. *Prog. Biophys. Mol. Biol.*, **59**, 23.
21. Bitner, A., Fialkowski, M., Smoukov, S.K. *et al.* (2005) Amplification of changes of a thin film's macromolecular structure into macroscopic reaction–diffusion patterns. *J. Am. Chem. Soc.*, **127**, 6936.
22. Bensemann, I.T., Fialkowski, M. and Grzybowski, B.A. (2005) Wet stamping of microscale periodic precipitation patterns. *J. Phys. Chem. B*, **109**, 2774.
23. Muller, S.C. and Ross, J. (2003) Spatial structure formation in precipitation reactions. *J. Phys. Chem. A*, **107**, 7997.
24. Antal, T., Droz, M., Magnin, J. *et al.* (1998) Derivation of the Matalon–Packter law for Liesegang patterns. *J. Chem. Phys.*, **109**, 9479.
25. Witt, D., Klajn, R., Barski, P. and Grzybowski, B.A. (2004) Applications, properties and synthesis of omega-functionalized *n*-alkanethiols and disulfides: the building blocks of self-assembled monolayers. *Curr. Org. Chem.*, **8**, 1763.
26. Love, J.C., Estroff, L.A., Kriebel, J.K. *et al.* (2005) Self-assembled monolayers of thiolates as a form of nanotechnology. *Chem. Rev.*, **105**, 1103.

27. Tour, J.M., Jones, L., Pearson, D.L. *et al.* (1995) Self-assembled monolayers and multilayers of conjugated thiols, alpha,omega-dithiols, and thioacetyl-containing adsorbates: understanding attachments between potential molecular wires and gold surfaces. *J. Am. Chem. Soc.*, **117**, 9529.
28. Weiss, E.A., Kriebel, J.K., Rampi, M.A. and Whitesides, G.M. (2007) The study of charge transport through organic thin films: mechanism, tools and applications. *Phil. Trans. R. Soc. A*, **365**, 1509.
29. Bishop, K.J.M. and Grzybowski, B.A. (2006) Localized chemical wave emission and mode switching in a patterned excitable medium. *Phys. Rev. Lett.*, **97**, 128702.
30. Winfree, A.T. (1974) Rotating chemical reactions. *Sci. Am.*, **230**, 82.
31. Field, R.J., Noyes, R.M. and Koros, E. (1972) Oscillations in chemical systems: II. Thorough analysis of temporal oscillation in the bromate–cerium–malonic acid system. *J. Am. Chem. Soc.*, **94**, 8649.
32. Tyson, J.J. (1985) *Oscillations and Traveling Waves in Chemical Systems*, (eds R.J., Field, and M. Burger), John Wiley & Sons, Ltd, New York. p. 93.
33. Reusser, E.J. and Field, R.J. (1979) Transition from phase waves to trigger waves in a model of the Zhabotinskii reaction. *J. Am. Chem. Soc.*, **101**, 1063.
34. Showalter, K., Noyes, R.M. and Turner, H. (1979) Chemical oscillations and instabilities: detailed studies of trigger wave initiation and detection. *J. Am. Chem. Soc.*, **101**, 7463.
35. Epstein, I. and Pojman, J. (1998) *An Introduction to Nonlinear Chemical Dynamics: Oscillations, Waves, Patterns, and Chaos*, Oxford University Press, New York.
36. Field, R.J. and Burger, M (eds). (1985) *Oscillations and Traveling Waves in Chemical Systems*, John Wiley & Sons, Ltd, New York.
37. Bishop, K.J.M., Fialkowski, M. and Grzybowski, B.A. (2005) Micropatterning chemical oscillations: waves, autofocusing, and symmetry breaking. *J. Am. Chem. Soc.*, **127**, 15943.
38. Briggs, T.S. and Rauscher, W.C. (1973) Oscillating iodine clock. *J. Chem. Ed.*, **50**, 496.
39. Furrow, S.D., Honer, K. and Cervellati, R. (2004) Inhibitory effects by ascorbic acid on the oscillations of the Briggs–Rauscher reaction. *Helv. Chim. Acta*, **87**, 735.
40. Cervellati, R., Innolcenti, G., Dall'Acqua, S. *et al.* (2004) Polyphenols from *Polygala* spp. and their antioxidant activity. *Chem. Biodivers.*, **1**, 415.
41. Cervellati, R., Renzulli, C., Guerra, M.C. and Speroni, E. (2002) Evaluation of antioxidant activity of some natural polyphenolic compounds using the Briggs–Rauscher reaction method. *J. Agric. Food Chem.*, **50**, 7504.
42. Cervellati, R., Honer, K. and Furrow, S.D. (2004) An experimental and mechanistic investigation of the complexities arising during inhibition of the Briggs–Rauscher reaction by antioxidants. *Helv. Chim. Acta*, **87**, 133.
43. Oravcova, J., Bohs, B. and Lindner, W. (1996) Drug–protein binding studies: new trends in analytical and experimental methodology. *J. Chromatogr. B*, **677**, 1.
44. Cuatrecasas, P. and Anfinsen, C.A. (1971) Affinity chromatography. *Annu. Rev. Biochem.*, **40**, 259.
45. Patton, W.F. (2002) Detection technologies in proteome analysis. *J. Chromatogr. B*, **771**, 3.
46. Poulsen, S.A., Davis, R.A. and Keys, T.G. (2006) Screening a natural product-based combinatorial library using FTICR mass spectrometry. *Bioorg. Med. Chem.*, **14**, 510.
47. Luong, J.H.T., Male, K.B. and Nguyen, A.L. (1998) Synthesis and characterization of a water-soluble affinity polymer for trypsin purification. *Biotech. Bioeng.*, **31**, 439.
48. Green, N.M. (1990) Avidin and streptavidin. *Meth. Enzymol.*, **184**, 51.

49. Diamandis, E.P. and Christopoulos, T.K. (1991) The biotin–(strept)avidin system: principles and applications in biotechnology. *Clin. Chem.*, **37**, 625.
50. Fishbein, W.N. (1972) Quantitative densitometry of 1–40 μg protein in acrylamide gel slabs with coomassie blue. *Anal. Biochem.*, **46**, 388.
51. Sedmak, J.J. and Grossberg, S.E. (1977) Rapid, sensitive and versatile assay for protein using coomassie brilliant blue G250. *Anal. Biochem.*, **79**, 544.
52. Press, W.H. (1992) *Numerical Recipes in C: The Art of Scientific Computing*, Cambridge University Press, New York.

10

Reaction–Diffusion in Three Dimensions and at the Nanoscale

The various modalities of wet stamping discussed so far can delineate the initial and/or boundary conditions for reaction–diffusion (RD) processes over a variety of *two-dimensional* substrates. Unfortunately, they are not easily extended to three dimensions, save some rather unimaginative tricks like stamping a three-dimensional (3D) substrate from different directions. In the absence of appropriate stamping methods, the way to perform RD fabrication in three dimensions is to make use of the boundary conditions specified by the shapes of the objects/particles, and to propagate RD inwards from the particle surfaces. The idea is that with appropriately chosen reaction and diffusion parameters, the inward-propagating RD will enable fabrication of structures *inside* of the particles. The procedure is loosely analogous to the ancient European craft of *Geduldflaschen* – in German, 'patience bottles' – whereby intricate structures (e.g., models of ships) are built inside closed containers (e.g., bottles) by assembling the components 'remotely' from outside (Figure 10.1). In our remote fabrication, it is the RD processes that will patiently put the molecular components into their proper places.

In this chapter, we will focus on two types of systems: (i) micro-/mesoscopic particles made from porous materials and (ii) nanoscopic ones made of metals, alloys, or semiconductors. In the former case, we will learn how to make structures within structures, and how to process them chemically after completion of RD. In the latter, we will discuss phenomena in which nanoscopic particle dimensions can offset extremely small diffusion coefficients in solids ($\sim 10^{-24}\,m^2\,s^{-1}$ at room temperature) and thus enable fabrication of nanostructures possessing some

Chemistry in Motion: Reaction–Diffusion Systems for Micro- and Nanotechnology Bartosz A. Grzybowski

Figure 10.1 Making models of ships inside bottles is a form of 'remote' fabrication – though not with RD and not at the microscale. The model shown is of the *Queen Margaret* barque (built 1890). (Picture reproduced with the kind permission of David S. Smith, http://seafarer.netfirms.com.)

unique and useful characteristics. As we proceed, we will develop theoretical models for these processes. In doing so, however, we will be able to implement analytical approaches only for the very symmetric geometries like spheres, for which only radial derivatives appear in the governing RD equations (Section 2.4). Other shapes like cubes or polygonal plates will require the use of finite element methods discussed in Section 4.3.2.

10.1 FABRICATION INSIDE POROUS PARTICLES

The desire to fabricate small core-and-shell particles (CSPs) is fueled by their various applications in delivery systems,[1,2] and as components of microstructured materials.[3,4] Although methods based on surface tension effects can produce spherical or spheroidal CSPs (e.g., using microfluidic encapsulation[5,6] or folding of prepolymer patches in the effective absence of gravity[7]), they cannot be used to prepare particles in which the symmetries of the entire particle and of its internal component(s) are different. For example, it is relatively easy to make spheres containing one or more smaller spheres,[1,7] but it is rather difficult to make a polyhedron containing a sphere. Here is where RD can help.

10.1.1 Making Spheres Inside of Cubes

Let us begin with the challenge of fabricating spherical metallic structures inside of small, transparent cubes (Figure 10.2). The cubes themselves are prepared either from agarose or polydimethylsiloxane (PDMS) by conventional molding[8] against photolithographic masters presenting cubical microwells. Agarose and PDMS are chosen because (i) in the absence of additives, they are optically transparent and

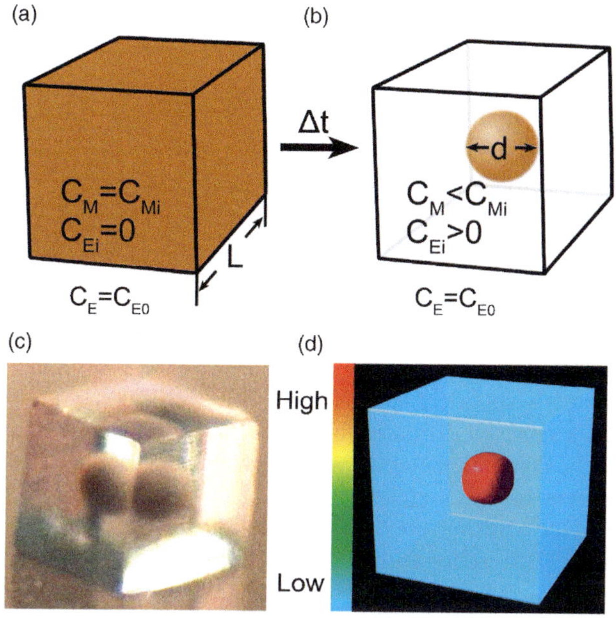

Figure 10.2 Fabrication by 3D RD. (a) A small particle (here, an agarose cube) is initially filled uniformly with copper colloids of concentration C_{M}^{0}. Upon immersion in an etchant solution (concentration C_{E}^{0}), a sharp RD front starts propagating into the particle and ultimately 'fabricates' a spherical core suspended in a clear agarose matrix. (c) Experimental image of a spherical copper core inside of a 1 mm agarose cube. (d) An example of a modeled core-and-shell structure. Red regions correspond to high concentration of unetched metal in the core

(ii) they allow for a range of chemistries to be performed in their bulk using either aqueous (for agarose) or organic (for PDMS) solvents. The initial state for the cubes is that of uniform loading with additives that are to participate later in the RD process (Figure 10.2(a)). For agarose cubes, these additives are small copper colloids (average diameter ~15 nm, concentration ~31 mM in terms of copper atoms) deposited electrolessly (see Chapter 8) in the already solidified cubes. For PDMS, the additives are gold nanoparticles (AuNPs; 5.6 nm, 38 mM in terms of gold atoms) stabilized with mixed monolayers of octanethiol and hexadecanethiol. These NPs are mixed with the PDMS prepolymer prior to molding and curing.

The RD fabrication commences when the cubes are placed in a solution of an appropriate chemical etchant, which oxidizes the metallic colloids/NPs into soluble metal salts that diffuse out of the cube and into solution. For agarose/copper cubes, the etchant is 100 mM HCl under aerobic conditions (i.e., HCl/O_2), which reacts with metallic copper to produce copper chloride: $2HCl + Cu + \frac{1}{2}O_2 \rightarrow CuCl_2 + H_2O$. For PDMS, a commercial, water-based Transene TFA gold etchant is mixed (1:1 v/v) with tetrahydrofuran (THF). The role of

THF is to swell the PDMS and allow the reactive I_3^- to penetrate into the particle where it oxidizes AuNPs into water-soluble $[AuI_2]^-$ according to $2Au + I^- + I_3^- \rightarrow 2[AuI_2]^-$.

Upon immersion in the etchant solution, the etching reaction propagates inwards as a sharp front leaving behind a region of clear agarose/PDMS. Figure 10.3(a) and 10.3(b) illustrate such propagation into cubical agarose/copper and PDMS/AuNP particles, respectively. In both cases, the sizes of the unetched regions ('the cores') decrease with the time of etching (front speed ~18.5 μm min^{-1} in agarose/copper and ~1.6 μm min^{-1} in PDMS/AuNP; Figure 10.3(c)) and their shapes gradually evolve from rounded-cubical to spherical. These shape changes can be quantified in the form of sphericity indices, Φ (Figure 10.3(d)), calculated from the two-dimensional projections of the cores and ranging from $\Phi = 1$ for a perfectly spherical core to $\Phi = 0$ for a cubical one. For both types of materials and for cubes of different sizes, the cores can be classified as spherical with $\Phi > 0.7$ when their diameters, d, are less than ~40% of the cube size, L. Also, statistics over sets of cubes etched for the same time reveal that the standard deviations, σ, of the core diameters at different times are 2.5–5% for agarose cubes and 2–6.7% for PDMS cubes with the upper values corresponding to the longest etching times. This polydispersity is minimal when the cubes are vigorously agitated during etching. In the absence of agitation, the cubes stick to one another and/or to the walls of the container, causing them to be etched to different extents and less symmetrically than cubes fully exposed to the etching solution.

10.1.2 Modeling of 3D RD

The key feature of the etching process is the sharpness of the RD front. To understand its origin, let us first, as usual, write the equations governing the RD process. Denoting the metal by the subscript M, and the etchant by E, we have

$$\begin{aligned} \partial C_E/\partial t &= D\nabla^2 C_E - \alpha k C_E C_M \\ \partial C_M/\partial t &= -\beta k C_E C_M \end{aligned} \tag{10.1}$$

where C_M is the concentration of metal colloids/NPs (in terms of atoms) immobilized in the cube, C_E is the concentration of etchant (limiting species), D is the diffusion coefficient of the etchant, k is the apparent reaction constant and α and β are the stoichiometric coefficients for the etching reaction: $\alpha E + \beta M \rightarrow$ soluble salt. These equations assume that metal particles do not diffuse through the agarose/PDMS matrix, and that the dissolved metal does not influence the reaction kinetics or the transport of the fresh etchant. The initial and boundary conditions are such that: (i) the concentration of metal is initially uniform throughout the cube, C_M^0, and (ii) the concentration of etchant is kept constant, C_E^0, at the surface of the cube (since the solution is well stirred). The RD equations can be further simplified by rescaling the variables: $\bar{C}_E = C_E/C_E^0$, $\bar{C}_M = C_M/C_M^0$, $\bar{\mathbf{x}} = \mathbf{x}/L$ and $\bar{t} = Dt/L^2$,

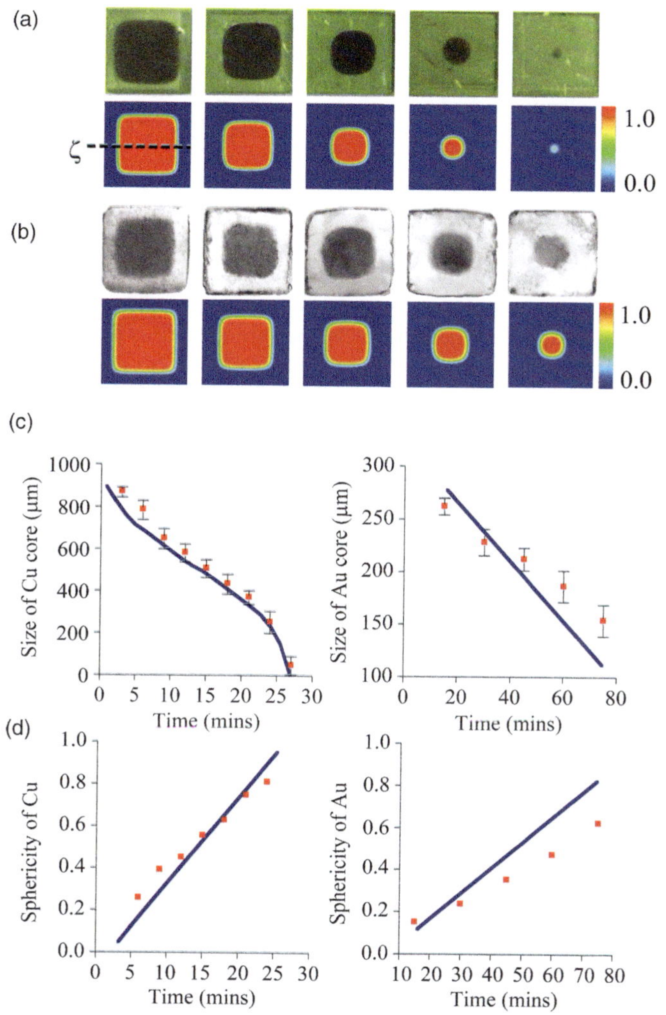

Figure 10.3 Evolution of cores inside of cubical particles. (a) The top row shows experimental, side-view images of 1 mm agarose/copper cubes etched for 6, 12, 18, 24 and 27 min. The bottom row shows the corresponding 2D projections of modeled 3D structures. (b) Analogous experimental and simulated structures in 400 μm PDMS/AuNP cubes at 15, 30, 45, 60 and 75 min. (c) Diameters of the copper and AuNP cores as a function of time (measured along the dashed line ζ shown in (a)). (d) Corresponding plots of the sphericity index Φ obtained from 2D core projections by (i) fitting a circle to an actual core shape and (ii) calculating the difference, A, in area between the two; and (iii) calculating the value of $\Phi = (A_{\square} - A)/A_{\square}$, where $A_{\square}$ is the area of a perfect square circumscribed about the core. In this way, $\Phi = 1$ for a perfectly circular core and $\Phi = 0$ for a square one. In (c) and (d), the markers correspond to experimental data and the blue lines to the simulations. Standard deviations are based on the averages over at least ten cubes for each time point.

where L is a characteristic length of the gel/polymer particle (e.g., the side of a cube). This procedure yields the nondimensional RD equations

$$\begin{aligned} \partial\bar{C}_{\mathrm{E}}/\partial\bar{t} &= \nabla^2\bar{C}_{\mathrm{E}} - \alpha \mathrm{Da}\bar{C}_{\mathrm{E}}\bar{C}_{\mathrm{M}} \\ \partial\bar{C}_{\mathrm{M}}/\partial\bar{t} &= -\beta\gamma\mathrm{Da}\bar{C}_{\mathrm{E}}\bar{C}_{\mathrm{M}} \end{aligned} \tag{10.2}$$

with initial conditions in the cube $\bar{C}_{\mathrm{M}}(\mathbf{x},0) = 1, \bar{C}_{\mathrm{E}}(\mathbf{x},0) = 0$, boundary condition for the etchant concentration at the cube's surface $\bar{C}_{\mathrm{E}}(\mathbf{x}_{\mathrm{S}},t) = 1$, and $\gamma = C_{\mathrm{E}}^0/C_{\mathrm{M}}^0$. The most important parameter in these equations is the dimensionless Damköhler number, $\mathrm{Da} = kL^2C_{\mathrm{M}}^0/D$, which here can be interpreted as a ratio of the characteristic width of the reaction zone,[9] $L_{\mathrm{RZ}} = (D/kC_{\mathrm{M}}^0)^{1/2}$, to the dimensions of the cube, L, as $\mathrm{Da} = (L/L_{\mathrm{RZ}})^2$. Thus, the 'sharpness' of the reaction zone, defined as L_{RZ}/L, may be expressed in terms of the Damköhler number as $\mathrm{Da}^{-1/2}$. We note, however, that this relation between sharpness and Da holds only asymptotically as the front moves far from the initial boundaries of the cube; thus, for intermediate times, the other dimensionless parameter, γ, may also influence the sharpness.

For the case of agarose/copper cubes, $\alpha = 1/2$, $\beta = 1$ and $C_{\mathrm{M}}^0 = 31$ mM. Since the etching process requires both HCl and O_2 (without oxygen, etching rates are much smaller), the effective etchant concentration is determined by the concentration of limiting reagent, O_2, dissolved in the etching solution (~8.4 mg L^{-1} at 25 °C) such that $C_{\mathrm{E}}^0 = 0.26$ mM and $\gamma \sim 0.01$. The diffusion coefficient of oxygen through the gel matrix is taken from the literature, $D = 2.1 \times 10^{-9}\,\mathrm{m^2\,s^{-1}}$, as is the rate constant of the surface reaction, $k_{\mathrm{S}} = 5.4 \times 10^{-4}\,\mathrm{m\,s^{-1}}$, determined previously for a planar copper surface etched with 100 mM HCl/O_2. For our system, however, this surface reaction rate has to be translated into a bulk etching rate accounting for the finite sizes of the colloidal particles (radii, $R \sim 5$–10 nm). The procedure for doing so is detailed in Example 10.1 and gives a Damköhler number for the process of $\mathrm{Da} \sim 10^4$.

Example 10.1 Transforming Surface Rates into Apparent Bulk Rates

Values for reaction rates describing metal etching are usually expressed in terms of the velocity of the receding surface of a bulk metal, $\mathrm{d}\delta/\mathrm{d}t$, with units of length per time. If the etching rate is first order with respect to the etchant of concentration C_{E}, the velocity of the receding surface can be written as $(1/\nu)(\mathrm{d}\delta/\mathrm{d}t) = k_{\mathrm{S}}C_{\mathrm{E}}$, where ν is the molar volume of the bulk metal and k_{S} is the surface rate constant of the etching reaction. Also, if the experimental values of this velocity are reported as $(\mathrm{d}\delta/\mathrm{d}t)_0$ for a given etchant concentration, C_{E}^0, the rate constant may be estimated as $k_{\mathrm{S}} = (\mathrm{d}\delta/\mathrm{d}t)_0/\nu C_{\mathrm{E}}^0$.

The question relevant to 3D fabrication – and instructive practice in chemical kinetics – is how these experimentally measured *surface* rate constants can be translated into the apparent *bulk* rate constants describing the dissolution of finite-sized colloidal particles immersed in the gel matrix.

To answer this question, we first consider the etching of a single spherical particle of radius R and containing N metal atoms, related to R as $N = \frac{4}{3}\pi R^3/\nu$. The etching rate may be defined as

$$\frac{1}{4\pi R^2}\frac{\mathrm{d}N}{\mathrm{d}t} = \frac{1}{\nu}\frac{\mathrm{d}R}{\mathrm{d}t} = -k_S C_E$$

Integrating this expression, we find that for $0 \leq t \leq R_0/k_S\nu C_E$, the radius of the particle evolves as $R = R_0 - k_S\nu C_E t$, where R_0 is the initial radius of the nanoparticle. Substituting this expression back into the equation above, we can write $\mathrm{d}N/\mathrm{d}t = -4\pi k_S C_E(R_0 - k_S\nu C_E t)^2$. Because this rate varies with t, we will integrate it over the period $R_0/k_S\nu C_E$ required to etch the entire particle. This procedure yields the *average* rate of particle etching:

$$\left\langle \frac{\mathrm{d}N}{\mathrm{d}t} \right\rangle = -4\pi R_0^2 k_S C_E/3$$

Now, for a collection of colloidal particles dispersed in some matrix, the rate at which metal atoms are consumed by the etching reaction is simply equal to the average rate for a single colloid times the total number of particles, N_P, in solution:

$$\frac{\mathrm{d}N_M}{\mathrm{d}t} = \left\langle \frac{\mathrm{d}N}{\mathrm{d}t} \right\rangle N_P$$

Here, the number of colloidal particles, N_P, is related to the number of metal atoms, N_M, as $N_M = (4\pi R_0^3/3\nu)N_P$, where $4\pi R_0^3/3\nu$ is simply the number of atoms in a single particle of radius R_0. Substituting this relation into the above equation and dividing by the total volume, V, we obtain

$$\frac{\mathrm{d}C_M}{\mathrm{d}t} = -\left(\frac{k_S\nu}{R_0}\right) C_M C_E$$

This rate equation, which describes how the total concentration of metal atoms, C_M, evolves in time, is identical to the second-order rate equations introduced in Equation (10.1). Thus, by inspection, we can identify $k = k_S\nu/R_0$ as the apparent bulk rate constant for the dissolution of metal atoms into soluble ions. Also, the Damköhler number for the RD process of core formation, defined in the main text as $\mathrm{Da} = kL^2C_M^0/D$, may now be related to the surface reaction rate as $\mathrm{Da} = k_S\nu L^2 C_M^0/DR_0$.

For the PDMS/AuNP cubes, I_3^- is the reagent limiting gold etching. In this case, $C_M^0 = 38$ mM and $C_E^0 = 8$ mM such that $\gamma \sim 0.2$; other parameters are $\alpha = 1/2$, $\beta = 1$ and $D = 5 \times 10^{-12}\,\mathrm{m^2\,s^{-1}}$. Using the experimentally estimated rate constant for AuNP etching, $k_S = 1.4 \times 10^{-6}\,\mathrm{m\,s^{-1}}$, the Damköhler number for $L = 400\,\mu$m cubes is then estimated at Da ~ 6200.

Figure 10.4 shows that when the nondimensional RD equations (10.2) are solved numerically (using the Fluent® finite volume method, which is a 3D version of the finite element method discussed in Section 4.3.2; for the source code see dysa.northwestern.edu) with the estimated Damköhler numbers, they give good quantitative agreement both in terms of the time evolution of the cores as well as their sphericities. Interestingly, the combination of core 'sharpness' (i.e., narrow reaction zone) and high sphericity is observed only for intermediate values of Da ($\sim 10^4$; Figure 10.4(b)). When Da is smaller

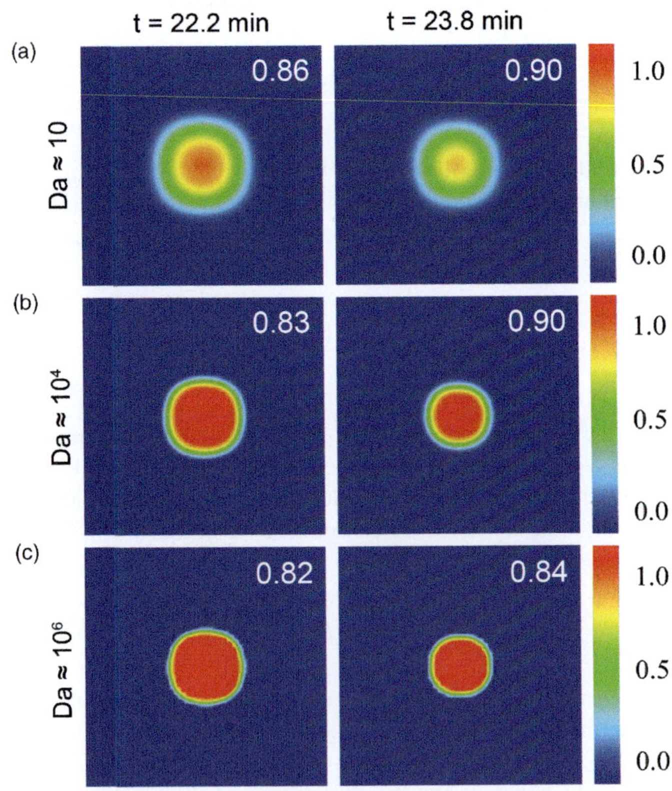

Figure 10.4 Effects of Damköhler number on the sharpness and the sphericity of metal cores formed in cubical particles at two different times ($t = 22.2$ and 23.8 min) and for $\gamma = 0.01$. (a) Da $= 10^3$; (b) Da $= 10^4$; (c) Da $= 10^6$. The intermediate values of Da give the best combination of core sharpness and sphericity. The numbers in the upper-right corners of the images give the sphericity indices, Φ

(e.g., of the order of 10^3; Figure 10.4(a)), the width of the reaction zone is roughly $Da^{-1/2}$ or 3% of the cube size, such that the contours of the cores – albeit more spherical – are significantly 'blurred'. In contrast, for high values of Da (e.g., 10^6; Figure 10.4(c)), etching becomes increasingly diffusion limited, and the etchant delivered to the reaction front is consumed immediately. As a result, the reaction front is sharp (~0.1% of L), but resembles more the initial, cubical contour.

10.1.3 Fabrication Inside of Complex-Shape Particles

For particles other than cubical, the RD process can generate cores of different shapes or even several cores per particle. This is shown in Figure 10.5, which shows experimental images (top row) and the corresponding calculated contours (bottom row) of CSPs of various shapes. The two leftmost examples illustrate a transformation of a curvilinear particle contour into a polygonal core and demonstrate that RD – with proper design of the curvature of the particle boundary – is not limited to etching rounded shapes. The third and fourth examples show that in particles with sharp internal corners, the RD fronts can 'break up' and produce multiple, separate cores in one particle. Finally, the two rightmost examples show particles having inward-pointing 'cuts' that control the angular orientation of the square cores. The fact that our simple RD model reproduces the experimentally developed cores indicates that in a well-stirred solution, any diffusion limitations on the delivery of fresh etchant into these cuts can be neglected. At this point it is probably obvious why the RD equations for these particles are best solved using commercial finite element solvers rather than in-house programs – imagine what Benedictine patience the coding of all these boundary conditions would take! With FEM software the source codes – available free of charge from dysa.northwestern.edu – are relatively simple to write and concise.

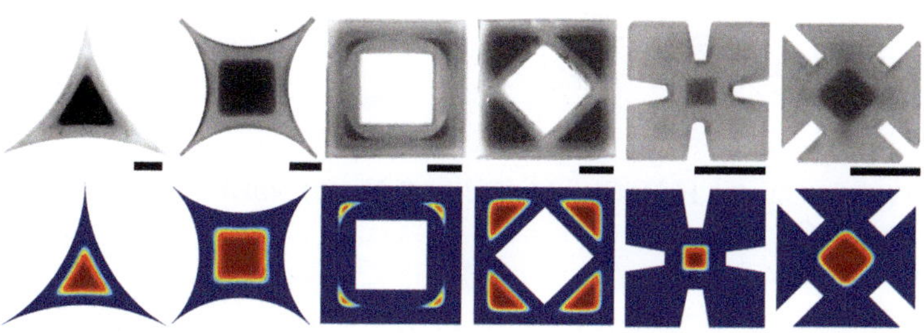

Figure 10.5 Examples of nonspherical cores generated by RD: triangular and square cores inside of particles with curved edges; multiple cores in 'frame'-shaped particles; square cores in cross-shaped particles. Scale bars = 500 μm. The bottom row has the corresponding structures modeled by FEM with parameters described in the text for agarose/copper system ($Da = 10^4$)

10.1.4 'Remote' Exchange of the Cores

Once fabricated, the cores can be further modified 'remotely' by galvanic replacement reactions (Figure 10.6). For example, when an agarose cube containing a copper core is immersed in 20 mM aqueous solution of $HAuCl_4{\cdot}3H_2O$, the following redox reaction takes place: $2AuCl_4^- + 3Cu \rightarrow 2Au + 3Cu^{2+} + 8Cl^-$. Since the difference in standard potentials, $E^0 = \varepsilon^0_{AuCl_4^-/Au} - \varepsilon^0_{Cu^{2+}/Cu} = 0.663$ V, is positive, this reaction spontaneously transforms the copper colloids in the cores into gold ones. Because the equilibrium constant for the reaction is very high

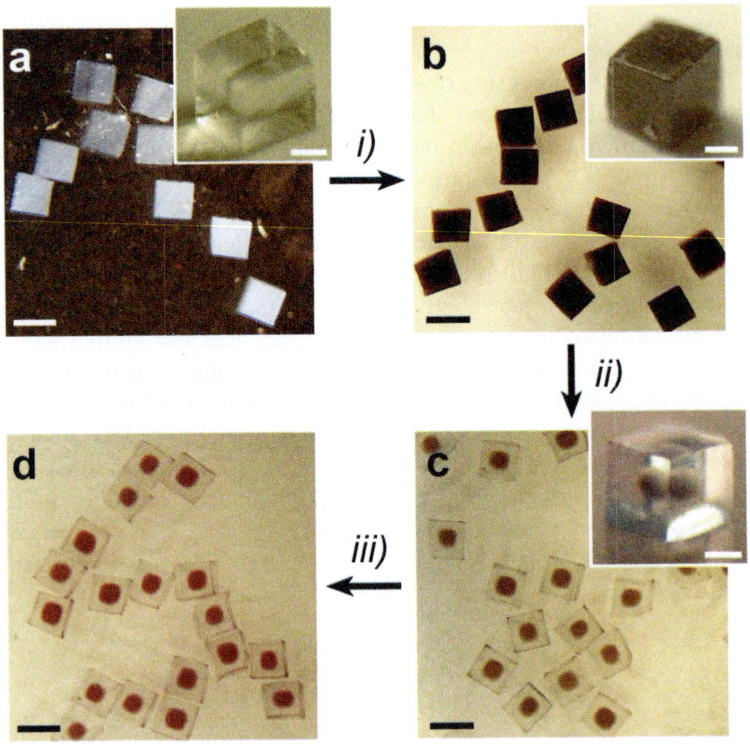

i) (a) $SnCl_2$, (b) H_2PdCl_4, (b) $[Cu(tart)_2]^{2-}$, (c) $NaBH_4$
ii) HCl, O_2
iii) $HAuCl_4$

Figure 10.6 Exchange of metal cores inside of agarose cubes. The images illustrate the fabrication sequence leading to gold cores inside of agarose cubes. (a) 1 mm pure agarose cubes are (b) filled uniformly with copper colloids. (c) Copper cores are etched using HCl/O_2 and (d) exchanged galvanically into gold cores. Galvanic reactions can exchange copper into other noble metals. Scale bars are 1 mm in the main images and 300 μm in the insets. The legend lists chemicals used ('tart' stands for tartrate)

($K_{eq} = e^{nFE^0/RT} = 10^{67}$, where $n = 6$ is the number of electrons exchanged in the reaction and F is the Faraday constant), the exchange reaction is quantitative (as verified by elemental analysis of the colloids comprising the cores). The process does not affect either the size or the shape of the core, and manifests itself by a color change of the core from brown (copper; Figure 10.6(c)) to red (gold; Figure 10.6 (d)). Also, the procedure can be extended to other metals having standard redox potentials higher than that of copper (i.e., $\varepsilon^0_{Me^{n+}/Me} > \varepsilon^0_{Cu^{2+}/Cu}$) and allows for fabrication of particles comprising silver ($\varepsilon^0_{Ag^+/Ag} = +0.800$ V), palladium ($\varepsilon^0_{Pd^{2+}/Pd} = +0.915$ V) and platinum ($\varepsilon^0_{Pt^{2+}/Pt} = +1.188$ V) cores. Other metals that could be used for the galvanic replacement include rhodium, mercury, iridium and others.

An interesting twist to this story is that under some conditions, the chemical modification of the cores can be only partial and can produce double-layer cores, as illustrated in Figure 10.8(c), where the outer regions of copper cores were oxidized to copper(I) chloride. Another possibility for multilayered particles is to etch the core of one metal and then deposit another metal in the particle. If the metals are catalytic (e.g., Pt, Pd, Au, etc.) such structures might have exciting uses for controlling *sequential* catalytic reactions (Figure 10.7). For example, if the cores contain PdNPs catalyzing hydrogenation reaction, while CuNPs catalyzing the so-called alkyne coupling reaction are dispersed throughout the entire particle volume, a phenylacetylene substrate entering the particle will be first coupled into 1,4-diphenylbutadiyne and only then, upon reaching the core, will undergo

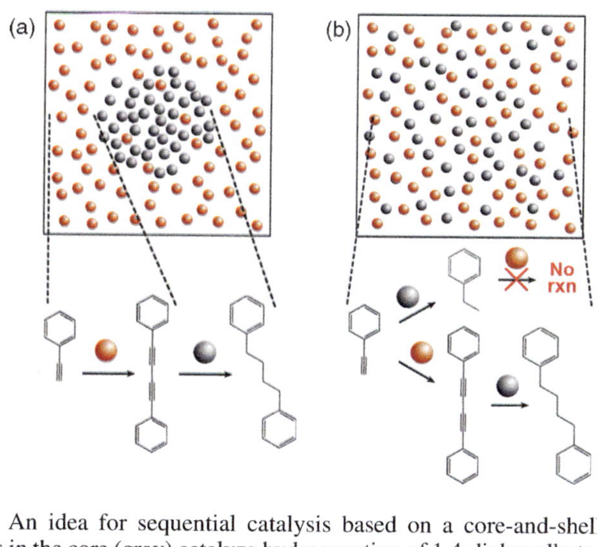

Figure 10.7 An idea for sequential catalysis based on a core-and-shell architecture. (a) The PdNPs in the core (gray) catalyze hydrogenation of 1,4-diphenylbutadiyne intermediate formed by the coupling of phenylacetylene substrates on CuNPs (orange) in the outer layer. This sequence of steps leads to one synthetic product, 1,4-diphenylbutane. (b) In contrast, if both types of NPs are distributed over the particle entire volume, the substrate can react with both types of NPs to give a mixture of products (ethylbenzene and 1,4-diphenylbutane)

hydrogenation to give 1,4-diphenylbutane (Figure 10.7(a)). Of course, to ensure the sequential nature of the process, the dimensions of the core and of the outer shell should be optimized such that the phenylacetylene substrate is fully converted into the 1,4-diphenylbutadiyne intermediate before reaching the core. Note that without the geometrical confinement of the catalysts, the substrates could react with both of them in an arbitrary sequence to give a mixture of products (Figure 10.7(b)).

10.1.5 Self-Assembly of Open-Lattice Crystals

Leaving other fabrication schemes and applications of individual particles to the creative reader, let us consider the collections of such CSPs. The opportunity here is to combine RD particle fabrication with self-assembly[10,11] – that is, the process by which discrete components organize without any human intervention into ordered and/or functional suprastructures. The unique feature of CSPs is that their self-assembly leads to structures in which the cores are separated from one another and 'suspended' in a transparent matrix (Figure 10.8). Such open-lattice crystals are interesting for applications in diffractive optics and are generally considered a challenging fabrication target requiring laborious layering of the matrix materials and careful placement of individual cores.

Self-assembly can make such structures without much effort by gentle agitation of the CSPs dispersed in a liquid phase. Though straightforward in concept, this approach entails some experimental nuances worth mentioning. First, there is an issue of controlling the particle surface properties. These properties are crucial to self-assembly and need to be adjusted carefully to give ordered assemblies. If the solvent in which the particles are dispersed is polar (e.g., water), the particles should be effectively nonpolar to attract one another; conversely, if the solvent is nonpolar (e.g., hexanes), particle surfaces should be polar. In addition, the difference in solvent–particle polarities (the so-called hydrophobic contrast[12]) also matters: if it

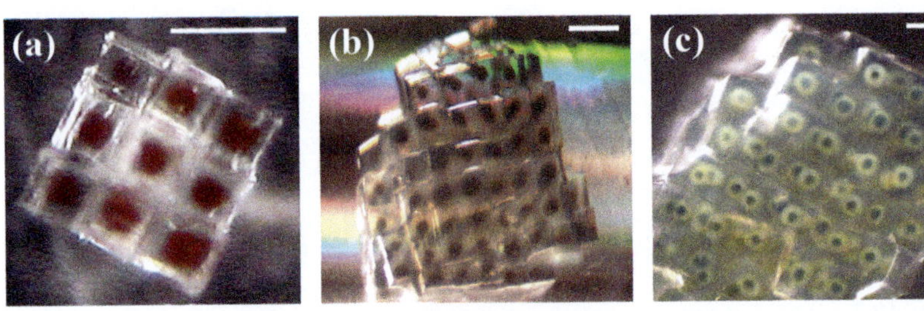

Figure 10.8 Three-dimensional assemblies of sphere-in-cube particles. (a) A $3 \times 3 \times 3$ 'supercube' made of 400 μm PDMS cubes with cores comprising 5.6 nm gold nanoparticles. (b) An assembly of about 250 agarose cubes (500 μm each) containing copper cores. (c) 1 mm agarose cubes with copper cores whose outer layers (greenish regions) were oxidized to CuCl. Scale bars = 1 mm

is too small, the interparticle attractions will be too weak to produce a stable assembly; if it is too large, the particles will clump rapidly into a structureless aggregate. With agarose particles containing mostly water, the area of exposed solid surface is too small to control the hydrophobic contrast directly via chemical modification. With PDMS pieces, on the other hand, the surfaces can be modified chemically (e.g., by surface oxidation), but the degree of such modification is hard to control. Fortunately, in both cases, hydrophobic contrast can be adjusted to a desired degree by changing the solvent composition, although finding an optimal solvent system can sometimes be time consuming. For the agarose/copper system, the fine-tuned procedure for self-assembly involves transferring the water-based pieces into acetone and then adding 5% v/v hexane while gently shaking the solution on an orbital shaker. For PDMS particles, self-assembly proceeds well under gentle agitation in a mixture of acetonitrile and 1-octadecene in 1250:1 v:v ratio (with pure, polar acetonitrile, the particles clump into orderless structures; with too much nonpolar 1-octadecene, the particles do not form large assemblies).

A second experimental complication concerns particle shapes. As a rule of thumb, the less symmetric the particles, the more difficult it is to achieve their self-assembly into ordered structures. In this spirit, assembling spheres or cubes is relatively straightforward, while assembling ellipsoids or crosses is usually much harder. Since low-symmetry particles can assume more orientations for which particle–particle attractions are of similar magnitudes, there are more possible ways in which these particles can form energetically similar – but not necessarily ordered – structures. In the language of thermodynamics, we would say that the system can be trapped in local energetic minima before reaching the global minimum corresponding to a perfectly ordered structure. The solution to this problem is to allow the particles to adjust their positions during self-assembly by coming together and then falling apart before finding perfect orientation. For such equilibration to be efficient, the particles should experience as many encounters as possible. With molecules or nanoparticles, this last condition is not an issue since the inherent thermal agitation (Brownian motion) results in as many as 10^{13}–10^{14} collisions per second. With mesoscopic (hundreds of micrometers to millimeters) particles, however, Brownian motions are negligible, and one must use external agitation (mechanical shaking, ultrasound waves, etc.) to make the particles collide. Although collision rates of hundreds per minute are realistic, they are still a far cry from the molecular rates, and mesoscale self-assembly can take a very long time to equilibrate. Thus, development of new agitation schemes is necessary to achieve efficient self-assembly of mesoscopic particles of shapes more complex than the cubes we used here.

The last remark we make in this section is about the practicality of the 3D RD/self-assembly structures. Although we have already mentioned the uses of supracrystals as diffractive elements, it must be stated very clearly that for this to become a reality, the particles must be smaller by at least few times (tens of micrometers). While there is nothing fundamental that prohibits such miniaturization, there are some technical challenges to be overcome (e.g., molding smaller particles, controlling reaction rates to etch very small cores). On the other hand, the soft materials we are working with

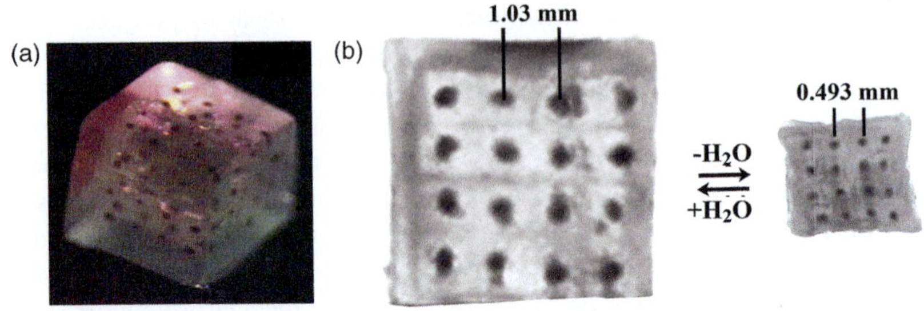

Figure 10.9 Shrinkable assemblies. (a) A $4 \times 4 \times 4$ supercube of 1 mm agarose cubes containing spherical copper cores. This cube is encased in PDMS. (b) The supercube shrinks when dried and expands when rehydrated. The change in linear dimensions is roughly a factor of two

present some unique opportunities – for instance, larger pieces can be first assembled and then shrunk by dehydrating the hydrogel matrix (Figure 10.9).

Leaving all these miniaturization issues for future investigations, we are now going to switch gears and talk about particles that are not porous but are extremely small. We are going to the nanoscale!

10.2 DIFFUSION IN SOLIDS: THE KIRKENDALL EFFECT AND FABRICATION OF CORE–SHELL NANOPARTICLES

Although our current repertoire of RD fabrication techniques is quite flexible in terms of the reacting chemicals, it is limited by the required porosity/permeability of the supporting medium (a gel, a polymer, or an elastomer). In this section, we will study – and later apply in nanofabrication – phenomena that overcome this limitation with the help of so-called solid interdiffusion.

Can atoms of crystalline solids such as metals or semiconductors diffuse into one another? Indeed, they can. The first experimental demonstration of such interdiffusion dates back to 1896 when Sir Roberts-Austen observed it in a couple of gold and lead metals. For a long time afterward, scientists believed that interdiffusion occurs by a direct exchange of atoms (Figure 10.10) or via a 'ring diffusion' mechanism, in which the atoms swap their positions in a concerted fashion.[13] It was not until 1947 when a young American scientist, Ernest Kirkendall, performed his famous experiment suggesting that interdiffusion is instead due to the swapping of atoms and crystal vacancies (Figure 10.10). Interestingly, this explanation had originally met with much disbelief or even hostility since the concentrations of vacancies (about one in a million lattice sites of a metal)[13] were historically considered too low for the atom–vacancy exchange to be the dominant effect. Nevertheless, Kirkendall stuck to his explanation tenaciously and after some bitter struggles (involving repeated rejection of his paper[13]) had the effect confirmed and, ultimately, approved by the community.[14]

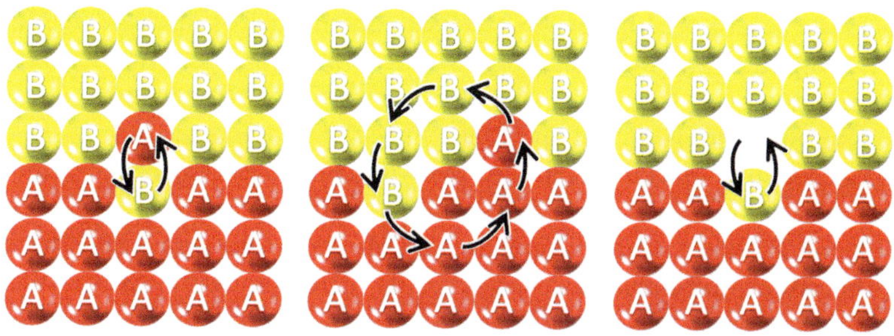

Figure 10.10 Plausible mechanisms of interdiffusion between solids A and B. Left: direct atom–atom exchange/'swapping'. Middle: ring diffusion. Right: Kirkendall atom–vacancy exchange

Let us briefly describe the 'canonical' Kirkendall experiment, in which metals/ alloys, A and B, are brought into contact, and a thin wire is placed carefully between them. The wire is attached to a laboratory bench and acts as a point of reference (Figure 10.11). Now the metals are welded together by placing them in

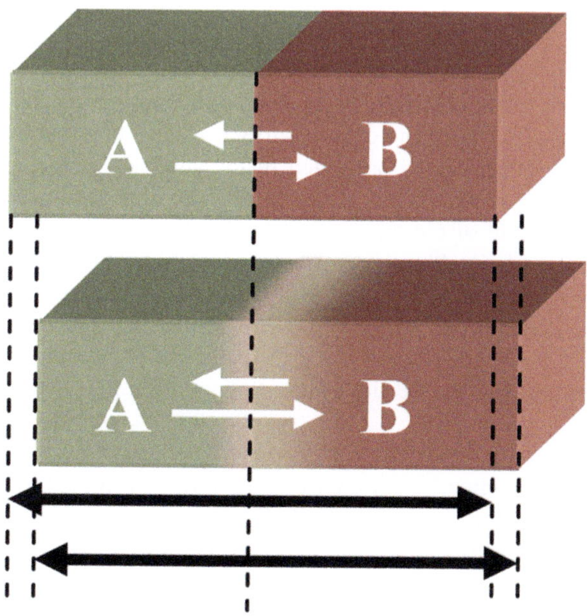

Figure 10.11 Scheme of Kirkendall's experiment, in which a heated AB diffusion couple translates with respect to a fixed point of reference

an oven at a temperature high enough to speed up interdiffusion but lower than the melting temperature (usually, one-third of the melting temperature).[15] When the specimen is removed from the oven, it is observed that the wire has 'traveled' laterally through the specimen, or rather, since the wire is fixed to the laboratory bench, the specimen has traveled past the wire. This net translation cannot be a result of either atom–atom exchange or the ring diffusion mechanism, for in such cases the metal slabs should have changed only their relative compositions, not their dimensions and locations. In the words of Kirkendall himself, 'diffusion formulas based on an equal interchange of solute and solvent [i.e., of A and B] atoms and a substantially stationary interface will be in error'.[13] The only explanation, with reference to Figure 10.10, is that metal A acquired a net amount of vacancies, or empty space, while metal B acquired a net amount of mass.

Since its discovery, the Kirkendall effect has been observed in materials other than the brass and copper used in the original experiments. The effect has also found numerous applications, most recently in nanotechnology where it can be cleverly applied to fabricate hollow and core-and-shell nanostructures.

Let us consider an illustrative experimental example (reported by the Berkeley-based group of Paul Alivisatos), in which a nanocrystal of cobalt is immersed in a boiling solution of sulfur and solvent (Figure 10.12).[16] At elevated temperatures, cobalt atoms diffuse rapidly into the solution leaving behind vacancies in the crystal. Since isolated vacancies on the crystal lattice are energetically unfavorable, they 'aggregate' near the crystal's center to ultimately produce the so-called Kirkendall void. Simultaneously, sulfur atoms from solution impinge onto the crystal surface where they react with cobalt to produce cobalt sulfide, Co_3S_4, arranged in the shell-like 'reaction zone'. The end result of this reaction-diffusion process is a hollow, compositionally homogeneous particle of Co_3S_4.

This process lends itself to an interesting theoretical RD analysis. Historically, models of the Kirkendall effect have considered the process to be strongly diffusion-limited. This assumption is justified for macroscopic systems, in which atoms diffuse over large distances; however, it may fail for nanoscale systems, where the time scales of reaction and diffusion are often commensurate.

In developing a zero-order RD model, we will make two simplifications. First, based on previous studies,[16–20] we will assume that the concentration profile of cobalt atoms within the cobalt sulfide shell adjusts rapidly relative to the speed of growth of the layer itself. Under such circumstances, we can use the quasi-steady approximation, in which the concentration profile of cobalt within the 'shell' obeys the *steady-state* diffusion equation. Second, we will assume that the diffusion of cobalt atoms into the liquid sulfur is much faster than the diffusion of sulfur atoms into the bulk cobalt. This assumption rests on the physical basis that there are more vacancies per unit volume in a liquid than in a solid. Hence, we neglect the diffusion of sulfur atoms into the cobalt nanocrystal entirely.

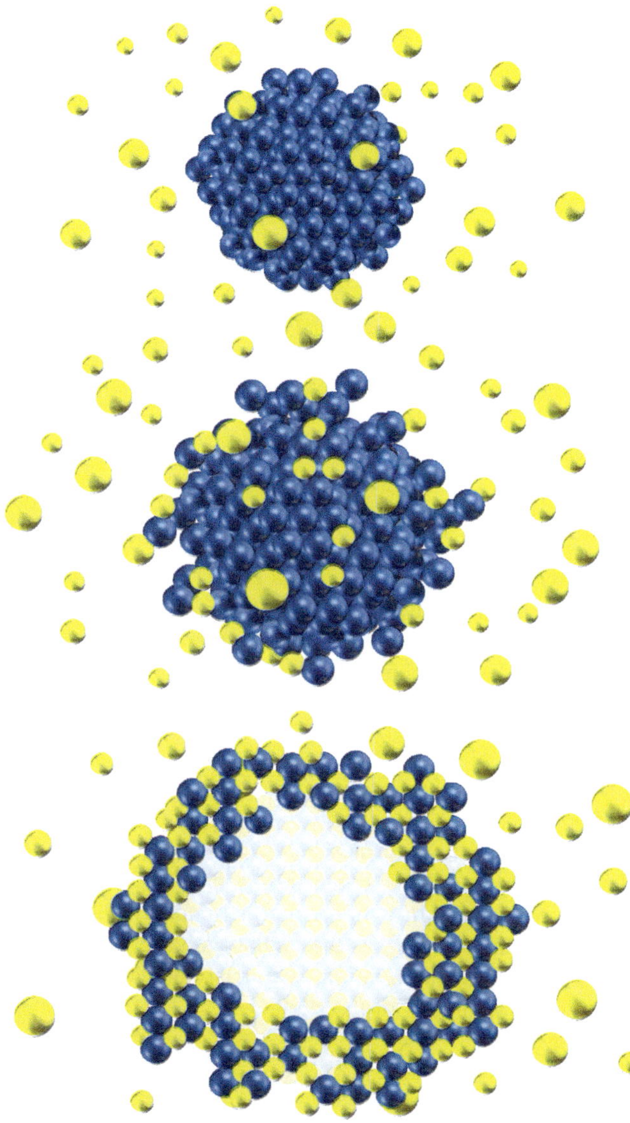

Figure 10.12 Top: a cobalt nanocrystal (represented as an aggregate of blue spheres) is immersed in a solution of sulfur (yellow spheres) and *ortho*-dichlorobenzene (not shown). Middle: when the temperature is raised to 400 K, cobalt atoms begin to diffuse into the solution leaving behind vacancies, which aggregate into one large void. Bottom: at the same time, the incoming sulfur atoms react with cobalt to give a hollow Co_3S_4 nanocrystal

With reference to Figure 10.13, the concentration of cobalt atoms, c, in the shell is governed by the steady-state diffusion equation $\nabla^2 c = 0$. At the interior boundary of the shell, the concentration of cobalt is equal to that of the bulk solid, $c(R_1) = 1/\nu_{Co}$, where ν_{Co} is the molar volume of cobalt; at the exterior boundary, conservation of

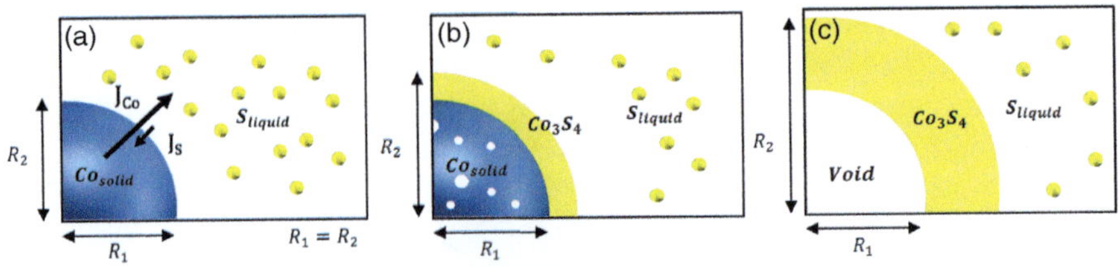

Figure 10.13 Growth of hollow Co_3S_4 nanocrystals modeled by RD. (a) Initially, the radius of a particle is equal to that of the cobalt crystal, $R_1 = R_2$. The diffusive fluxes of cobalt, J_{Co} (dominant), and sulfur, J_S (negligible), are also indicated by the arrows. (b) As cobalt atoms diffuse outwards, they leave behind crystal vacancies and react with sulfur to create a shell of outer radius R_2, which increases with time. (c) Ultimately, the vacancies aggregate into a Kirkendall void surrounded by a shell of Co_3S_4

cobalt atoms requires that $-D(\partial c/\partial r)_{R_2} = kc(R_2)$ (Section 8.1), where D is the diffusion coefficient and k is the surface reaction rate constant of the reaction producing Co_3S_4. To simplify the solution, we introduce the following dimensionless variables: $\bar{r} = r/R_1$, $\bar{c} = c/c_b$ and $\bar{R} = R_2/R_1$. Thus, our dimensionless problem becomes $\nabla^2\bar{c} = 0$ with boundary conditions $\bar{c}(1) = 1$ and $-(\partial\bar{c}/\partial\bar{r})_{\bar{R}} = \mathrm{Da}\bar{c}(\bar{R})$, where $\mathrm{Da} = kR_1/D$ is the familiar Damköhler number characterizing the ratio of the diffusive time scale, $\tau_{\mathrm{diff}} = R_1^2/D$, to the reaction time scale, $\tau_{\mathrm{rxn}} = R_1/k$. Solving the above equation in the relevant spherical coordinate system (Section 2.4) gives the following concentration profile for cobalt within the shell ($1 \leq \bar{r} \leq \bar{R}$):

$$\bar{c}(\bar{r}) = \frac{1 + \mathrm{Da}(\bar{R}/\bar{r} - 1)\bar{R}}{1 + \mathrm{Da}(\bar{R} - 1)\bar{R}} \tag{10.3}$$

Now, to determine how the interface between the Co_3S_4 shell and the sulfur solution evolves in time, we first equate the number of cobalt atoms diffusing through the interface, $J_{\mathrm{Co}}4\pi R_2^2\mathrm{d}t$, with the number deposited in the product layer, $(3/\nu_{\mathrm{Co_3S_4}})4\pi R_2^2\mathrm{d}R_2$, where $\nu_{\mathrm{Co_3S_4}}$ is the molar volume of Co_3S_4 and 3 is a stoichiometric factor. From this relation, we arrive at the following differential equation for the outer radius, R_2, of the shell:

$$\frac{\mathrm{d}R_2}{\mathrm{d}t} = \frac{1}{3}\nu_{\mathrm{Co_3S_4}}J_{\mathrm{Co}} \tag{10.4}$$

Applying Fick's law for the diffusive flux, $J_{\mathrm{Co}} = -D(\partial c/\partial r)_{R_2}$, and reintroducing dimensionless quantities, we find that the evolution of $\bar{R}$ obeys

$$\frac{\mathrm{d}\bar{R}}{\mathrm{d}t} = \frac{\nu_{\mathrm{Co_3S_4}}}{3\nu_{\mathrm{Co}}[\tau_{\mathrm{rxn}} + \tau_{\mathrm{diff}}(\bar{R}^2 - \bar{R})]} \tag{10.5}$$

Here, we see explicitly how the system's dynamic depends on the relevant reactive and diffusive time scales, τ_{rxn} and τ_{diff}, respectively. Scaling time by the reaction time scale (of course, we could just as easily scale by the diffusive time scale – i.e., both are acceptable!), we find

$$\frac{d\bar{R}}{d\bar{t}} = \frac{\nu_{Co_3S_4}}{3\nu_{Co}[1 + Da(\bar{R}^2 - \bar{R})]} \tag{10.6}$$

This equation may be readily solved analytically to give the outer radius of the shell $\bar{R}(\bar{t})$ as a function of time; in Figure 10.14, this solution is plotted for several representative values of Da.

Having taken the time to derive this equation in a most general fashion (i.e., for any Damköhler number, Da), let us briefly examine the limiting cases of reaction control and diffusion control. For example, when the characteristic reaction time scale is much larger than that of diffusion (i.e., $\tau_{rxn} \gg \tau_{diff}$, corresponding to reaction-limited process), Equation (10.5) simplifies to give a constant rate of growth for the Co_3S_4 shell (cf. Figure 10.14; Da = 0):

$$\frac{d\bar{R}}{dt} = \frac{\nu_{Co_3S_4}}{3\nu_{Co}\tau_{rxn}} = \frac{k\nu_{Co_3S_4}}{3\nu_{Co}R_1} \quad \text{for} \quad Da \ll 1 \tag{10.7}$$

Conversely, for diffusion control (i.e., $\tau_{diff} \gg \tau_{rxn}$), the growth of the shell becomes independent of the reaction rate and slows in time as the shell becomes thicker (due to increasingly shallow concentration gradients):

$$\frac{d\bar{R}}{dt} = \frac{\nu_{Co_3S_4}}{3\nu_{Co}\tau_{diff}(\bar{R}^2 - \bar{R})} = \frac{D\nu_{Co_3S_4}}{3\nu_{Co}R_1^2(\bar{R}^2 - \bar{R})} \quad \text{for} \quad Da \gg 1 \tag{10.8}$$

These limiting cases are of practical importance when one would like to control the thickness of the alloyed shell by tuning the experimental soaking times. In particular, the reaction-limited behavior is ideal for such control since the shell thickness increases as a simple linear function of time.

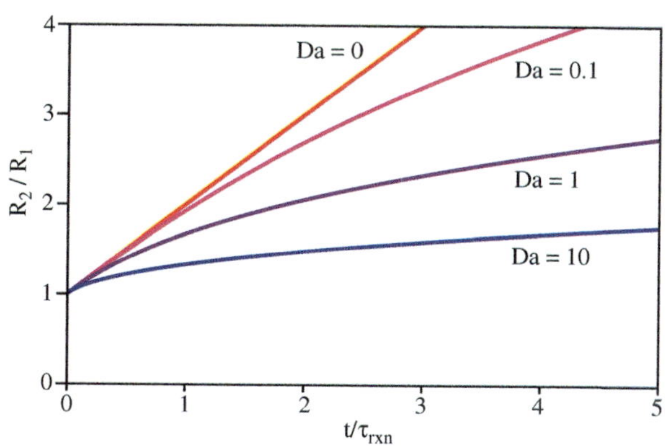

Figure 10.14 Growth of Co_3S_4 product layer versus time. The four lines correspond to different values of Da

Table 10.1 Examples of nanostructures synthesized with the help of the Kirkendall effect. From Fan *et al.* Formation of Nanotubes and Hollow Nanoparticles Based on Kirkendall and Diffusion Processes: A Review, *Small*, 3,10, 1660–1671 copyright (2007) Wiley-VCH

Material	Morphology[a]	Growth process[b]	Ref.	Year of publication
Co_3S_4, CoO, CoSe	Hollow nanocrystals	Wet sulfidation or oxidation of Co nanocrystals	17	2004
Co_3S_4, $CoSe_2$, CoTe	Hollow nanochains	Solution reaction of Co nanonecklace	26	2006
ZnO	Microcages	Dry oxidation of Zn polyhedra	27[b]	2004
ZnO	Dandelion	Hydrothermal reaction	28	2004
Cu_2O	Hollow NPs	Low-temperature dry oxidation	29	2007
ZnS	Hollow nanospheres	Wet sulfidation of ZnO nanospheres	30	2005
PbS	Hollow nanocrystals	Reaction of Pb NPs with vapor S	31	2005
CuS	Octahedral cages	Sulfidation of Cu_2O octahedra	32	2006
Cu_7S_4	Polyhedron nanocages	Sulfidation of Cu_2O nanocube	33	2005
Fe_xO_y	Porous thin film	Hydrothermal reaction	34	2005
Fe_xO_y	Hollow NPs	Room-temperature oxidation of <8 nm particles	35	2005
ZnO	Hollow NPs	Low-temperature oxidation of <20 nm particles	36	2007
Al_xO_y	Amorphous hollow NPs	Low-temperature oxidation of <8 nm Al particles	29	2007
AuPt	Hollow NPs	Solution reaction	37	2004
MoS_2	Cubic microcages	Solution reaction	38	2006
MoO_2	Hollow microspheres	Hydrothermal reaction	39	2006
Ni_2P, Co_2P	Hollow NPs	Wet phosphidation of Ni NPs	40	2007
FePt@CoS_2	Yolk–shell NPs	Wet sulfidation of FePt@Co NPs	41	2007
Pt–Cu	Core–shell NPs	Solution reaction	42	2005
AlN	Hollow nanospheres	Reaction of Al NPs with NH_3CH_4 gas	43	2006
AlN	Hollow nanospheres	Annealing of Al NPs in ammonia	44	2007
SiO_2	Hollow nanospheres	Water oxidation of Si NPs	45	2004
Co_3O_4	Porous nanowires	Oxidation of $Co(OH)_2$ nanowires	46	2006
$SrTiO_3$, $BaTiO_3$	Porous spheres	Hydrothermal reaction of TiO_2 spheres	47	2006
CdS	Polycrystal nanoshell	Reaction of Cd nanowire with H_2S	48	2005
$ZnAl_2O_4$	Crystalline nanotubes	Solid-state reaction of core–shell nanowires	49	2006
Ag_2Se	Nanotubes	Photodissociation of adsorbed CSe_2 on Ag nanowires	50,51	2006
Zn_2SiO_4	Monocrystal nanotubes	Solid-state reaction of core–shell nanowires	20[b]	2007
Co_3S_4	Quasi-monocrystal nanotubes	Reaction of $Co(CO_3)_{0.35}Cl_{0.20}(OH)_{1.10}$ nanowires in solution with H_2S	52	2007
CuO	Polycrystal nanotubes	Dry oxidation of Cu nanowires	53	2005
CuS	Monocrystal nanotubes	Reaction of CuCl nanorod with H_2S	54	2007

[a]NPs = nanoparticles.
[b]The Kirkendall effect was not pointed out by the authors but was most likely one of the growth mechanisms.

While the model is general in nature and, in principle, applicable to different materials, it can only be treated as a first-order approximation.[20] For one, we do not know the exact diffusion coefficient and reaction rates on/in the nanoscopic crystals, and we cannot be even sure that the diffusion is Fickian at this scale. This leaves plenty of room for future theoretical work including discrete atomic simulations. Even without such quantitative models in place, however, the Kirkendall effect has already proven to be a very useful strategy for nanofabrication. Table 10.1, adapted from a review in the journal *Small*, lists the types of particles that have been synthesized using the Kirkendall effect: spherical shells (Figure 10.15(a,b)), hollow polyhedra, yolk–shell particles (Figure 10.15(c)),

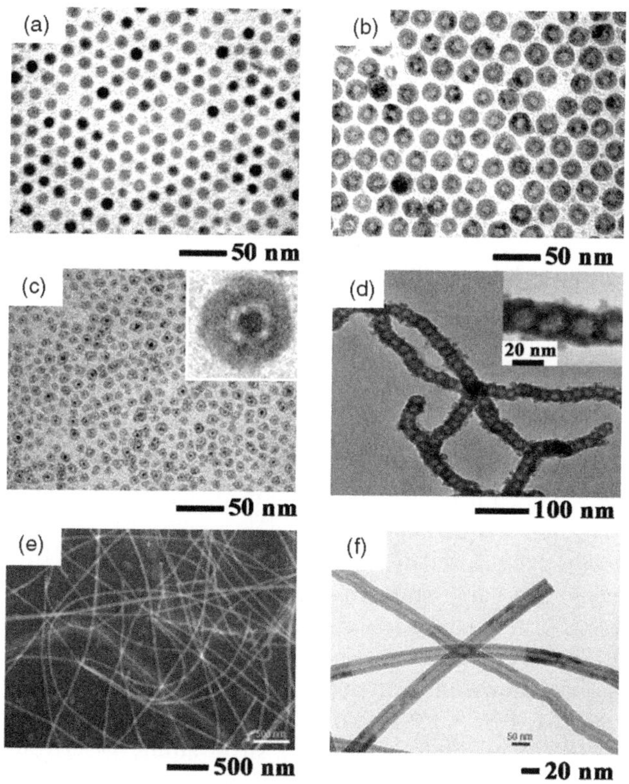

Figure 10.15 Examples of nanostructures synthesized via the Kirkendall effect. (a) Cobalt nanocrystals used as precursors for (b) cobalt sulfide, Co_3S_4, nanoshells. (c) Yolk–shell 'nanoreactors' comprising a platinum core enclosed in a spherical shell of cobalt oxide, CoO. The particles were synthesized by first depositing cobalt onto platinum nanocrystal 'seeds' and then voiding the shell by the Kirkendall effect. (d) Chain of cobalt selenide, $CoSe_2$, nanocrystals made by voiding a chain of 'wired' cobalt nanocrystals using selenium. The cobalt particles were initially assembled into a necklace structure by the interactions between their magnetic dipoles. (e, f) Images of $ZnAl_2O_4$ nanotubes prepared from ZnO–Al_2O_3 core–shell nanowires. (a–c) reproduced from reference 17 with permission from AAAS, (d) reproduced from reference 26 with permission, copyright (2006) Wiley-VCH, (f) reproduced, with permission, from reference 49, copyright (2006), Nature Publishing Group

particle chains (Figure 10.15(d)), dandelions and nanotubes (Figure 10.15(e,f)). The interest in these nanostructures is motivated by their potential uses in biological sensing (e.g., based on optical detection using metallic shells and boxes), in lightweight materials, in drug delivery ('nanocapsules'), and in catalysis as nanoreactors, in which a catalytic core is positioned inside of a noncatalytic shell (Figure 10.15(c)).

10.3 GALVANIC REPLACEMENT AND DE-ALLOYING REACTIONS AT THE NANOSCALE: SYNTHESIS OF NANOCAGES

As mentioned at the end of the previous section, one of the major uses of hollow metal nanoparticles is in optical detection.[21] Due to their small sizes, metal nanoparticles have optical properties very different than those of the corresponding bulk metals – for instance, ~5 nm particles of gold are red/violet whereas silver particles appear yellow/orange. These colors result from the confinement of the electrons within the metal NPs and from collective electron motions – known as surface plasmon resonance (SPR)[22,23] – excited by the impinging light. The wavelength of light that excites SPR depends on the NP size and shapes and shifts when analyte molecules (i.e., molecules one wishes to detect) come close to the nanoparticle surfaces. In the latter context, nanoparticles having sharp corners are usually more sensitive (i.e., they give larger SPR shifts)[23] than more spherical ones. The group led by Younan Xia at the Washington University at Saint Louis has recently demonstrated[21,24,25] how a very unique class of such hollow, sharp-edge particles – the so-called nanocages – can be made by RD combining a galvanic replacement reaction and de-alloying.

Nanoframes are synthesized by a two-stage process (Figure 10.16). First, 50 nm silver nanocubes (gray) are immersed in an aqueous solution of a predetermined amount of $HAuCl_4$. A galvanic replacement reaction that takes place deposits a layer of gold (green) on the cube's surface and creates a hollow void within the cube (starting with a small hole in one of the cube faces; step 1). The resulting 'nanobox', composed of ~15% gold and ~85% silver, is then etched using $Fe(NO_3)_3$ or NH_4OH. Etching removes the silver atoms selectively and creates a frame-like structure composed of almost pure gold – that is, the 'nanoframe' (Figure 10.16; steps 2 and 3). RD processes are found to play a significant role in (i) the deposition of gold and dissolution of silver during the replacement reaction, (ii) the diffusive mixing of gold and silver within the nanobox and (iii) the etching of the box to form the final nanoframe.

In the first process, the silver cube reacts with $AuCl_4^-$ according to the following galvanic replacement reaction: $3Ag\ (s) + AuCl_4^-(aq) \rightarrow 3Ag^+(aq) + Au\ (s) + 4Cl^-\ (aq)$. The net result of this process is that metallic gold is deposited onto the cube surface while the oxidized silver dissolves into the surrounding solution.

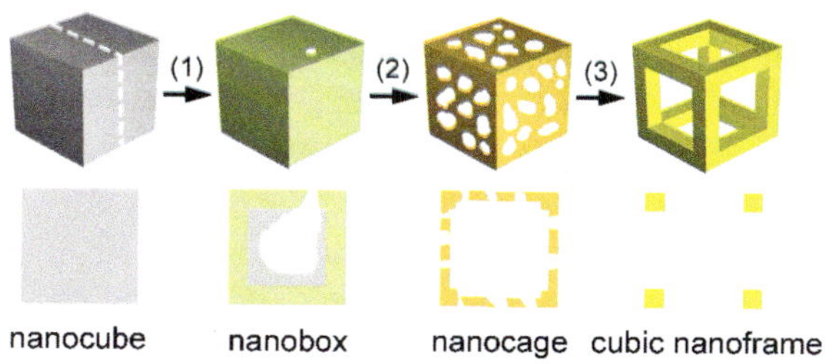

Figure 10.16 Synthesis of gold nanocages from silver nanocubes. A nanocube is transformed into a hollow nanobox by a reaction between silver and $HAuCl_4$ (step 1). Voiding of the cube proceeds through a small hole defect through which Ag^+ escapes. When silver is de-alloyed using $Fe(NO_3)_3$ or NH_4OH, the nanobox first evolves into a porous nanocage (step 2) and then, as more etchant is added, into a cubic gold nanocage (step 3). The drawings in the lower panel correspond to the cross-sections of 3D structures. (Image courtesy of Prof. Younan Xia, University of Washington, St Louis. Reprinted from reference 25 with permission. Copyright (2007), American Chemical Society.)

However, as gold is deposited onto the cube surface, it inhibits the escape of Ag^+ ions – therefore, silver ions are released predominantly from a location where the rate of gold deposition is smallest. Experimental evidence (cf. Figure 10.17(b) and the equations we will see shortly) indicates that this location corresponds to a concave defect that forms on one of the cube's faces. As silver ions diffuse from such a defect, they create a hole that grows in size and becomes even more concave. The increased concavity further inhibits the diffusive deposition of gold and thus speeds up the release of silver ions. This feedback process results in the formation of a large, hollow void in the cube's center. Importantly, the deposition (onto the cube's outer surface) and dissolution (from the cube's inner surface) can proceed at different locations owing to the high electric conductivity of Ag/Au, which permits the transfer of electrons from the cube's center ($Ag \rightarrow Ag^+ + e^-$) to its outer surface ($AuCl_4^- + 3e^- \rightarrow Au + 4Cl^-$).

To illustrate the effects of diffusion on the deposition of gold in a small, hole-like defect, consider the simplified, 2D model (Figure 10.18(a)). Outside of the cube, the concentration of $AuCl_4^-$, c, is governed by the steady-state diffusion equation, $0 = D\nabla^2 c$, with a typical diffusion coefficient ($D = 10^{-5}\,\text{cm}^2\,\text{s}^{-1}$). Here, we have neglected the time dependence of the concentration profile, which approaches its steady-state values after only $L^2/D \sim 1\,\mu\text{s}$ for $L = 50\,\text{nm}$. At the cube surface, we have the usual surface reaction boundary condition, $D(\nabla c \cdot \boldsymbol{n}) = kc$ (where $\boldsymbol{n}$ is the unit vector normal to the surface and directed into solution), characterizing the conservation of gold atoms at the surface. Far from the cube, the $AuCl_4^-$ concentration approaches a constant value of c_0. Scaling the above equations in the usual manner, $\bar{\boldsymbol{x}} = \boldsymbol{x}/L$ and $\bar{c} = c/c_0$, the boundary conditions become $\nabla\bar{c} \cdot \boldsymbol{n} = \text{Da}\bar{c}$ and $\bar{c}(\infty) \rightarrow \bar{c}_0$, where $\text{Da} = kL/D$ is the Damköhler number. In the present system, the rate of galvanic reaction is approximated as $k = 2\,\text{m}\,\text{s}^{-1}$ (estimated from a similar platinum replacement reaction), such that $\text{Da} \sim 100$.

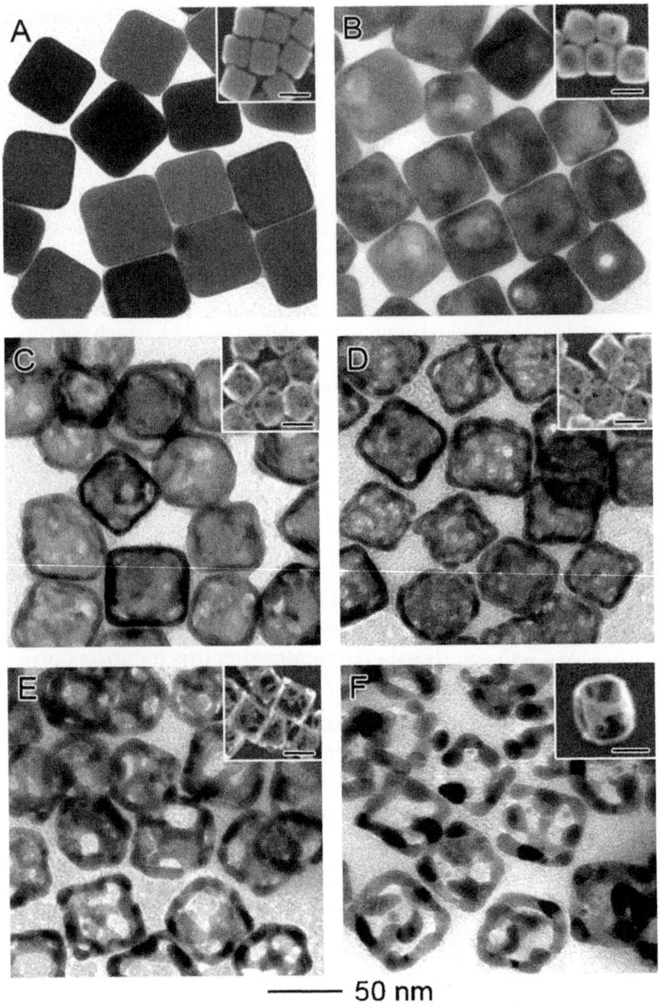

Figure 10.17 Transmission electron microscopy and scanning electron microscopy (insets) images of (a) 50 nm silver nanocubes; (b) Au/Ag alloy nanoboxes obtained by reacting the nanocubes with 4.0 ml of 0.2 mM $HAuCl_4$ aqueous solution. Note holes formed in the cube faces. (c–f) Nanocages and nanoframes obtained by etching nanoboxes with 5, 10, 15 and 20 μl of 50 mM aqueous $Fe(NO_3)_3$ solution. The inset in (f) was obtained at a tilting angle of 45°, clearly showing the 3D structure of a nanoframe. The scale bars in all insets are 50 nm. (Images courtesy of Prof. Younan Xia, University of Washington at St Loius. Reprinted from reference 25, with permission. Copyright (2007), American Chemical Society.)

The relatively large Damköhler number implies that the rate of diffusion is much slower than that of reaction; consequently, the process is diffusion-limited. Figure 10.18(b) shows the steady-state concentration profiles for $AuCl_4^-$ near the circular defect as obtained via numerical solution of the above equation. Note

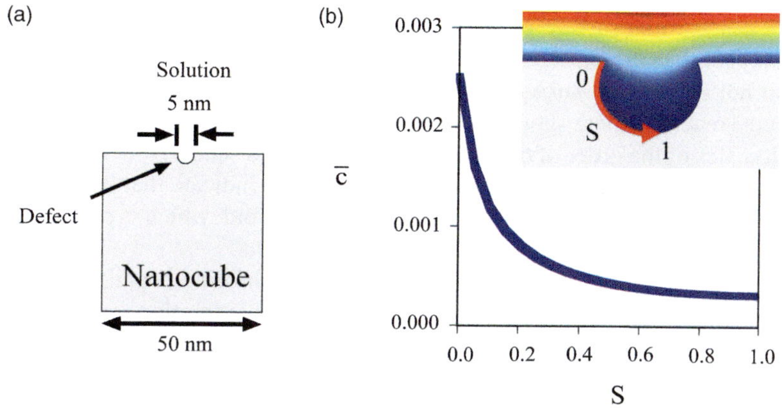

Figure 10.18 Deposition of gold onto a silver nanocube presenting a small, concave defect. (a) Scheme of the cube indicates typical dimensions. (b) The calculated, normalized (with respect to the initial concentration, c_0) concentration of Au^{3+} ions along the length of the defect, denoted by the curvilinear axis, S (red line in the inset). This plot demonstrates that the concentration of Au^{3+} ions decreases rapidly inside the defect and a much smaller amount of gold is deposited therein than on the cube's flat faces. Conversely, the release of Ag^+ is faster from the defect than from other locations on the cube's surface

that the concentration of gold is minimal within the concave defect, and the deposition reaction proceeds slowest therein. Consequently, the dissolution of silver ions occurs fastest in the defect, which deepens increasingly rapidly by the feedback mechanism described above.

The end result of this process is the formation of a hollow silver cube surrounded by a thin gold shell. Experimentally, it was determined (by energy-dispersive X-ray spectroscopy, EDX) that the composition of the final nanoboxes is ~15% gold on a molar basis. Furthermore, because both gold and silver have the same face-centered cubic crystal structure and similar lattice constants (4.0786 Å versus 4.0862 Å), the volume fraction of gold is approximately equal to the mole fraction. From this information, we can estimate the size of the roughly spherical void within the cube as well as the thickness of the gold layer. Denoting the final dimension of the box as L_f, the final volume of gold is $V_{Au} = L_f^3 - L^3$ and that of silver is $V_{Ag} = L^3 - 3V_{Au}$, where the factor of three accounts for the stoichiometry of the replacement reaction. Thus, from the equation $V_{Au}/(V_{Au} + V_{Ag}) = 0.15$, we find that $L_f \approx 1.037L$. The thickness of the gold layer is then given by $\delta = L_f - L$ or $\delta \sim 1.9$ nm, and the radius of the void is $R = (9V_{Au}/4\pi)^{1/3}$ or $R \sim 22$ nm.

The above analysis addresses only the diffusion of Au^{3+} ions in solution surrounding the nanocube. In addition, diffusion also occurs – albeit slowly – within the cube due to the gradients in the gold and silver composition – e.g., between gold-rich surface and silver-rich center of the cube. For the experimental temperature of 100 °C, the binary diffusion coefficient for Ag/Au is of the order of $D_{AuAg} \sim 10^{-20}\ m^2\ s^{-1}$. Thus, the characteristic time for atoms to diffuse from the cube's center to its surface is $\tau_{diff} \sim L^2/D_{AuAg} \sim 100$ min, which is about twenty times longer than the experimental time scale (a few minutes). Therefore, while we should not expect the Au/Ag box to be compositionally homogeneous after the exchange reaction, there should be an appreciable degree of diffusional mixing on length scales of the order of δ, the thickness of the gold shell. (Note that the rather long time scales of solid interdiffusion in this system indicate that the Kirkendall effect cannot be responsible for the formation of the void, which is created within only a few minutes.)

Quantitatively, this mixing process is described by the time-dependent diffusion equation $\partial x_{Au}/\partial t = D_{AuAg}\nabla^2 x_{Au}$, where x_{Au} is the mole fraction (or equivalently the volume fraction) of gold atoms ($x_{Au} + x_{Ag} = 1$). Initially, silver and gold are distributed within the core and shell geometry as described above and, at the boundaries, there is no flux of either metal atoms into solution. This equation can be integrated numerically (using a finite volume method; see Section 4.3.2) to give the gold composition after a specified time of diffusion (Figure 10.19(a,b)). Note that the walls of the cube show significant mixing, while the corners maintain a higher fraction of gold atoms due to the initial geometry.

In the final step of nanoframe formation, the nanoboxes are placed in an etching solution of $Fe(NO_3)_3$ or NH_4OH, which react selectively with silver to erode the cube center and its walls, leaving only a frame of gold. Unlike the previous step that required elevated temperature, etching takes place at room temperature, at which the binary diffusion coefficient for Au/Ag decreases

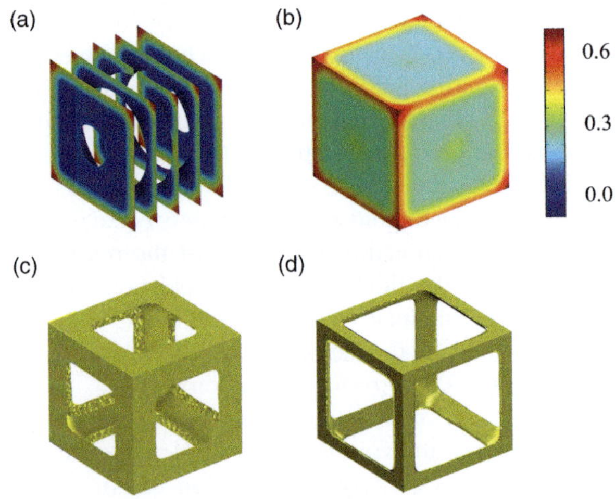

Figure 10.19 Evolution of nanoboxes into nanocages modeled by the finite volume method. (a) Concentration profile of gold at five cross-sections of the box; (b) surface concentration. In both cases, concentrations are normalized with respect to the initial concentration of gold deposited onto the nanocube. (c) A nanobox at an intermediate stage of etching and (d) the final nanocage

dramatically to $\sim 10^{-24}\,m^2\,s^{-1}$. Consequently, intermetallic diffusion at this step can be neglected and only the silver removal reaction needs to be considered. This removal process occurs most rapidly at surface regions characterized by the highest silver content, x_{Ag} – that is, at the interior of the box and at the centers of the faces. When silver atoms are being selectively removed, sites of vacancy are created in the crystalline lattice of the metallic structure. With vacancies, the remaining gold atoms have fewer neighboring atoms to bind to and become more unstable. Importantly, the local number-density of metallic atoms may drop to a critical low, ρ_{min}, such that the remaining gold atoms may have practically no neighbors to bind to (in physical parlance, the structure then percolates). When this happens, gold atoms are lost to the solution. Figure 10.19(c,d) illustrates this process modeled for $\rho_{min} = 1/3\rho$, where ρ denotes the number-density of metallic atoms in the original nanobox. While this particular value of ρ_{min} gives the best agreement with the experimentally measured etching times, other values predict qualitatively similar structural evolution and ultimately lead to well-defined nanocages such as the one shown in Figure 10.19(d).

The example of nanocages illustrates vividly the opportunities that 3D RD fabrication at the nanoscale offers. With just a few physical phenomena it can give structures that are topologically more complex than the staring templates, and would be exceedingly difficult, if not impossible, to make in any other way. Given what we can already do with a few basic 'ingredients' (the Kirkendall effect, galvanic replacement or de-alloying) acting on monocomponent particles, one can easily imagine what could be done if we were able to apply RD to particles composed of multiple domains and/or address the particles in site-specific ways (e.g., by etching from different crystalline faces with different speeds). The progress of nanotechnology is so rapid these days that it will probably be sooner rather than later when these pipe dreams become an experimental reality.

REFERENCES

1. Chu, L.Y., Utada, A.S., Shah, R.K. *et al.* (2007) Controllable monodisperse multiple emulsions. *Angew. Chem. Int. Ed.*, **46**, 8970.
2. Pekarek, K.J., Jacob, J.S. and Mathiowitz, E. (1994) Double-walled polymer microspheres for controlled drug-release. *Nature*, **367**, 258.
3. Kim, S.H., Jeon, S.J. and Yang, S.M. (2008) Optofluidic encapsulation of crystalline colloidal arrays into spherical membrane. *J. Am. Chem. Soc.*, **130**, 6040.
4. Nguyen, D., Chambon, P., Rosselgong, J. *et al.* (2008) Simple route to get very hydrophobic surfaces of fibrous materials with core–shell latex particles. *J. Appl. Polym. Sci.*, **108**, 2772.
5. Utada, A.S., Lorenceau, E., Link, D.R. *et al.* (2005) Monodisperse double emulsions generated from a microcapillary device. *Science*, **308**, 537.
6. Kim, J.W., Utada, A.S., Fernandez-Nieves, A. *et al.* (2007) Fabrication of monodisperse gel shells and functional microgels in microfluidic devices. *Angew. Chem. Int. Ed.*, **46**, 1819.
7. Fialkowski, M., Bitner, A. and Grzybowski, B.A. (2005) Self-assembly of polymeric microspheres of complex internal structures. *Nature Mater.*, **4**, 93.

8. Xia, Y.N., McClelland, J.J., Gupta, R. *et al.* (1997) Replica molding using polymeric materials: a practical step toward nanomanufacturing. *Adv. Mater.*, **9**, 147.
9. Koza, Z. (1996) Asymptotic behaviour of initially separated A + B(static) → 0 reaction–diffusion systems. *Physica A*, **240**, 622.
10. Grzybowski, B.A. and Whitesides, G.M. (2002) Self-assembly at all scales. *Science*, **295**, 2418.
11. Fialkowski, M., Bishop, K.J.M., Klajn, R. *et al.* (2006) Principles and implementations of dissipative (dynamic) self-assembly. *J. Phys. Chem. B*, **110**, 2482.
12. Tanford, C. (1980) *The Hydrophobic Effect*, John Wiley & Sons, Ltd, New York.
13. Nakajima, H. (1997) The discovery and acceptance of the Kirkendall effect: the result of a short research career. *JOM*, **49**, 15.
14. Kirkendall, E.O. and Smigelskas, A.D. (1947) Zinc diffusion in alpha brass. *Trans. Am. Inst. Mining Metall. Peterol. Eng.*, **171**, 130.
15. Ertl, G., Knözinger, H. and Weitkamp, J. (1999) *Preparation of Solid Catalysts*, Wiley-VCH, Berlin.
16. Yin, Y., Erdonmez, C.K., Cabot, A. *et al.* (2006) Colloidal synthesis of hollow cobalt sulfide nanocrystals. *Adv. Funct. Mater.*, **16**, 1389.
17. Yin, Y., Rioux, R.M., Erdonmez, C.K. *et al.* (2004) Formation of hollow nanocrystals through the nanoscale Kirkendall effect. *Science*, **304**, 711.
18. Dybkov, V.I. (1986) Reaction diffusion in heterogeneous binary systems. *J. Mater. Sci.*, **21**, 3078.
19. Paul, A. (2004) The Kirkendall effect in solid state diffusion. PhD thesis, Technische Universiteit Eindhoven.
20. Zhou, J., Liu, J., Wang, X.D. *et al.* (2007) Vertically aligned Zn_2SiO_4 nanotube/ZnO nanowire heterojunction arrays. *Small*, **3**, 622.
21. Skrabalak, S.E., Chen, J., Au, L. *et al.* (2007) Gold nanocages for biomedical applications. *Adv. Mater.*, **19**, 3177.
22. Jain, P.K., Huang, X., El-Sayed, I.H. and El-Savad, M.A. (2007) Review of some interesting surface plasmon resonance-enhanced properties of noble metal nanoparticles and their applications to biosystems. *Plasmonics*, **2**, 107.
23. Anker, J.N., Hall, W.P., Lyandres, O. *et al.* (2008) Biosensing with plasmonic nanosensors. *Nature Mater.*, **7**, 442.
24. Sun, Y.G. and Xia, Y.N. (2004) Mechanistic study on the replacement reaction between silver nanostructures and chloroauric acid in aqueous medium. *J. Am. Chem. Soc.*, **126**, 3892.
25. Lu, X.M., Au, L., McLellan, J. *et al.* (2007) Fabrication of cubic nanocages and nanoframes by dealloying Au/Ag alloy nanoboxes with an aqueous etchant based on $Fe(NO_3)_3$ or NH_4OH. *Nano Lett.*, **7**, 1764.
26. Gao, J.H., Zhang, B., Zhang, X.X. and Xu, B. (2006) Magnetic-dipolar-interaction-induced self-assembly affords wires of hollow nanocrystals of cobalt selenide. *Angew. Chem. Int. Ed.*, **45**, 1220.
27. Fan, H.J., Scholz, R., Kolb, F.M. *et al.* (2004) Growth mechanism and characterization of zinc oxide microcages. *Solid State Commun.*, **130**, 517.
28. Liu, B. and Zeng, H.C. (2004) Fabrication of ZnO 'dandelions' via a modified Kirkendall process. *J. Am. Chem. Soc.*, **126**, 16744.
29. Nakamura, R., Tokozakura, D., Nakajima, H. *et al.* (2007) Hollow oxide formation by oxidation of Al and Cu nanoparticles. *J. Appl. Phys.*, **101**, 074303.

30. Shao, H.F., Qian, X.F. and Zhu, Z.K. (2005) The synthesis of ZnS hollow nanospheres with nanoporous shell. *J. Solid State Chem.*, **178**, 3522.
31. Wang, Y.L., Cai, L. and Xia, Y.N. (2005) Monodisperse spherical colloids of Pb and their use as chemical templates to produce hollow particles. *Adv. Mater.*, **17**, 473.
32. Xu, H.L. (2006) Synthesis of octahedral CuS nanocages via a solid–liquid reaction. *Nanotechnology*, **17**, 3649.
33. Cao, H.L., Qian, X.F., Wang, C. *et al.* (2005) High symmetric 18-facet polyhedron nanocrystals of Cu_7S_4 with a hollow nanocage. *J. Am. Chem. Soc.*, **127**, 16024.
34. Zhang, L.Z., Yu, J.C., Zheng, Z. and Leung, C.W. (2005) Fabrication of hierarchical porous iron oxide films utilizing the Kirkendall effect. *Chem. Commun.*, 21 2683.
35. Wang, C.M., Baer, D.R., Thomas, L.E. *et al.* (2005) Void formation during early stages of passivation: initial oxidation of iron nanoparticles at room temperature. *J. Appl. Phys.*, **98**, 094308.
36. Nakamura, R., Lee, J.G., Tokozakura, D. *et al.* (2007) Formation of hollow ZnO through low-temperature oxidation of Zn nanoparticles. *Mater. Lett.*, **61**, 1060.
37. Liang, H.P., Guo, Y.G., Zhang, H.M. *et al.* (2004) Controllable AuPt bimetallic hollow nanostructures. *Chem. Commun.*, 1496.
38. Ye, L.N., Wu, C.Z., Guo, W. and Xie, Y. (2006) MoS_2 hierarchical hollow cubic cages assembled by bilayers: one-step synthesis and their electrochemical hydrogen storage properties. *Chem. Commun.*, 45 4738.
39. Guo, C.H., Zhang, G.J., Shen, Z.R. *et al.* (2006) Hydrothermal synthesis and formation mechanism of micrometer-sized MoO_2 hollow spheres. *Chinese J. Chem. Phys.*, **19**, 543.
40. Chiang, P.Y., Lin, T.W., Dai, J.H. *et al.* (2007) Flux synthesis, crystal structure, and luminescence properties of a new europium fluoride-silicate: $K_5Eu_2FSi_4O_{13}$. *Inorg. Chem.*, **46**, 3619.
41. Gao, J.H., Liang, G.L., Zhang, B. *et al.* (2007) FePt@CoS_2 yolk-shell nanocrystals as a potent agent to kill HeLa cells. *J. Am. Chem. Soc.*, **129**, 1428.
42. Zhou, S.H., Varughese, B., Eichhorn, B. *et al.* (2005) Pt–Cu core–shell and alloy nanoparticles for heterogeneous NOx, reduction: anomalous stability and reactivity of a core--shell nanostructure. *Angew. Chem. Int. Ed.*, **44**, 4539.
43. Ma, Y.W., Huo, K.F., Wu, Q. *et al.* (2006) Self-templated synthesis of polycrystalline hollow aluminium nitride nanospheres. *J. Mater. Chem.*, **16**, 2834.
44. Zheng, J., Song, X.B., Zhang, Y.H. *et al.* (2007) Nanosized aluminum nitride hollow spheres formed through a self-templating solid–gas interface reaction. *J. Solid State Chem.*, **180**, 276.
45. Colder, A., Huisken, F., Trave, E. *et al.* (2004) Strong visible photoluminescence from hollow silica nanoparticles. *Nanotechnology*, **15**, L1.
46. Li, Y.G., Tan, B. and Wu, Y.Y. (2006) Freestanding mesoporous quasi-single-crystalline Co_3O_4 nanowire arrays. *J. Am. Chem. Soc.*, **128**, 14258.
47. Wang, Y.W., Xu, H., Wang, X.B. *et al.* (2006) A general approach to porous crystalline TiO_2, $SrTiO_3$, and $BaTiO_3$ spheres. *J. Phys. Chem. B*, **110**, 13835.
48. Li, Q.G. and Penner, R.M. (2005) Photoconductive cadmium sulfide hemicylindrical shell nanowire ensembles. *Nano Lett.*, **5**, 1720.
49. Fan, H.J., Knez, M., Scholz, R. *et al.* (2006) Monocrystalline spinel nanotube fabrication based on the Kirkendall effect. *Nature Mater.*, **5**, 627.
50. Tan, H., Li, S.P. and Fan, W.Y. (2006) Core–shell and hollow nanocrystal formation via small molecule surface photodissociation: Ag@Ag_2Se as an example. *J. Phys. Chem. B*, **110**, 15812.

51. Ng, C.H.B., Tan, H. and Fan, W.Y. (2006) Formation of Ag_2Se nanotubes and dendrite-like structures from UV irradiation of a CSe_2/Ag colloidal solution. *Langmuir*, **22**, 9712.
52. Chen, X.Y., Zhang, Z.J., Qiu, Z.G. *et al.* (2007) Hydrothermal fabrication and characterization of polycrystalline linneite (Co_3S_4) nanotubes based on the Kirkendall effect. *J. Colloid Interface Sci.*, **308**, 271.
53. Chang, Y., Lye, M.L. and Zeng, H.C. (2005) Large-scale synthesis of high-quality ultralong copper nanowires. *Langmuir*, **21**, 3746.
54. Wang, Q., Li, J.X., Li, G.D. *et al.* (2007) Formation of CuS nanotube arrays from CuCl nanorods through a gas–solid reaction route. *J. Cryst. Growth*, **299**, 386.

11

Epilogue: Challenges and Opportunities for the Future

And so our journey through the world of reaction–diffusion (RD) comes to an end. While the selection of systems and problems we covered in this book is certainly arbitrary or even biased, we hope it is comprehensive enough to illustrate one, general thought – namely that if 'programmed' properly, RD systems can become a unique tool with which to manipulate matter, build structures and sense events at small scales. There is much promise for RD beyond what we managed to discuss! But there are also many hurdles to be overcome that require more creativity, experimental skill and theoretical expertise. We will therefore conclude this book with a short 'laundry list' of what we consider the most challenging but also most promising issues for future RD research.

1. Parallel chemistries. Most of the examples we covered involved a single chemical reaction. If we are serious about using RD in bio-inspired ways, we should be able to encode multiple RD processes in the same system and make them perform fabrication or sensing tasks in parallel. One way to do this is by choosing reactions that are mutually 'orthogonal' and do not interfere with one another. Figure 11.1(a) shows one example of such a RD system, in which two precipitation reactions are initiated from the same wet stamp delivering a mixture of $CuCl_2$ and $FeCl_3$ into a gel containing $K_4[Fe(CN)_6]$. A subtle interplay between the solubility products of the two forming salts ($K_{sp}^{(1)} = [Fe^{3+}]^4[[Fe(CN)_6]^{4-}]^3 = 3.3 \times 10^{-41}$ versus $K_{sp}^{(2)} = [Cu^{2+}]^2[Fe(CN)_6]^{4-} = 1.3 \times 10^{-16}$) and the diffusion constants of the ions ($D_{Cu^{2+}} : D_{Fe^{3+}} : D_{[Fe(CN)_6]^{4-}} \approx 2 : 1 : 0.3$) makes the process sequential in the sense that the brown copper salt precipitates first near the stamped features, and only when it is consumed does the blue iron precipitate deposits away from the features.[1] Although reactions placing up to three different metals to selective locations are

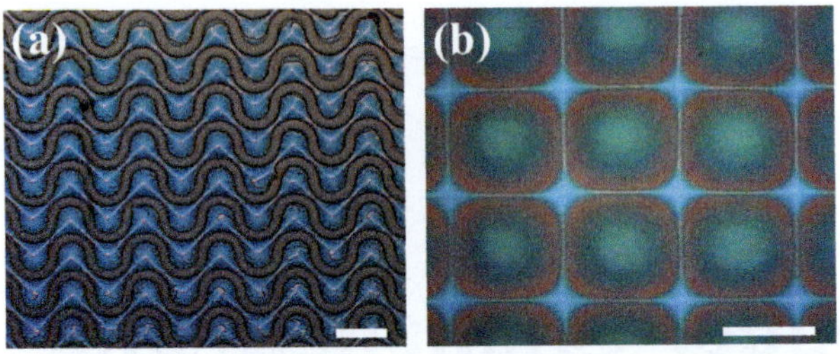

Figure 11.1 Multicolor patterns via sequential reactions. (a) Pattern formed by wet stamping an array of wiggly lines delivering a mixture of $FeCl_3$ and $CuCl_2$ (0.5 M each). (b) A three-color pattern from an array of circles delivering a mixture of $CuCl_2$, $FeCl_3$ and $CoCl_2$ (0.014 M:0.31 M:0.67 M)

possible (Figure 11.1(b)), this approach is limited by the relative scarcity of inorganic reactions for which the values of K_{sp} are suitable and cross-precipitation of the salts and common-ion effects are absent. A much more general and powerful approach could be to combine RD with methods such as microfluidics/electrophoresis that would first spread out the reagents in the stamp and then use this stamp to initiate different reactions over different regions of the substrate (Figure 11.2).

2. Gradients. An issue closely related to parallel chemistries is to be able to realize multiple experimental conditions within the same RD process. One of the possibilities here is to use RD to set up spatial and temporal gradients that control fluxes of chemicals and chemical processes at different locations. Figure 11.3 illustrates this idea in a system in which RD sets up and controls a flux of NaCl which, it turn, controls the growth of protein crystals (here, lysozyme) immobilized in an array of microwells.[2] RD gradients combined with arrayed microreactors can be a valuable tool for screening studies, including optimization of crystal growth conditions or biochemical binding assays.

3. Complex chemistries. Applying RD only to simple inorganic reactions is like using a supercomputer to solve quadratic equations. To realize its true potential, RD should be extended to more complex and important chemistries, involving organic and bioorganic molecules. This is especially true for the sensing and amplification schemes, which should be applied not to ions or small molecules (see Chapter 9) but to enzymes,[3] antibodies or nucleic acids. The RD approach would be even more powerful if multiple analytes could be screened in one experiment – for instance by the methods mentioned in (1) and (2) above.

4. RD in the solid state. Can RD in solids be extended beyond the Kirkendall effect and intermetallic diffusion (Chapter 10) to organic materials? Although high temperatures cannot usually be used to speed up diffusion in organics (cf. Chapter 10), there are already some useful examples of RD processes that transform organic frameworks at room temperature and without compromising material properties. For example, a Danish team led by Niels Larsen used RD initiated by wet stamping to control the oxidation state and to pattern electrically conductive/nonconductive regions in thin films of poly(3,4-ethylenedioxythiophene)

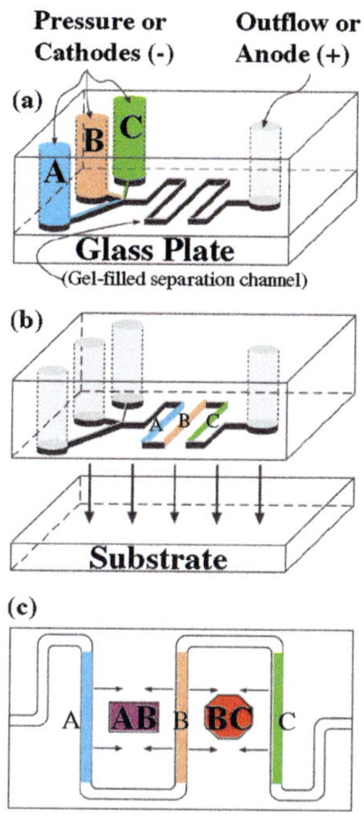

Figure 11.2 Electrophoretic separation of multiple reagents (here, A, B, C) in a microchannel fabricated on a surface of a polydimethylsiloxane block and filled with a hydrogel could be used to initiate different RD processes (A + B, B + C) over different locations of the substrate

(PEDOT).[4] Using RD in this context can be a very promising strategy for fabricating organic/flexible electronic circuits. In another example (Figure 11.4), the group of Chad Mirkin at Northwestern University used RD to exchange metal cations inside of microbeads made of so-called infinite coordination polymers (materials used in hydrogen storage and catalysis).[5] In the future, it would be interesting to explore whether similar exchange reactions can be run in organic crystals (at least those that have columnar structures), and whether sequences of RD can produce more complex micro-/nanoarchitectures.

5. Manipulation of individual (macro)molecules. This is not a joke. We have already seen that RD processes can create well-defined structures with dimensions down to low tens of nanometers (for instance, at the end of Chapter 7). This is only a few times larger than the size of a single protein! If we managed to miniaturize RD systems a bit more – by using smaller wet-stamped features and media with lower diffusion coefficients – and to adapt the pertinent bioorganic chemistries, we could conceivably use RD to resolve structures composed of individual macromolecules. Our personal favorites (hypothetical!) are Liesegang rings made by co-precipitation of inorganic salts with highly charged histone proteins or DNA strands. In the latter case, could consecutive bands separate strands of different length/charge?

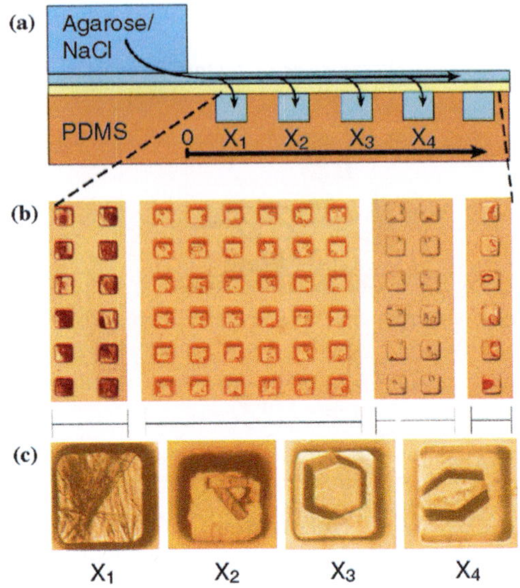

Figure 11.3 Reaction–diffusion gradients control the growth of protein microcrystals. (a) Side view scheme of the experimental setup. Protein precipitant (here, NaCl) diffuses from an agarose stamp through the gel transport layer (turquoise) and into gel-filled microwells (light blue) containing protein solution. A dialysis membrane (yellow) allows delivery of NaCl into the wells while preventing the escape of proteins. (b) An optical micrograph showing different regions of an array of wells containing lysozyme microcrystals. (c) Magnified images of typical crystals grown in the regions shown in panel (b). Crystal morphologies are controlled by the flux of precipitant what varies with spatial location and with time. For details, see Ref. 2 (Reproduced by permission from the American Chemical Society, 2008.)

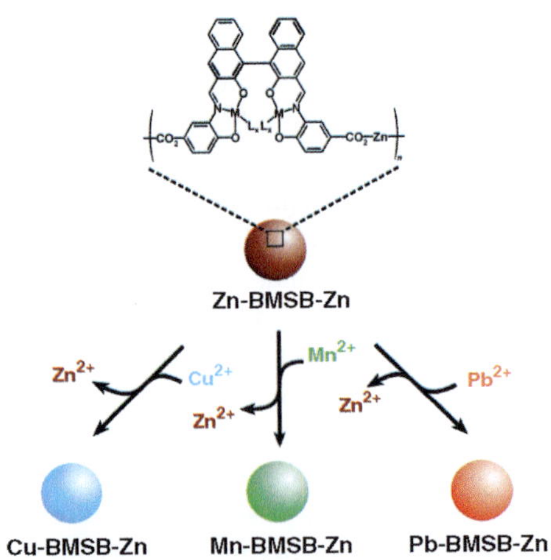

Figure 11.4 Infinite coordination polymer particles consisting of metalloligand (bis-metallo-tridentate Schiff base, BMSB) building blocks and metal interconnecting nodes. Zn^{2+} used in the original synthesis can be readily displaced from the BMSB tridentate pockets to give particles held together by Cu^{2+}, Mn^{2+} or Pd^{2+} cations. (Scheme adapted from Ref. 5)

6. Three-dimensional fabrication. RD in three dimensions has the potential to become a very powerful method for the fabrication of nanoscopic and microscopic particles of complex shapes and internal architectures. In the demonstrations of 'remote fabrication' we covered in Chapter 10, reaction fronts propagated uniformly from all directions and always led to structures at least as symmetric as the original templates (e.g., squares inside of cross-shaped particles; Section 10.1.3). With the existing methods for the site-selective modification/protection of micro- and nanoparticles (e.g., via the use of self-assembled monolayers), some locations on the surfaces of template particles could be 'passivated' and RD could be initiated only from the unprotected areas. Such processes should lead to structures less symmetric than the template – e.g., cubic particles having one face 'blocked' for diffusion should yield curvilinear 'pyramids' (Figure 11.5). Finding relationship(s) between the shape of the initial template and that of the final particle might also present interesting modeling/theoretical challenges (try this: how should a template look for RD to produce a Möbius strip?).

7. Combination with self-assembly. When cells – the ultimate RD nanosystems – self-assemble into larger structures and communicate with one another through the exchange of chemicals they carry, they create functional tissues and organisms. Learning from biology, one could envision artificial systems in which the building blocks carrying chemical reagents would first self-assemble in a programmed fashion, and would then exchange and react their contents to build structures and/or perform desired tasks. Such self-assembling RD (SA/RD) ensembles would rely on the use of soft/porous materials and on surface chemistries that would enable selective recognition of various types of pieces. The left part of Figure 11.6 shows a very primitive illustration of the SA/RD concept, where five agarose cubes containing a tartrate complex stabilized with basic solution of Cu^{2+} (blue) and four clear cubes soaked in a basic solution of formaldehyde (a reductant for the complex used for electroless deposition of Cu^0) assemble into a 3×3 checkerboard structure. When the two types of cubes come into contact, the rapidly diffusing formaldehyde solution produces structures of copper colloids inside the cubes containing copper solution. How to get from such toy models to 'artificial cells' (right part of Figure 11.6) depends only on our creativity – for, in principle, it is possible!

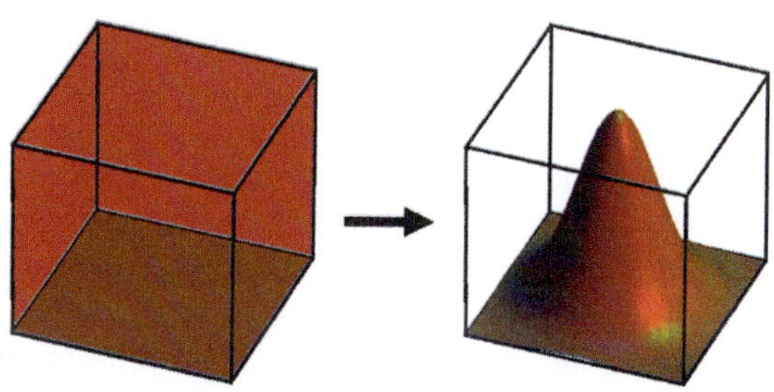

Figure 11.5 Fabrication in three dimensions. Etching of a metal cube having the lower face protected/'passivated' should yield a curvilinear pyramid shown in the right-hand panel

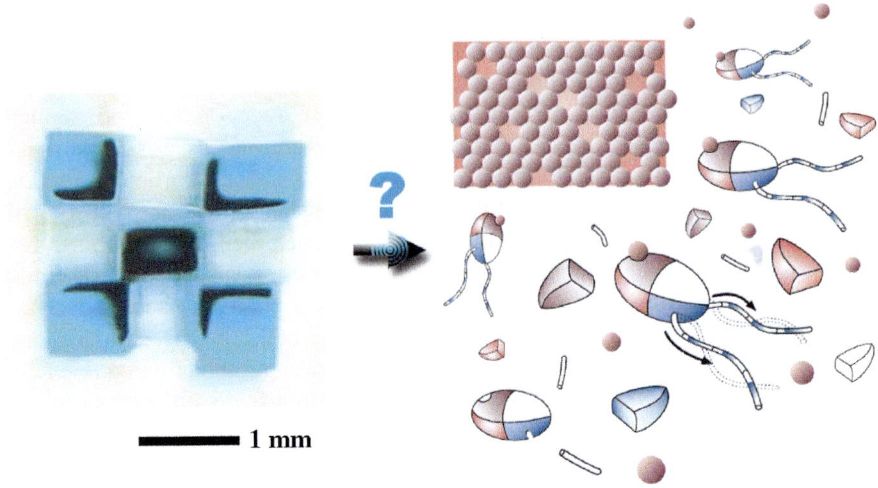

Figure 11.6 Fact and science fiction. The left panel shows an assembly of nine 0.9 mm cubes exchanging chemicals and building copper colloid structures in the tartrate/Cu^{2+} particles. The right panel shows an artistic vision of dynamic, self-assembling nanomachines, each carrying a microsphere cargo to be deposited onto a growing colloidal array. Each nanomachine is powered by reaction–diffusion waves (blue bands) running synchronously along its 'legs' (in the direction of the arrows), and is guided onto the array by chemotactic attraction of its red 'belly' towards the array's background (emitting 'red attractants'). The 'design' of these organisms was inspired by *Chlamydia algi.* The question marks next to the arrow express our current uncertainty of how to progress from simple 'toy' models to truly intelligent assemblies based on a synergistic combination of self-assembly and reaction–diffusion processes

8. Theoretical issues. Finally, there are some theoretical challenges. First, there is an issue of reverse engineering of RD – that is, finding the right reactions, concentrations and initial conditions that would create a desired pattern/structure. Since only a few RD processes may be treated analytically, the challenge of reverse engineering must, in general, be tackled numerically. One way is to use 'brute force' and optimize system parameters in an iterative manner by repeated numerical integration of the RD equations and subsequent tuning of parameters (cf. Example 6.3). With the rapid growth of computational power available to most researchers, such procedures are a viable possibility, especially in systems where only one or a few RD processes are operative. In systems in which many reactions occur in parallel, some heuristics could be implemented based on the 'additivity' of the outcomes of *orthogonal/independent* RD processes (e.g., periodic precipitation and gel swelling discussed in Section 7.8). Establishing which types of processes/reactions could be treated as independent and additive would be a useful albeit time-consuming thing to do. Development of rapid yet accurate algorithms to simulate RD is another outstanding issue.

The second challenge has to do with the scale at which RD occurs. When dealing with nanoscopic RD phenomena, we should question and then probe the validity of the continuum description based on partial differential equations. At very small scales, RD processes might need to be described by stochastic differential equations (Langevin equations), which account for the thermal motion of dissolved species and their interconversion through reactive collisions. For small systems, containing only few hundred molecules, concentration fluctuations become significant relative to the mean concentration values, especially when feedback processes act to amplify stochastic, microscale events into larger scale macroscopic changes (such was the case in the context of certain periodic precipitation processes described in Chapter 7). In general, when are microscale dynamics such as fluctuations important, how do we account for them and what are the consequences? These are all challenging questions that must be explored.

All in all, was not Richard Feynman correct in stating that 'there's plenty of room at the bottom' – not only for the sophisticated nanoparticles and nanotubes and but also for the apparently simple reaction–diffusion?

REFERENCES

1. Klajn, R., Fialkowski, M., Bensemann, I.T. *et al.* (2004) Multi-color micropatterning of thin films of dry gels. *Nature Mater.*, **3**, 729.
2. Mahmud, G., Bishop, K.J.M., Chegel, Y. *et al.* (2008) Wet-stamped precipitant gradients control the growth of protein microcrystals in an array of nanoliter wells. *J. Am. Chem. Soc.*, **130**, 2146.
3. Mayer, M., Yang, J., Gitlin, I. *et al.* (2004) Micropatterned agarose gels for stamping arrays of proteins and gradients of proteins. *Proteomics*, **4**, 2366.
4. Hansen, T.S., West, K., Hassager, O. and Larsen, N.B. (2007) Direct fast patterning of conductive polymers using agarose stamping. *Adv. Mater.*, **19**, 3261.
5. Oh, M. and Mirkin, C.A. (2006) Ion exchange as a way of controlling the chemical compositions of nano- and microparticles made from infinite coordination polymers. *Angew. Chem. Int. Ed.*, **45**, 5492.

Appendix A: Nature's Art

For centuries, the Western art of painting focused on realistic and figurative representations of the surrounding world. Correspondingly, the techniques the artists used put premium on the ability to control even the most minute details on the canvas. The advent of photography made these strides to some extent irrelevant, and new artistic movements emerged that gradually shifted from the figurative to the abstract. The techniques also evolved to meet the demands of new artistic trends – from Seurat's pointillism (still fully under the artist's control) to Pollock's drip painting, where canvas becomes a manifold of fractal-like structures the artist controls only marginally.

Reaction–diffusion (RD) initiated by wet stamping can be regarded as a micro-scale painting technique situated somewhere between these two extrema – we can control the initial conditions, but from there on the images evolve on their own. Also, since fluctuations are always present in small-scale RD, the process creates unique paintings and no two applications of the same stamp are ever the same. In this respect, the RD creations can be regarded as a form of artistic expression – not our expression, to be sure, but that of nature. Figures A.1–A.5 show some illustrative examples varying both in the types of chemical inks used as well as in the degree of human control and regularity.

Chemistry in Motion: Reaction–Diffusion Systems for Micro- and Nanotechnology Bartosz A. Grzybowski

(a)
(b)

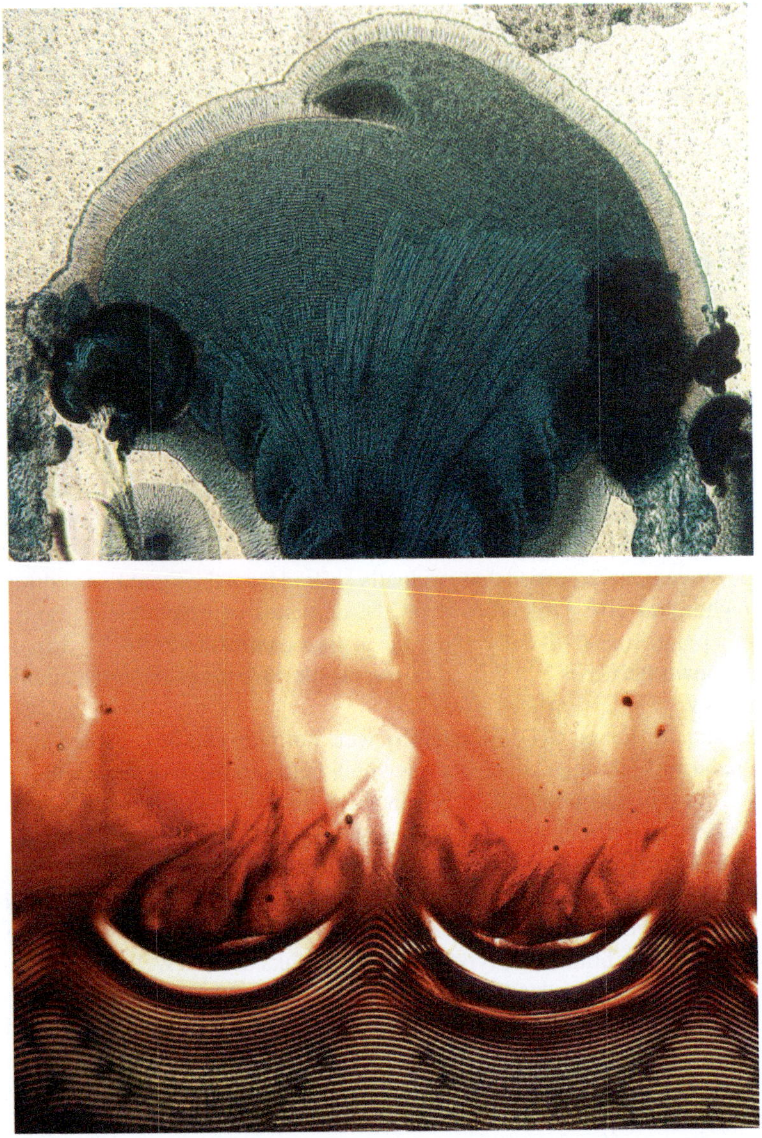

Figure A.1 Random and semi-random. The top image is created by a concentrated $FeCl_3$ ink sprinkled onto and then spreading randomly in a thin gelatin 'canvas' soaked with $K_4[Fe(CN)_6]$. No wet stamp is applied to initiate the process. The bottom image is based on the Liesegang ring chemistry ($AgNO_3$ ink in the stamp, $K_2Cr_2O_7$ 'developer' in the gelatin film) initiated from two crescent-like features. The Liesegang rings propagate downwards in an expected, regular fashion. The irregular 'fire' above the features is created by accidental contact between the stamp and the gel, which renders RD random. Both 'paintings' are about 1 mm across. The images were taken on a standard optical microscope

(a)

(b)

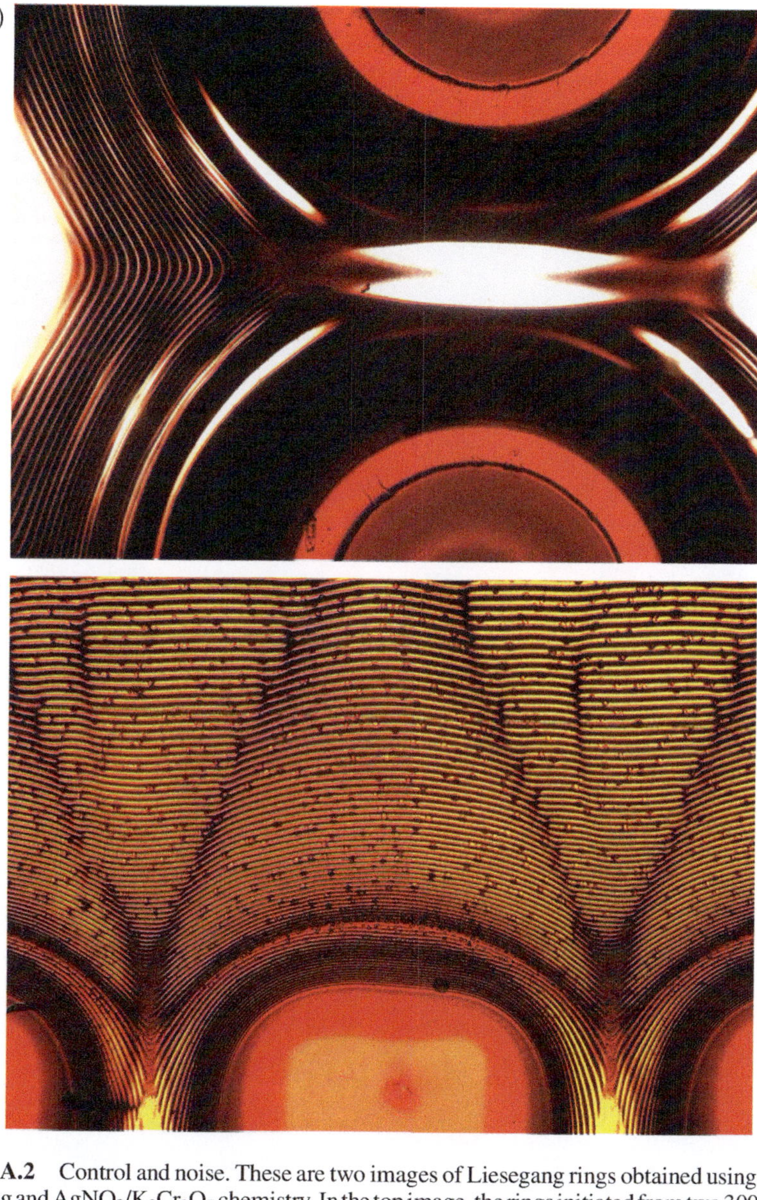

Figure A.2 Control and noise. These are two images of Liesegang rings obtained using wet stamping and $AgNO_3/K_2Cr_2O_7$ chemistry. In the top image, the rings initiated from two 300 μm rings and propagating from right to left are continuous and never split or cross one another. In the bottom image, the precipitation bands propagating from bottom to top (away from the stamped 500 μm squares) bifurcate/split along the apparent, inclined lines – this is an interesting example when an inherently probabilistic/noisy event (i.e., the bifurcations) is brought under experimental control. The effects of noise are still manifest, however, in the random placement of the little 'beads' on the rings. These beads are small crystals of $Ag_2Cr_2O_7$ that form because the concentrations of 'paints' used were high. Overall, the structure comprises both deterministic and stochastic elements. The presence of the latter indicates that every time we stamp the squares, we are going to obtain a slightly different 'micropainting'!

(a)
(b)

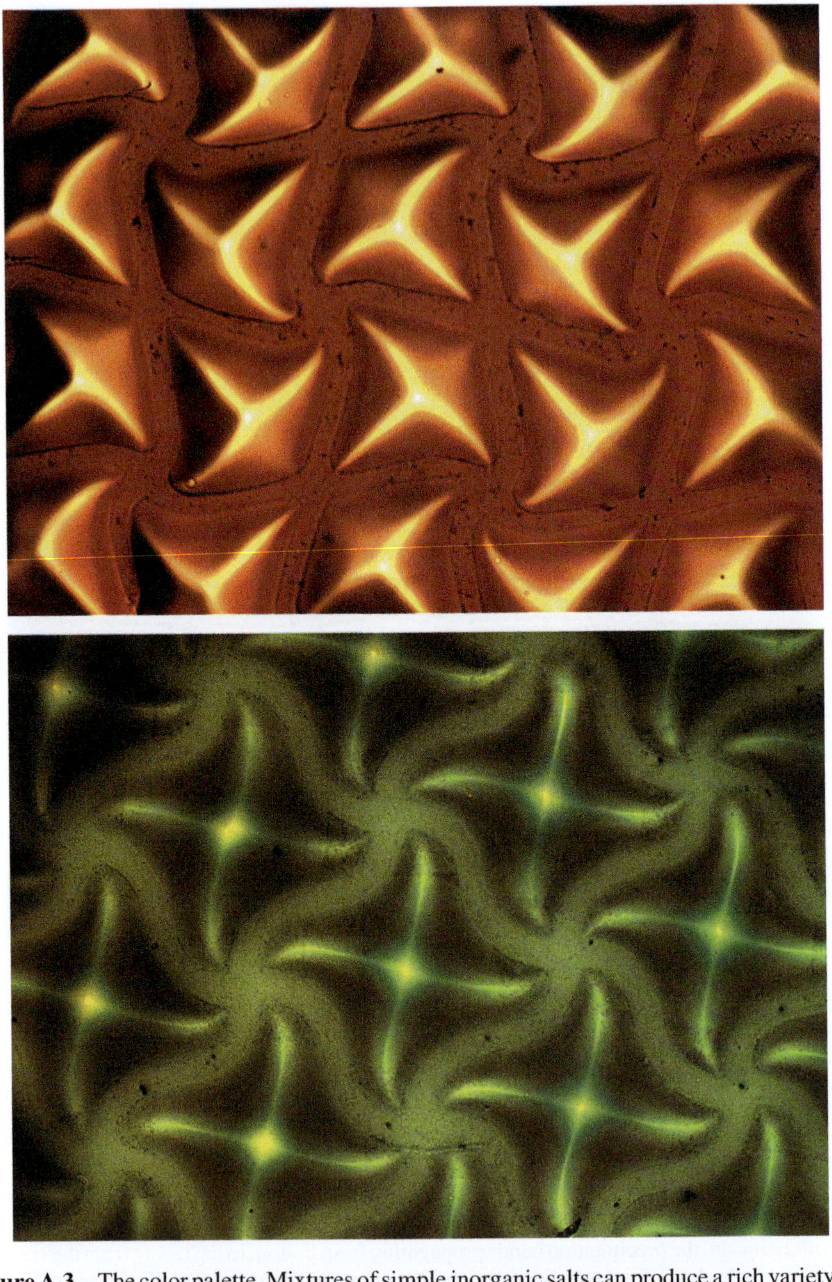

Figure A.3 The color palette. Mixtures of simple inorganic salts can produce a rich variety of colors and hues. In the upper image, the wet stamp delivers a mixture of $CoCl_2$ and $CuCl_2$ (5%:5% w/w) to a gel containing 1 wt% of $K_4[Fe(CN)_6]$. In the lower image, a mixture of three salts, $CoCl_2/FeCl_3/CuCl_2$ (5%:5%:5% w/w) is wet stamped onto the same canvas. The gradients of color develop because the salts have different solubility products and diffusion coefficients. As a result, the compositions of colored precipitates vary from one location to another. In reality, each image has an area of about 4 mm^2

(a)

(b)

Figure A.4 Evolution and complexity. The top image captures an intermediate stage of pattern formation. Note that the $FeCl_3$ ink invades the gelatin/$K_4[Fe(CN)_6]$ canvas only from the nodes of the stamped micronetwork. We have explained this effect in Chapter 9 when discussing RD micro-chameleons. In the bottom image, the stamped relief is more complex and so is the emerging rosette pattern. In both images, the canvas is about 1.5 mm across

(a)

(b)

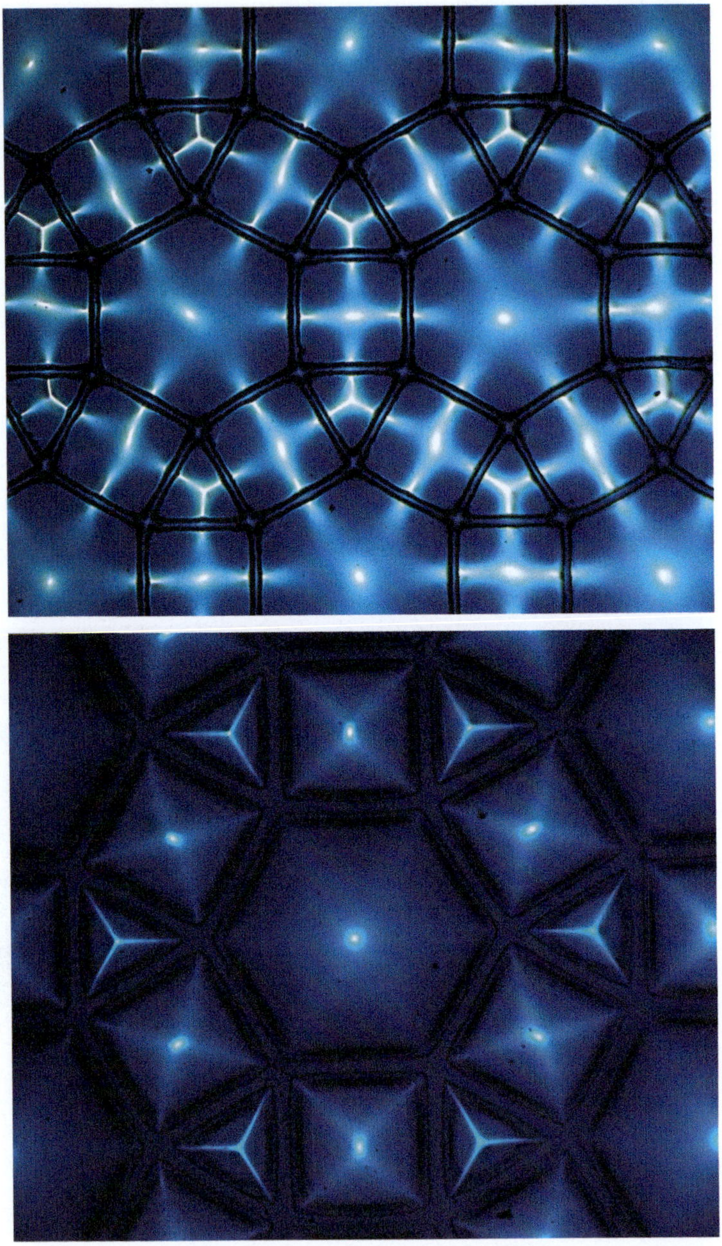

Figure A.5 Thick and thin. The same stamp applied on a thicker (~40 μm, top) and thinner (~10 μm) gelatin/$K_4[Fe(CN)_6]$ produces drastically different images. On a thick gel, the $FeCl_3$ ink flows into the canvas only from the nodes of the stamped pattern; on a thin gel, the ink invades from all stamped locations. The scientific origin of this bimodality is explained in Section 9.1. The images are about 1.5×1.5 mm

Appendix B: Matlab Code for the Minotaur (Example 4.1)

```
clc; tic; X = []; X = imread ('maze2.bmp', 'bmp');
                                   %Read the initial Maze from
                                   %a bitmap file.
                                   %The outer edge of the
                                   %bitmap must consist
                                   %of only sources, sinks, or
                                   %walls.
map = [1,1,1;0,0,0;0,1,0;1,0,0];
                                   %Set the color map
for i = 5:256
  map(i,:) = [1,1-(i-5)/251,1-(i-5)/251];
                                   %These values produce
                                   %shades of red
end                                %for the diffusing aroma.
C = zeros(size(X,1),size(X,2), 2);
                                   %Set the maze to be
                                   %initially empty
tmax = 100000; alpha = 0.2; sink = 0; source = 1;
                                   %Define the concentrations
                                   %and α
for i = 1:size(X,1)                %Parse the input maze,
                                   %changing the
  for j = 1:size(X,2)              %colors as necessary. Note
                                   %that these
```

Chemistry in Motion: Reaction–Diffusion Systems for Micro- and Nanotechnology Bartosz A. Grzybowski

```
  if X(i,j) == 255        %Diffusion space
                              %correspond to a 256 color
                              %bitmap.
  X(i,j) = 0;
elseif X(i,j) == 0        %Wall
                              %Also, reset the colors to
                              %the above color map.
  X(i,j) = 1; C(i,j,:) = 2;
  elseif X(i,j) == 250  %Source (Lair)
                              %Colors are from the standard
                              %pallet in
  X(i,j) = 2; C(i,j,:) = source;
                              %Microsoft Paint, with the
                              %wallscolored
  elseif X(i,j) == 249  %Sink (Exit)
                              %black, the diffusion areas
                              %white, the source
    X(i,j) = 3; C(i,j,:)= sink;
                              %area(s) green, and the sink
                              %areas red.
      end
    end
  end
imwrite(X,map,int2str(0),'bmp')
                              %Save the initialstate as '0'
for t = 1:tmax
  if mod(t,100) == 0          %Output the concentrations
    [toc, t/100]              %every 100 iterations
    for i = 1:size(X,1)
      for j = 1:size(X,2)
        if X(i,j) == 0        %Translate actual
                              %concentrations
          O(i,j) = floor(C(i,j,2)*251+5);
                              %into the color map used
                              %above
      else
          O(i,j) = X(i,j)+1;
        end
      end
    end
    imwrite(O,map,int2str(t),'bmp')
                              %Write the file, with the name
end                           %corresponding to the time
                              %step
```

```
C(:,:,1)=C(:,:,2);                  %Reset the concentration
                                    %matrix
for i = 1:size(X,1)
  for j = 1:size(X,2)
    if X(i,j) == 0                  %Perform diffusion only at
                                    %nodes
  switch (X(i+1,j)*10)+(X(i-1,j))
                                    %that allow diffusion, also
                                    %read:
  case {0, 2, 3, 20, 22, 23,30, 32, 33}
                                    %Treatment of individual
                                    %nodes.
     C(i,j,2) = C(i,j,1)+alpha*(C(i-1,j,1)+C(i+1,j,1)-
                2*C(i,j,1));
    case {1,21,31}
     C(i,j,2) = C(i,j,1)+alpha*(C(i+1,j,1)+C(i+1,j,1)-
                2*C(i,j,1));
    case {10,12,13}
     C(i,j,2) = C(i,j,1)+alpha*(C(i-1,j,1)+C(i-1,j,1)-
                2*C(i,j,1));
    case 11
     C(i,j,2) = C(i,j,1);
    end
    switch X(i,j+1)*10+(X(i,j-1))
                                    %Abbreviations defined below
    case {0, 2, 3, 20, 22, 23, 30, 32, 33}
                                    %DD
     C(i,j,2) = C(i,j,2)+alpha*(C(i,j-1,1)+C(i,j+1,1)-
                2*C(i,j,1));
    case {1,21,31}                  %DW
     C(i,j,2) = C(i,j,2)+alpha*(C(i,j+1,1)+C(i,j+1,1)-
                2*C(i,j,1));
    case {10,12,13}                 %WD
     C(i,j,2) = C(i,j,2)+alpha*(C(i,j-1,1)+C(i,j-1,1)-
                2*C(i,j,1));
    case 11                         %WW
     C(i,j,2) = C(i,j,2);
        end
      end
    end
  end
end

%{
```

Treatment of individual nodes. In a square lattice with four types of nodes (diffusion, source, sink or wall), each node has four independent nearest neighbors (NNs). Each of the four NNs can have any one of those four values, resulting in a maximum of 256 unique combinations of NNs in a given calculation. Writing code to address each of these unique combinations is both tedious and error-prone due to there being many copies of very similar code. However, the 256 combinations of NNs can be divided into two sets of four mathematically unique cases by: (i) separating the diffusion operation into two discrete steps and (ii) grouping mathematically identical nodes. In (i), by separating the diffusion operation into two discrete steps, one along the *i* axis and one along the *j* axis, the four NNs of each node are treated as two independent sets of two NNs. This reduces the 256 unique combinations of four NNs to two sets of 16 unique pairs of NNs. In (ii), the forward time centered space (FTCS) method for constant concentration NNs (either sources or sinks) is mathematically identical to the method for NNs that allow diffusion, leaving only two types of NN – diffusing (D) and wall (W). When only two mathematically different NNs exist, the number of pairs of NNs on each axis is reduced from 16 to 4. The four pairs of NNs correspond to: both NNs allow diffusion (DD), the left or upper neighbor allows diffusion and the right or lower neighbor is a wall (DW), the left or upper neighbor is a wall and the right or lower neighbor allows diffusion (WD) and both NNs are walls (WW). The concentration change at a particular node is calculated by encoding the type of NN on each axis, then applying the appropriate concentration changes (DD, DW, WD or WW) for each axis.

The code can be made significantly more efficient by writing the diffusion algorithm twice – using an explicit step from C(:,:,1) to C(:,:,2), followed by an explicit step from C(:,:,2) to C(:,:,1) – instead of copying the entire matrix during each iteration.

Appendix C: C++ Code for the Zebra (Example 4.3)

```
#include <stdio.h>
#include <stdlib.h>
#include <time.h>
using namespace std;

#define UP (0x00000001)          //Define the velocity vectors,
                                 //and some constants
#define RI (0x00000002)          //for adding and removing
                                 //particles.
#define DN (0x00000004)
#define LE (0x00000008)          //NX(x) counts the number of
                                 //particles at a node
#define RU (0x0000000E)          //RotL(x) rotates x π/2
                                 //radians counterclockwise
#define RR (0x0000000D)          //RotR(x) rotates x π/2
                                 //radians clockwise
#define RD (0x0000000B)
#define RL (0x00000007)
#define MA (0x0000000F)
#define EMPTY (0x00000000)
#define NX(x) ((x&UP)+((x&RI)>>1)+((x&DN)>>2)+((x&LE)>>3))
#define RotL(x) (x&UP)?(LE+(x>>1)):(x>>1)
#define RotR(x) (x&LE)?(UP+(MA&(x<<1))):(MA&(x>>1))

int addone(int);                 //Two functions for adding
                                 //and removing
```

Chemistry in Motion: Reaction–Diffusion Systems for Micro- and Nanotechnology Bartosz A. Grzybowski

```
int remone(int);            //particles from nodes.
                            //Included after the main()

int main(void){
  int size = 250;
  const double dX = .1;     //Define the probabilities for
                            //diffusion

  const double dY = .3;
  const double r1 = 0;      //Define the rate constants.
  const double r2 = 1;
  const double r3 = 11;
  const double r4 = 7;
  const double r5 = 6;
  const double r6 = 6;
  const double rmax = 150;
  const int maxstep = 5000;//Simulation stopping point

  FILE *fid;
  int X[size][size], Y[size][size], C[size][size],PrX
    [size][size], PrY[size][size];
  int i,j,k,l,cn, addx, addy, x,y;
  char fname[50];
  double p,paddA,pAtoB,pkillA,paddB,pkillB;

 srand48(time(0));

 for(i=0; i < size; i++){   //Initialize the matrices.
   for(j = 0; j<size; j++){
     X[i][j] = EMPTY;
     Y[i][j] = EMPTY;
     C[i][j] = EMPTY;
     PrX[i][j] = EMPTY;
     PrY[i][j] = EMPTY;
   }
 }

 for(k = 0; k<5; k++){      //Set up the initial
                            //distribution of X
   for(i = 0;i<size; i++){
     for(j = 0; j < 5; j++){
     X[i][j+50*k+24] |= addone(X[i][j+50*k+24]);
     X[i][j+50*k+24] |= addone(X[i][j+50*k+24]);
     }
   }
 }
```

```
size-;                              //Decrement size here
                                    //to set it for indexing
                                    //the last row or column.
for(k = 0; k <= maxstep; k++){
  if( drand48() < dX){              //Propagate and randomize
    for(i = 0; i <= size; i++){     //PrA is used for temporary
                                    //concentration storage
    for(j = 0; j <= size; j++){
      if((X[i][j]&UP) == UP){       //Move up
        if(i == 0){    PrX[size][j] |= UP;
        }else{         PrX[i-1][j] |= UP;}}
      if((X[i][j]&RI) == RI){       //Move right
        if(j == size){ PrX[i][1] |= RI;
        }else{         PrX[i][j+1] |= RI;}}
    if((X[i][j]&DN) == DN){         //Move down
      if(i == size){ PrX[1][j] |= DN;
      }else{         PrX[i+1][j] |= DN;}}
  if((X[i][j]&LE) == LE){           //Move left
    if(j == 0){        PrX[i][size] |= LE;
    }else{         PrX[i][j-1] |= LE;}}
}
}
for(i = 0; i <= size; i++){         //Rotate. Note that EMPTY
                                    //and MA are
for(j = 0; j <= size; j++){         //symmetrical on rotation
  if(PrX[i][j] != EMPTY && PrX[i][j] != MA){
    p = drand48();                  //Perform rotations with
                                    //equal probability
    if(p<1/4) X[i][j] = RotR(PrX[i][j]);
      else if(p<1/2) X[i][j] = RotL(PrX[i][j]);
      else if(p<3/4) X[i][j] = RotR(RotR(PrX[i][j]));
      else X[i][j] = PrX[i][j];
    }else{X[i][j] = PrX[i][j];}
    PrX[i][j] = EMPTY;
  }
  }
}
if(drand48()<dY){                   //Identical to the
                                    //propagation/rotationofX
  for(i = 0; i <= size; i++){       //See comments above
  for(j = 0; j <= size; j++){
    if((Y[i][j]&UP) == UP){
      if(i == 0){ PrY[size][j] |= UP;
      }else{ PrY[i-1][j] |= UP;}}
```

```
    if((Y[i][j]&RI) == RI){
      if(j == size){ PrY[i][1] |= RI;
      }else{ PrY[i][j+1] |= RI;}}
    if((Y[i][j]&DN) == DN){
      if(i == size){ PrY[1][j] |= DN;
      }else{ PrY[i+1][j] |= DN;}}
    if((Y[i][j]&LE) == LE){
      if(j == 0){ PrY[i][size] |= LE;
      }else{ PrY[i][j-1] |= LE;}}
}
}
for(i = 0; i <= size; i++){
for(j = 0; j <= size; j++){
  if(PrY[i][j] != EMPTY && PrY[i][j] != MA){
    p = drand48();
    if(p<1/4) Y[i][j] = RotR(PrY[i][j]);
    else if(p<1/2) Y[i][j] = RotL(PrY[i][j]);
    else if(p<3/4) Y[i][j] = RotR(RotR(PrY[i][j]));
    else Y[i][j] = PrY[i][j];
  }else{Y[i][j] = PrY[i][j];}
  PrY[i][j] = EMPTY;
  }
  }
}
for(i = 0; i <= size; i++){           //Reactions
  for(j = 0; j <= size; j++){
  x = NX(X[i][j]);                    //Store these two numbers
                                      //to increase speed.
  y = NX(Y[i][j]);                    //Calculate reaction
                                      //probabilities for each
                                      //node
paddA = (r1+r3*x)/rmax;               //0->X
pAtoB = r5*x/rmax;                    //X->Y
pkillA = r6*y/rmax;                   //X->0
paddB = r2/rmax;                      //0->Y
pkillB = r4*y/rmax;                   //Y->0
addx = 0;                             //Then determine if the
                                      // reactions happen
addy = 0;                             //and how much X and Y to
                                      //add or subtract
if(drand48() < paddA) addx++;
if(drand48() < pAtoB){addx--; addy++;}
if(drand48() < pkillA){addx--;}
if(drand48() < paddB){addy++;}
```

```
if(drand48() < pkillB){addy--;}
if(addx > 0) X[i][j] |= addone(X[i][j]);
                                        //Perform the
                                        //necessary
                                        //additions or
if(addx <= -1) X[i][j] = remone(X[i][j]);
                                        //subtractions.
if(addx == -2) X[i][j] = remone(X[i][j]);
if(addy > 0) Y[i][j] |= addone(Y[i][j]);
if(addy > 1) Y[i][j] |= addone(Y[i][j]);
if(addy <= -1) Y[i][j] = remone(Y[i][j]);
if(addy == -2) Y[i][j] = remone(Y[i][j]);
C[i][j] = C[i][j] + X[i][j];            //Integrate the
                                        //concentration of X
                                        //for processing
}
}
if(k%100 == 0) {printf("%6d\n",k);} //Output to screen
if(k%1000 == 0){                        //Write a file with the
                                        //integrated X
                                        //concentrations
     sprintf(fname,"%d.dat", k);
     fid = fopen(fname,"w+");
     for(i = 0; i <=size; i++){
     for(j = 0; j<=size; j++){
       fprintf(fid,"%2d ", C[i][j]);
     }
     fprintf(fid,"\n");
     }
     fclose(fid);
   }
 }
 return 0;
}                                       //EndofMain function

int addone(int A){                      //Take a given
                                        //velocityvector,
                                        //and add one more
 double q;                              //particle to it.
 int c1, c2, c3;
 if(NX(A) >= 3){ return MA;}            //Trivial cases
 if(NX(A) == 2){                        //The general scheme
                                        //is to determine
                                        //which
```

```
  if(A&UP == UP){                            //directions are
                                             //open, then add
                                             //randomly to
    if (A&RI == RI){c1 = DN; c2 = LE;}       //one of those
                                             //directions
    else if (A&DN == DN){c1 = RI; c2 = LE;}
    else{c1 = RI; c2 = DN;}
  }else if(A&RI == RI){
   if (A&DN == DN){c1 = UP; c2 = LE;}
   else{c1 = UP; c2 = DN;}
  }else {c1 = UP; c2 = RI;}
  if(drand48() < 0.5){return c1;}
  else {return c2;}
}
if(NX(A) == 1){                              //Same process,
                                             //except for
                                             //NX(A) ==1
  if(A&UP == UP){c1 = RI; c2 = DN; c3 = LE;}
  else if(A&RI == RI){c1 = UP; c2 = DN; c3 = LE;}
  else if(A&DN == DN){c1 = UP; c2 = RI; c3 = LE;}
  else {c1 = UP; c2 = RI; c3 = DN;}
  q = drand48();
  if(q< 1/3){return c1;}
  else if(q<2/3){return c2;}
  else {return c3;}
}
if(NX(A) == 0){                              //Another trivial
                                             //case. Randomly
                                             //select a
                                             //direction.
q = drand48();
  if(q< 1/4){return UP;}
  else if(q<2/4){return RI;}
  else if(q<3/4){return DN;}
  else {return LE;}
  }
}

int remone(int A){                           //This function
                                             //will remove one
                                             // velocity from
  double q;                                  //a velocity vector.
  int c1, c2, c3;
```

```
  if(NX(A) <= 1){return EMPTY;}      //Trivial cases
  if(NX(A) == 2){return (((addone(A^MA))^MA)^A);}
                                     //The ^ operator is
                                     // XOR, and is used
  if(NX(A) == 3){return ((addone(A^MA))^MA);}
                                     //to remove a particle
                                     //by adding one
  if(NX(A) == 4){                    //to the inverse
                                     // vector, then
    q = drand48();                   //re-inverting to
                                     //get the new vector
    if(q< 1/4){return RU;}           //The opposite trivial
                                     //case from
    else if(q<2/4){return RR;}       //above. Randomly
                                     //select an empty
    else if(q<3/4){return RD;}       //path, and return the
                                     //inverse.
    else {return RL;}
  }
}
```

Index

References to figures are given in italic type. Preferences to tables are given in bold type.

acetylcholine 197
ADP (adenosine diphosphate) 6–7
aerogels 97
agarose gel 33–34, 95–96
 in 3D RD fabrication 228–230
 mold preparation 101–102
 see also WETS
alternating direction implicit (ADI) method 82–83
m-aminobenzamidine 221
AMP (adenosine monophosphate) 55
amplification (of materials properties) 195–197
 RD micronetworks 197–202
angelfish 8
animate systems, overview 5–8
applications, overview 9–12
aspirin 53
ATP (adenosine triphosphate) 6–7, 55
backward time centered space (BTCS) differencing 81–82

Belousov, Boris 5
Belousov-Zabotinsky oscillator 208
 diffusive coupling 212–213
 kinetics 211–212
 wave emission 213–215
 WETS patterning 210–211
Bessel functions 36–37
binding constants *222*
 proteins 219–221
blood 52
Briggs-Rauscher reaction 56–57, 216
Brownian motion 17, 42–43
Brusselator 55–56, *57*
Buckle Finder software 120
buckled surfaces 118–121, *119*, 152–160
 applications 155–158
tert-butyl chloride 48

Cahn-Hilliard equation 132
calcium phosphate 51
calcium signalling (in cell) 5–6
calcium sulfate 51
cAMP 195–196
catalysis 9–11
 autocatalytic reactions 52–55
 packed-bed reactors 9–10
 using core-and-shell particles 237–238
caterpillar micromixer 123

Chemistry in Motion: Reaction-Diffusion Systems for Micro- and Nanotechnology Bartosz A. Grzybowski

cells (living)
  cultured on SAMs 183
  feedback systems 5–6
  grown on wrinkled substrates 156–157
  motility mechanism 185
  optical imaging 184
  regulatory processes and RD 5–8
chameleons 196–197, *196*
chemical equilibrium 50–51
chemical plating *see* electroless plating
chemical reactions
  autocatalytic 52–54
  autoinhibiting 54–55
  equilibrium 50–51
  galvanic replacement 248–253
  ionic 51–52
  oscillating 55–57
  rate 45–49
cobalt, interdiffusion with sulfur 242–246
computing times, Crank-Nicholson modeling 137
COMSOL software 70, *71*
concentration profiles *21*, *29*
continuous random time walk 40–42, 43
convolution, Laplace transforms **27**
cooperativity 53–54
coordinate systems
  cylindrical and spherical 34–38
  rectangular 20–34
copper, eletroless plating 166–167
core-and-shell particles (CSP)
  core exchange 236–238
  formation of crystals from 238–240
  spherical, inside cubes 228–230
Crank-Nicholson modeling 84, 220
  periodic precipitation 137
crystals, from CSPs 238–240
CSP *see* core-and-shell particles
CTRW (continuous random time walk) formalism 40–42, 43

Damköhler number 64, 232
  and metal core properties in 3D etching 233, *234*
dentistry 216
dermatology 216
Dictyostelium discodeum 8
differential equations *see* partial differential equations
differentiation, Laplace transforms **27**
diffraction gratings 145
diffraction structures 145–152, *146*
  pattern calculation 149–152
  substrate patterning 146
diffusion
  diffusive flux 18
  'drift' 41
  governing equation 17–20
  governing equations
    boundary conditions 24–26
    solution
      by Laplace transforms 26–29
      by separation of variables 21–26
    Stokes-Einstein equation 176
  in nonhomogeneous media 38–43
  in solids *see* Kirkendall effect
  symmetry in solution of governing equations 31–34
  in a thin tube *18*
    finite 31–33
    infinite 28–29, 30–31
  varying constants 258
drag 176
drugs, delivery systems 37–38

electroless plating 166–167
  RD in gels 172–178
  RD in plating solution 167–172
electrolytes
  concentration, and periodic precipitation pattern geometry 142–144
  for periodic precipitation **128**
  and PP buckling 155
equilibrium reactions 50–51
etching 178–180
  governing equations 178–180
  in 3 dimensions 230
  for cell biology 184–186
  conductive oxides 187–188
  glass 189–192

three-dimensional 230–235
reaction rates 232–233

fabrication
3D structures
nanocages 248–253
spheres inside cubes 228–235
structures inside non-cubical particles 235
using Kirkendall effect **246**
advantages of RD 11–12
disjoint features 117–121
gel stamps 98–101
microlenses 105–109, 111–117
micromixers 122–124
optical diffraction structures 145–152
feedback reactions 55
FEM 70–80
Fick, Adolf 18
Fick's law 18–19, 39
finite difference (FD) analysis 66–70
finite element analysis 70–80
Galerkin method 74
focal adhesions 185
Fokker-Planck diffusion 43
formaldehyde 213–214
fortifications 7
forward time centered space (FTCS) differencing 81
Fourier series 22–23
Fresnel zone plate (FZP) **147**, 147–149
with buckled surfaces 157–158
Fresnel-Kirchoff modeling 150

Galerkin finite element analysis 73–75
gallium arsenide, microetching 188–189
geduldflaschen 225
gel electrophoresis 96
gelatin 94–95
gels 93
capsules 37–38
choice for WETS 94–97
crosslinking 147, 155
and periodic precipitation 142
definition 57
in electroless metal deposition 173–177
for periodic precipitation **128**
reaction rates 57–59
giraffes 8
glass
gel etching 181
microetching 189–192
glycolysis 55
gold
gel etching 180, *181*
nanoframe fabrication 248–251
as self-assembled monolayer (SAM) substrate *182*
gradient control 258, *260*

Heaviside step function 86–87
hemoglobin 53–54
hexacyanoferrate 105–109
hydrofluoric acid 181, 189
hydrogel stamps 100–101

IBM Power6 processor *121*
indium-tin oxide 187
integration, Laplace transforms **27**
interpolating functions 71
iodide 180
ionic reactions 51–52
IP3 messenger 5–6
iron(III) chloride, reaction-diffusion 197–202

KFN model 211–212
Kipling, Rudyard 89–90
Kirkendall, Ernest 240
Kirkendall effect 240–248
basic mechanism 241–242

laminar flow 122
Laplace equation 21
Laplace transforms 26–29, **27**
lattice gas automata (LGA) 88–89
microlens fabrication 111–117
and Monte Carlo modeling 116–117
le Duc, Stephane 2
lenses *see* microlenses
Liesegang, Raphael 2–3
Liesegang rings *2*

linear superposition 34
lithium chloride 51
lymphocytes 197

Matalon-Packter law 129
 and PP feature geometry 139
Matlab 23
metal foils
 microstructuring 181–186
 overview 165–172
 reaction-diffusion
  in plating solution 167–172
  in substrate 172–178
 standard fabrication 165–166
methanol 215
microfabrication *see* fabrication
microfluidic devices 121–124, 189
microlenses *104*, 189, *190*
 fabrication 105–109
  by lattice gas modeling 111–117
 Fresnel zone plate (FZP) 147–149, *148*
 shape optimization using Monte Carlo techniques 116–117
micromixers 189
 fabrication 122–124
microprocessors *104*
microtubules 7, 185
migraine headaches 7
minerals 9
miniaturization 103–104
Minotaur 68–70
mixers *see* micromixers
molds 100, 101–102
molecular manipulation 259–260
Monte Carlo simulation 112–113
multilevel surfaces *119*
myocardiac tissue 7

nanocages 248–253
nanowrinkles 152–160
noise 135
 and time-reversibility 114–115
nonlinear amplification 12, 195–196
 applications 215–222
 by RD micronetworks 197–202
 using low-symmetry networks 203–205

optics
 analogy to periodic precipitation mechanisms 144–145
 diffraction structures 145–152
 see also microlenses
oscillatory reactions 5–6, 55–57
 see also Belousov-Zhabotinsky oscillator
Ostwald, Wilhelm 129
Ostwald-Liesegang mechanisms 9

PAAm (polyacrylamide) 96–97
packed-bed reactors 9–10
parallel chemistries 257–258
parallel-ridge micromixer 121–122
partial differential equations
 solution by Laplace transforms 26–29
 solution by separation of variables 20–26
PDMS (polydimethylsiloxane)
 in 3D RD fabrication 229–230
 WETS 100, 101
periodic precipitation (PP)
 analogy with optics 144–145
 at nanoscale 160
 diffraction pattern calculation 149–152
 explanatory models 129–130
 fabrication of optical diffraction structures 145–152
 gel crosslinking 142
 gel thickness 140–141
 governing equations 130–137
 immobile precipitate 135
 integration with other microfabrication techniques 158–160
 overview 127
 patterns in two dimensions
  overview 137–139
  feature dimensions and spacing 139–140
  and gel crosslinking 142
 patterns in two-dimension *143*
 phenomenology 128–130

and SAM detection 205–208
simulation time 137
stacked 159–160
stochastic effects 135–137
three-dimensional patterns 152–160
via spinodal decomposition 131–134
and WETS 137–138
photolithography
limitations 103–104
and wet stamp fabrication 98–100
photomask preparation, WETS 98–99
polyacrylamide 96–97
polydimethylsiloxane *see* PDMS
PP *see* periodic precipitation
precipitation
bands, scaling laws 129
time discretization 86–87
see also periodic precipitation
proteins
binding constants 219–221
crystal growth control 260

Rat2 fibroblast cells *156*
rational design 109–111
RD *see* reaction-diffusion
reaction rates 45–49
autocatalysis 52–55
in gels 57–59
non-apparent reaction orders 46–47
oscillating reactions 55–57
sequential reactions 49
three-dimensional etching 232–233
reaction-diffusion (RD)
applications, future prospects 258–263
governing equations
general form 61–62
susceptible to analytic solution 62–66
finite difference methods 66–70
finite element analysis 70–80
for gel etching 178–180
mesoscopic models 87–90, 87
precipitation reactions 86–87
in 3 dimensions 230–235
time discretization 80–87
backward time centered space (BTCS) differencing 81–82
method of lines 84–85
operator splitting 83–84
see also diffusion
electroless deposition
in gels 172–178
in plating solution 167–172
history 1–3
in inanimate systems 9–12
initiation 93–94
and metal film deposition 167–172
in nature 4–9, *4*
and nonlinear amplification mechanisms 197–202
solid-state 259
theoretical challenges 262–263
see also periodic precipitation
refraction 144
roughness, of gel surface 172–178
Runga-Kutta method 84

seashells 9
self-assembled monolayers (SAM) 182–184, 205–208
cell cultures 183
self-assembly 261
aggregates of CSPs 238–240
sensors
formaldehyde and methanol 208–210
as measurement devices 216–217
optical 148–149
outlook 215–222
protein-ligand binding 217–219
for self-assembled monolayers 205–208, *207*
shape functions 71
silica gels 97
silicon, gel etching 181
silver nitrate 105–109
skin patterns 8
Snell's law 144
solids, diffusion 240–248
solubility, ionic compounds 51–52
spacing law, precipitation bands 129
spatial discretization
finite difference analysis 66–70
finite element analysis 70–77
spinodal decomposition 131–134

SPR spectroscopy 248
square features 119–121
squid *196*
stalactites 9
stochastic effects, periodic precipitation 135
stoichiometry 46
Stokes-Einstein equation 175, 176
substrates
 buckling parameters 153–155
 for cell culture 156–157
 electroless plating 172–178
 for SAMs *182*
sulfur, interdiffusion with cobalt 242–246
superposition 34, 186
supersaturation 129
surface plasmon resonance (SPR) 248
Sylgaard 184 (PDMS) 100, 101
symmetry, and diffusion modeling 31–34

temporal discretization 80–87
tessellations 198–202
tetrahydrofuran (THF) 229–230
TFA 182
TGFβ (transforming growth factor) 8
THF 229–230
thiols, detection 205–208
tigers 8
tile-centred tessellation 198, *199*
tilings 198–202
time discretization, Crank-Nicholson method 82–83
time-reversibility 112, 114–115
tin, electroless plating, reaction-diffusion mechanisms 166
translation, Laplace transforms **27**
Turing, Alan 5, 89
Tyson scaling 212

Voronoi tesselation 198

WETS (wet stamping) 94
 Belousov-Zabotinsky oscillators 208, 210–211
 diffusion 106–107
 and metal oxide etching 187
 and periodic precipitation 137–139
 stamp fabrication 98–101, **99**
width law
 periodic precipitation 129
 precipitation bands 129
Winfree solution 210

xerogels 97

zebras 8, 89–90
zinc oxide 187

Printed and bound by CPI Group (UK) Ltd, Croydon, CR0 4YY

16/09/2026

14963382-0001